AF358479

Latest Advancements in Semiconductor Materials, Devices, and Systems

Latest Advancements in Semiconductor Materials, Devices, and Systems

Guest Editors

Zeheng Wang
Jing-Kai Huang

Basel • Beijing • Wuhan • Barcelona • Belgrade • Novi Sad • Cluj • Manchester

Guest Editors

Zeheng Wang
DATA61
CSIRO
Sydney
Australia

Jing-Kai Huang
Department of Systems
Engineering
City University of Hong Kong
Hong Kong
China

Editorial Office
MDPI AG
Grosspeteranlage 5
4052 Basel, Switzerland

This is a reprint of the Special Issue, published open access by the journal *Micromachines* (ISSN 2072-666X), freely accessible at: www.mdpi.com/journal/micromachines/special_issues/DEL4N73D24.

For citation purposes, cite each article independently as indicated on the article page online and using the guide below:

Lastname, A.A.; Lastname, B.B. Article Title. *Journal Name* **Year**, *Volume Number*, Page Range.

ISBN 978-3-7258-2808-1 (Hbk)
ISBN 978-3-7258-2807-4 (PDF)
https://doi.org/10.3390/books978-3-7258-2807-4

Contents

About the Editors

Zeheng Wang

Dr. Zeheng Wang is a CERC Fellow at CSIRO, specializing in quantum artificial intelligence, AI for science, and semiconductor devices. He earned his PhD in silicon quantum computing devices from the University of New South Wales and has extensive experience in semiconductor fabrication, physics, simulation, and AI applications in medicine.

Jing-Kai Huang

Dr. Huang is an experienced researcher with a wide-ranging background in semiconductor R&D. Prior to working at City University of Hong Kong, he contributed significantly to the field at King Abdullah University of Science and Technology (2015–2019, KSA), TSMC (2019, TW), and the National Nano Device Laboratories (2010–2011, TW). His expertise lies in the nanofabrication of ultra-scaled electronic devices and the scalable manufacture of low-dimensional materials for advanced electronics. Outstanding achievements led to Dr. Huang being awarded the prestigious Scientia Ph.D. scholar award for the "Beyond the Silicon Technology" project at UNSW. His current research focuses on integrating emerging materials such as 2D semiconductors, complex oxides (high-κ, ferroelectric, memory), and porous coordination polymers (low-κ, sensor) to advance future semiconductor technology nodes.

Preface

The rapid evolution of semiconductor technologies continues to drive breakthroughs across a multitude of industries, from computing and telecommunications to healthcare and renewable energy. Recognizing the importance of fostering knowledge exchange in this transformative field, we are pleased to present this reprint volume, "Latest Advancements in Semiconductor Materials, Devices, and Systems", a collection of articles that were originally published in a Special Issue of *Micromachines*.

This reprint aims to provide readers with a comprehensive overview of cutting-edge research on semiconductor materials and devices, emphasizing novel approaches to material synthesis, advanced device architectures, numerical simulations, and practical applications. The scope of this collection spans both traditional silicon-based systems and emerging technologies, such as two-dimensional materials and wide-bandgap semiconductors, reflecting the dynamic landscape of semiconductor research.

The motivation behind curating this reprint is to offer an accessible resource for researchers, engineers, and students, bridging the gap between fundamental studies and applied advancements. By compiling these exemplary works, we aim to inspire innovative thinking and promote collaborative efforts in the pursuit of future semiconductor breakthroughs.

The contributions in this volume are authored by a diverse group of scientists and engineers from around the world, whose expertise and dedication have enriched this collection. Each article represents a valuable piece of the broader mosaic of semiconductor science, offering readers both theoretical insights and practical perspectives.

We extend our deepest gratitude to the authors for their outstanding contributions and commitment to advancing knowledge in this field. Special thanks are due to the editorial team at *Micromachines* for their guidance and meticulous review process, as well as to the reviewers, whose constructive feedback ensured the high quality of the published articles.

This volume is dedicated to the global semiconductor research community, whose efforts continue to illuminate the path toward technological progress. We hope that this book will serve as a valuable resource, sparking new ideas and fostering collaborations across disciplines.

Zeheng Wang and Jing-Kai Huang
Guest Editors

Editorial

Editorial for the Special Issue on the Latest Advancements in Semiconductor Materials, Devices and Systems

Xinghuan Chen [1], Fangzhou Wang [2,*], Zirui Wang [3], Zeheng Wang [4] and Jing-Kai Huang [5,6]

1   The China Electronic Product Reliability and Environmental Testing Research Institute,
     Guangzhou 510610, China
2   Songshan Lake Materials Laboratory, Dongguan 523808, China
3   School of Integrated Circuits, Peking University, Beijing 100871, China
4   Manufacturing, CSIRO, Lindfield, NSW 2070, Australia
5   Department of Systems Engineering, The City University of Hong Kong, Kowloon, Hong Kong 999077, China
6   School of Materials Science and Engineering, University of New South Wales (UNSW),
     Sydney, NSW 2052, Australia
*   Correspondence: fz_wang@foxmail.com

Citation: Chen, X.; Wang, F.; Wang, Z.; Wang, Z.; Huang, J.-K. Editorial for the Special Issue on the Latest Advancements in Semiconductor Materials, Devices and Systems. *Micromachines* **2024**, *15*, 1422. https://doi.org/10.3390/mi15121422

Received: 25 November 2024
Accepted: 26 November 2024
Published: 27 November 2024

The field of semiconductor research is experiencing a paradigm shift as the boundaries of Moore's Law are being approached [1]. While the exploration of next-generation semiconductors, such as 2D materials and wide-bandgap semiconductors, continues to reveal new horizons for electronic devices, advancements in traditional semiconductors, such as silicon (Si) and germanium (Ge), remain crucial for current and emerging technologies [2–5]. These advancements, in conjunction with sophisticated modeling, simulation, and fabrication techniques, are paving the way for remarkable innovations not only in materials and devices but also in circuits and systems, thus enriching the landscape of semiconductor research. In this Editorial, we summarize 10 cutting-edge papers that highlight the wide range of innovations within the design, optimization, and modeling of devices and systems based on various materials, as well as their applications. These contributions can be broadly categorized into five key areas: (1) revolutionary materials, (2) the optimization of devices, (3) advanced modeling methods, (4) the techniques used for characterizing traps, and (5) circuits and systems.

The creation of wide- and ultrawide-bandgap semiconductor materials, such as silicon carbide (SiC), gallium nitride (GaN), and diamond, has significantly improved the performance of power electronics [6–8]. Rafin et al. provided a comprehensive review of wide- and ultrawide-bandgap semiconductor power devices, comparing Si, SiC, GaN, and emerging diamond technology devices [9]. Wide- and ultrawide-bandgap semiconductor power devices exhibit significant superiority over Si in terms of their voltage blocking, switching speeds, efficiency, and thermal performance. SiC and GaN devices are becoming increasingly prevalent, particularly in electric vehicles, renewable energy, aerospace, and high-frequency applications. Diamond's exceptionally wide bandgap could enable unprecedented power densities and the high-temperature operation of power devices.

GaN-based high-electron-mobility transistors (HEMTs) are considered promising candidates for next-generation high-efficiency power conversion applications [10–13]. The optimization of these devices is critical to improving their performance [14]. Chen et al. present a novel enhancement-type GaN HEMT with a high power transmission capability [15]. In this transistor, a graded Al mole fraction is utilized to broaden the conduction band and create a three-dimensional electron sea (3DES) so that a coherent channel can be formed. Benefiting from the high electron density of the 3DES, this device exhibits an outstanding high-power performance. Deng et al. systematically compared the differences between MOCVD-SiNx and LPCVD-SiNx in terms of their Ohmic contact and related interfaces [16]. The growth interface of LPCVD-SiNx can suppress leakage effectively. Furthermore, it was discovered that LPCVD-SiNx devices can improve their RF output performance by creating a lower Ohmic contact resistance. Therefore, LPCVD-SiNx devices exhibit an excellent performance in small-sized modules working in low-voltage applications, which highlights their potential uses in small 5G terminals. Yang et al. report on the optimization of ultraviolet (UV) photodetectors with a TiO_2 nanorod (NR)-containing active layer and a solid–liquid heterojunction (SLHJ) via the atomic layer deposition (ALD) of an Al_2O_3 passivation layer [17]. The oxygen vacancies in the TiO_2 are effectively filled by the diffusing of Al, Ti, and O atoms between the Al_2O_3 passivation layer and TiO_2 NRs during the annealing treatment. A difference of four orders of magnitude is observed in the photocurrent-to-dark-current ratio, demonstrating the advantages of an ALD-Al_2O_3 passivation layer in UV photodetectors.

Advanced modeling methods are vital to optimizing the design of devices. Recently, many researchers have become interested in artificial neural network (ANN) methods, using them in areas such as the subthreshold swing modeling of GaN HEMTs [18], GaN Ohmic contacts used for fabrication processes [19,20], and predicting the effect of statistical variability on Si junctionless nanowire transistors [21]. Within this area, Zhao et al. have proposed a compact ANN model generation methodology for GAA nanosheet FETs (NS-FETs) used at advanced technology nodes [22]. The DC and AC characteristics of NSFETs can be reproduced by the optimized ANN model with a fitting error (MSE) of 0.01.

The characterization of traps is useful for improving device reliability, and the CV test, pulse IV, deep-level transient spectroscopy, simulation model, etc., are techniques frequently used to identify the state of traps in semiconductor devices [23–27]. Zou et al. reviewed the characterization techniques used for detecting bulk traps and interface traps in GaN HEMTs [28]. Electrical, optical, and junction capacitance methods have been widely used to probe the traps in GaN HEMTs. However, their low sensitivity, poor spatial resolution, and frequency range limit trap characterization, so further optimizations and innovations in their characterization techniques are needed.

Circuits and systems play a critical role in the explosive development of information technology, which has been regarded as a powerful engine in contemporary human civilization [29–31]. Interested in this topic, Jiang et al. proposed a lumped circuit based on a 3DIC physical structure and calculated the values of all the lumped elements in the circuit model and the transmission line model [32]. The 3DIC jitters, integrating DRAM logic and 3DIC designs into the simulation environment, were analyzed by the proposed CPSIA method, which determined that the timing uncertainty introduced by the 3DIC crosstalk ranged from 31 ps to 62 ps. Chen et al. proposed a novel frequency-domain broadband model (Sensi-Freq-Model) of the conduction susceptibility of integrated circuits, which accurately quantifies the conduction immunity of components in the frequency domain and builds a model of integrated circuits based on their quantized data [33]. The "Sensi-Freq-Model" can reduce the broadband modeling time by about 90% compared to the traditional ICIM-CI method, with a normalized mean square error (NMSE) of 18.5 dB. Li et al. designed a high-efficiency internally matched power amplifier with a 2.5 μm GaN HEMT [34]. An output power of 43.75 dBm, large-signal gain over 15.75 dBm, and PAE of 78.5% at 2.45 GHz were obtained from the proposed power amplifier, which is 13.4×13.5 mm^2 in size. Xiao et al. presented a bandgap reference (BGR) source capable of operating over a wide input range [35]. The high-order curvature compensation method was used and a pre-regulation circuit was incorporated into the BGR. Then, a temperature coefficient (T_C) of 0.88 ppm/°C and stable operation with variations in the power supply voltage were achieved over a temperature range of −40 °C to 130 °C.

In conclusion, this Special Issue showcases the dynamic landscape of semiconductor materials, devices, and systems, highlighting the innovations seen across a broad spectrum of research areas. From the exploration of wide- and ultrawide-bandgap materials to advancements in device optimization, modeling techniques, and characterization methods, these contributions underscore the field's diversity and its pivotal role in technological progress. These studies not only push the boundaries of current semiconductor capabilities but also lay a strong foundation for future breakthroughs in energy efficiency, computational power, and miniaturization, driving us closer to the next era of electronic innovation.

Acknowledgments: This work was not supported by any internal or external funding associated with the authors' affiliations.

Conflicts of Interest: The authors declare no conflict of interest.

References

1. Cao, W.; Bu, H.; Vinet, M.; Cao, M.; Takagi, S.; Hwang, S.; Ghani, T.; Banerjee, K. The Future Transistors. *Nature* **2023**, *620*, 501–515. [CrossRef] [PubMed]
2. Spencer, J.A.; Mock, A.L.; Jacobs, A.G.; Schubert, M.; Zhang, Y.; Tadjer, M.J. A Review of Band Structure and Material Properties of Transparent Conducting and Semiconducting Oxides: Ga_2O_3, Al_2O_3, In_2O_3, ZnO, SnO_2, CdO, NiO, CuO, and Sc_2O_3. *Appl. Phys. Rev.* **2022**, *9*, 011315. [CrossRef]
3. Fiori, G.; Bonaccorso, F.; Iannaccone, G.; Palacios, T.; Neumaier, D.; Seabaugh, A.; Banerjee, S.K.; Colombo, L. Electronics Based on Two-Dimensional Materials. *Nat. Nanotechnol.* **2014**, *9*, 768–779. [CrossRef] [PubMed]
4. Zeng, S.; Tang, Z.; Liu, C.; Zhou, P. Electronics Based on Two-Dimensional Materials: Status and Outlook. *Nano Res.* **2021**, *14*, 1752–1767. [CrossRef]
5. Huang, J.-K.; Wan, Y.; Shi, J.; Zhang, J.; Wang, Z.; Wang, W.; Yang, N.; Liu, Y.; Lin, C.-H.; Guan, X.; et al. High-κ Perovskite Membranes as Insulators for Two-Dimensional Transistors. *Nature* **2022**, *605*, 262–267. [CrossRef]
6. Buffolo, M.; Favero, D.; Marcuzzi, A.; De Santi, C.; Meneghesso, G.; Zanoni, E.; Meneghini, M. Review and Outlook on GaN and SiC Power Devices: Industrial State-of-the-Art, Applications, and Perspectives. *IEEE Trans. Electron Devices* **2024**, *71*, 1344–1355. [CrossRef]
7. Varley, J.B.; Shen, B.; Higashiwaki, M. Wide Bandgap Semiconductor Materials and Devices. *J. Appl. Phys.* **2022**, *131*, 230401. [CrossRef]
8. Tsao, J.Y.; Chowdhury, S.; Hollis, M.A.; Jena, D.; Johnson, N.M.; Jones, K.A.; Kaplar, R.J.; Rajan, S.; Van De Walle, C.G.; Bellotti, E.; et al. Ultrawide-Bandgap Semiconductors: Research Opportunities and Challenges. *Adv. Electron. Mater.* **2018**, *4*, 1600501. [CrossRef]
9. Rafin, S.M.S.H.; Ahmed, R.; Haque, M.A.; Hossain, M.K.; Haque, M.A.; Mohammed, O.A. Power Electronics Revolutionized: A Comprehensive Analysis of Emerging Wide and Ultrawide Bandgap Devices. *Micromachines* **2023**, *14*, 2045. [CrossRef]
10. Liu, C.; Chen, X.; Sun, R.; Lai, J.; Chen, W.; Xin, Y.; Wang, F.; Wang, X.; Li, Z.; Zhang, B. On the Abnormal Reduction and Recovery of Dynamic R_{ON} Under UIS Stress in Schottky p-GaN Gate HEMTs. *IEEE Trans. Power Electron.* **2023**, *38*, 9347–9350. [CrossRef]
11. Chen, K.J.; Häberlen, O.; Lidow, A.; Tsai, C.L.; Ueda, T.; Uemoto, Y.; Wu, Y. GaN-on-Si Power Technology: Devices and Applications. *IEEE Trans. Electron Devices* **2017**, *64*, 779–795. [CrossRef]
12. Chen, X.; He, Z.; Shi, Y.; Wang, Z.; Wang, F.; Sun, R.; Chen, Y.; Chen, Y.; He, L.; Lu, G.; et al. Physical Insight into the Abnormal VTH Instability of Schottky P-GaN HEMTs under High-Frequency Operation. *Appl. Phys. Lett.* **2024**, *124*, 173501. [CrossRef]
13. Wang, X.; Chen, W.; Sun, R.; Liu, C.; Xia, Y.; Xin, Y.; Xu, X.; Wang, F.; Chen, X.; Chen, Y.; et al. Degradation Behavior and Mechanism of GaN HEMTs With P-Type Gate in the Third Quadrant Under Repetitive Surge Current Stress. *IEEE Trans. Electron Devices* **2022**, *69*, 5733–5741. [CrossRef]
14. Wang, Z.; Feng, X.; Yang, D.; Chen, C.; Wang, Z.; Chen, X.; Wang, C.; Wang, S.; Cao, J.; Yao, Y. A Novel High Performance Lateral AlGaN/GaN Schottky Barrier Diode Using Highly Effective Field Plate with Polarization Enhanced Channel. In Proceedings of the 2019 Electron Devices Technology and Manufacturing Conference (EDTM), Singapore, 12–15 March 2019; pp. 377–379.
15. Chen, X.; Wang, F.; Wang, Z.; Huang, J.-K. Exploring the Potential of GaN-Based Power HEMTs with Coherent Channel. *Micromachines* **2023**, *14*, 2041. [CrossRef] [PubMed]
16. Deng, L.; Zhou, L.; Lu, H.; Yang, L.; Yu, Q.; Zhang, M.; Wu, M.; Hou, B.; Ma, X.; Hao, Y. Comprehensive Comparison of MOCVD- and LPCVD-SiNx Surface Passivation for AlGaN/GaN HEMTs for 5G RF Applications. *Micromachines* **2023**, *14*, 2104. [CrossRef]
17. Yang, Y.-T.; Lin, S.-C.; Wang, C.-C.; Ho, Y.-R.; Chen, J.-Z.; Huang, J.-J. Performance Improvement of TiO2 Ultraviolet Photodetectors by Using Atomic Layer Deposited Al2O3 Passivation Layer. *Micromachines* **2024**, *15*, 1402. [CrossRef]
18. Yao, Y.; Wang, Z.; Li, L.; Yang, D.; Wang, S.; Chen, X. Modelling on GaN Power HEMT with Condideration of Subthreshold Swing Using Artificial Intelligence Technology. In Proceedings of the 2019 International Conference on IC Design and Technology (ICICDT), Suzhou, China, 17–19 June 2019; pp. 1–3.
19. Wang, Z.; Li, L.; Leon, R.C.C.; Yang, J.; Shi, J.; van der Laan, T.; Usman, M. Improving Semiconductor Device Modeling for Electronic Design Automation by Machine Learning Techniques. *IEEE Trans. Electron Devices* **2024**, *71*, 263–271. [CrossRef]
20. Wang, Z.; Li, L.; Yao, Y. A Machine Learning-Assisted Model for GaN Ohmic Contacts Regarding the Fabrication Processes. *IEEE Trans. Electron Devices* **2021**, *68*, 2212–2219. [CrossRef]
21. Carrillo-Nuñez, H.; Dimitrova, N.; Asenov, A.; Georgiev, V. Machine Learning Approach for Predicting the Effect of Statistical Variability in Si Junctionless Nanowire Transistors. *IEEE Electron Device Lett.* **2019**, *40*, 1366–1369. [CrossRef]
22. Zhao, Y.; Xu, Z.; Tang, H.; Zhao, Y.; Tang, P.; Ding, R.; Zhu, X.; Zhang, D.W.; Yu, S. Compact Modeling of Advanced Gate-All-Around Nanosheet FETs Using Artificial Neural Network. *Micromachines* **2024**, *15*, 218. [CrossRef]

23. Yang, S.; Huang, S.; Wei, J.; Zheng, Z.; Wang, Y.; He, J.; Chen, K.J. Identification of Trap States in P-GaN Layer of a p-GaN/AlGaN/GaN Power HEMT Structure by Deep-Level Transient Spectroscopy. *IEEE Electron Device Lett.* **2020**, *41*, 685–688. [CrossRef]

24. Wang, F.; Chen, W.; Wang, Y.; Sun, R.; Ding, G.; Yu, P.; Wang, X.; Feng, Q.; Xu, W.; Xu, X.; et al. Analytically Modeling the Effect of Buffer Charge on the 2DEG Density in AlGaN/GaN HEMT. *IEEE Trans. Electron Devices* **2024**, *71*, 1654–1661. [CrossRef]

25. Sun, R.; Chen, X.; Liu, C.; Chen, W.; Zhang, B. Degradation Mechanism of Schottky P-GaN Gate Stack in GaN Power Devices under Neutron Irradiation. *Appl. Phys. Lett.* **2021**, *119*, 133503. [CrossRef]

26. Wang, Z.; Wang, H.; Wang, Y.; Sun, Z.; Zeng, L.; Wang, R.; Huang, R. New Insights into the Random Telegraph Noise (RTN) in FinFETs at Cryogenic Temperature. In Proceedings of the 2024 IEEE International Reliability Physics Symposium (IRPS), Grapevine, TX, USA, 14–18 April 2024; pp. 6A.3-1–6A.3-7.

27. Wang, Z.; Lu, H.; Sun, Z.; Shen, C.; Peng, B.; Li, W.-F.; Xue, Y.; Wang, D.; Ji, Z.; Zhang, L.; et al. New Insights into the Interface Trap Generation during Hot Carrier Degradation: Impacts of Full-Band Electronic Resonance, (100) vs (110), and nMOS vs pMOS. In Proceedings of the 2023 International Electron Devices Meeting (IEDM), San Francisco, CA, USA, 9–13 December 2023; pp. 1–4.

28. Zou, X.; Yang, J.; Qiao, Q.; Zou, X.; Chen, J.; Shi, Y.; Ren, K. Trap Characterization Techniques for GaN-Based HEMTs: A Critical Review. *Micromachines* **2023**, *14*, 2044. [CrossRef] [PubMed]

29. Loeckx, J.; Gielen, G. Assessment of the DPI Standard for Immunity Simulation of Integrated Circuits. In Proceedings of the 2007 IEEE International Symposium on Electromagnetic Compatibility, Honolulu, HI, USA, 8–13 July 2007; pp. 1–5.

30. Masetti, G.; Graffi, S.; Golzio, D.; Kovács-V, Z.M. Failures Induced on Analog Integrated Circuits by Conveyed Electromagnetic Interferences: A Review. *Microelectron. Reliab.* **1996**, *36*, 955–972. [CrossRef]

31. Sebastian, A.; Le Gallo, M.; Khaddam-Aljameh, R.; Eleftheriou, E. Memory Devices and Applications for In-Memory Computing. *Nat. Nanotechnol.* **2020**, *15*, 529–544. [CrossRef]

32. Jiang, X.; Jia, X.; Wang, S.; Guo, Y.; Guo, F.; Long, X.; Geng, L.; Yang, J.; Liu, M. A Cross-Process Signal Integrity Analysis (CPSIA) Method and Design Optimization for Wafer-on-Wafer Stacked DRAM. *Micromachines* **2024**, *15*, 557. [CrossRef]

33. Chen, X.; Xie, S.; Wei, M.; Yang, Y. A Study on the Frequency-Domain Black-Box Modeling Method for the Nonlinear Behavioral Level Conduction Immunity of Integrated Circuits Based on X-Parameter Theory. *Micromachines* **2024**, *15*, 658. [CrossRef]

34. Li, C.; Zhang, Z.; Pei, Y.; Chen, C.; Feng, G.; Xu, Y. Internally Harmonic Matched Compact GaN Power Amplifier with 78.5% PAE for 2.45 GHz Wireless Power Transfer Systems. *Micromachines* **2024**, *15*, 1354. [CrossRef]

35. Xiao, Y.; Wang, C.; Hou, H.; Han, W. A Sub-1 Ppm/°C Reference Voltage Source with a Wide Input Range. *Micromachines* **2024**, *15*, 1273. [CrossRef]

 *micromachines*

Article

Exploring the Potential of GaN-Based Power HEMTs with Coherent Channel [†]

Xinghuan Chen [1], Fangzhou Wang [2,*], Zeheng Wang [3,*] and Jing-Kai Huang [4,5]

1   The China Electronic Product Reliability and Environmental Testing Research Institute, Guangzhou 510610, China
2   Songshan Lake Materials Laboratory, Dongguan 523808, China
3   Manufacturing, CSIRO, Lindfield, NSW 2070, Australia
4   Department of Systems Engineering, City University of Hong Kong, Hong Kong, China
5   School of Materials Science and Engineering, University of New South Wales (UNSW), Sydney, NSW 2052, Australia
*   Correspondence: wangfangzhou@sslab.org.cn (F.W.); zenwang@outlook.com (Z.W.)
†   The paper is an expanded version of the paper presented at the ICSICT, Kunming, China, 3–6 November 2020.

Abstract: The GaN industry always demands further improvement in the power transport capability of GaN-based high-energy mobility transistors (HEMT). This paper presents a novel enhancement-type GaN HEMT with high power transmission capability, which utilizes a coherent channel that can form a three-dimensional electron sea. The proposed device is investigated using the Silvaco simulation tool, which has been calibrated against experimental data. Numerical simulations prove that the proposed device has a very high on-state current above 3 A/mm, while the breakdown voltage (above 800 V) is not significantly affected. The calculated Johnson's and Baliga's figure-of-merits highlight the promise of using such a coherent channel for enhancing the performance of GaN HEMTs in power electronics applications.

Keywords: GaN; HEMT; figure of merit; coherent channel

Citation: Chen, X.;
Wang, F.; Wang, Z.; Huang, J.-K.
Exploring the Potential of GaN-Based
Power HEMTs with Coherent
Channel. *Micromachines* **2023**, *14*,
2041. https://doi.org/10.3390/
mi14112041

Academic Editors: Ha Duong Ngo
and Aiqun Liu

Received: 7 August 2023
Revised: 18 October 2023
Accepted: 25 October 2023
Published: 31 October 2023

1. Introduction

In recent years, wide bandgap semiconductor devices have gained significant attention in power electronics due to their superior performance compared to their silicon-based counterparts [1–4]. Among these, Gallium Nitride (GaN) and its alloys with Indium and Aluminum have been shown to be particularly promising for high-frequency applications, thanks to the high mobility two-dimensional electron gas (2DEG) that forms in their heterojunctions [5–10]. The AlGaN/GaN High Electron Mobility Transistor (HEMT) is one such device that takes advantage of this high mobility 2DEG, with electron mobilities that can exceed 1800 cm^2/V·s [11,12].

However, natural GaN HEMTs are depletion-mode devices due to their intrinsically continuous 2DEG, which can lead to standby leakage and increase the complexity of driver circuits [13–15], therefore, the enhancement-mode devives are desired in power electronic applications. Various device structures, such as fluoride ion implantation, pillar structure, and field coupling gate, have been proposed and explored to address this issue [16–18]. Additionally, extensive studies from our group and others have demonstrated that p-GaN on HEMT is a charming strategy for the realization of the enhancement-type GaN HEMTs based on the process in contemporary foundries [19–23]. Despite these efforts, the low concentration of the thin 2DEG still limits the current transportation, preventing the device from reaching its full potential [24].

To overcome this limitation, we hereby explore the use of a coherent channel for realizing a novel AlGaN/GaN HEMT with an enhancement-type functionality and very high-power transmission capability based on p-GaN-on-HEMT architecture [25]. Our device uses a coherent channel consisting of a GaN cap channel with n-type doping and an AlGaN layer channel with graded Al mole fraction. The graded Al mole fraction broadens the conduction band and creates a three-dimensional electron sea (3DES), which is different from the electron slab induced by buck doping. This new structure allows for a significantly higher current density and a higher breakdown voltage (BV) compared to traditional HEMTs. A numerical analysis of our proposed device shows that it has the potential to boost the performance of GaN-based power applications significantly.

2. Structure and Mechanism

The proposed Coherent Channel High Electron Mobility field-effect Transistor (CC-HEMT) structure with three-dimensional electron sea (3DES) heterojunction can be fabricated on an AlGaN(graded)/GaN wafer, which can be realized by the typical MOCVD process. In this structure, the Al mole fraction of the AlGaN layer (15 nm) linearly increases from 0 at the heterojunction to 0.3 at the AlGaN surface, as shown in Figure 1. A p-type GaN layer is then deposited on top of the heterojunction, followed by a highly n-type-doped thin GaN layer of 10 nm, which serves as the active region of the source to ensure that the Ohmic contact is based on n-GaN for source and drain so that the source and drain Ohmic contact can be formed simultaneously. An etching and passivation process between the p-GaN pillar and drain electrode should be employed to prevent p-type leakage from the source to the drain by RIE/ICP-RIE and PECVD, and the passivation dielectric is Silicon Nitride (SiN). The gate is designed in a trench form and is covered with a 10 nm HfO_2 dielectric layer, which can be conducted through ALD growth to form high quality gate dielectric. The gate metal extends over the source and 3DES regions, creating a continuous current channel perpendicular to the heterojunction. The standard fabrication process can be referred to as shown in Figure 2. The aforementioned processes have been well-developed and have already shown potential for achieving the superior performance of various GaN-based devices [26–31].

The working mechanisms of the proposed device under different biasing conditions can be seen in Figure 3. During off-state, the inversion channel is not formed in the p-GaN layer, so the Ohmic source and 3DES channel are separated by the p-GaN layer, and the device blocks high drain voltage by the reverse-biased p-GaN/3DES junction. When the device turns on, an inversion channel is formed in the p-GaN layer, connecting the 3DES and the Ohmic source, and current will flow though the coherent channel (combined by n-type GaN cap and graded AlGaN) and the inversion channel formed in p-GaN layer. The channel length is defined by the inversion layer, which should not be too short and located at a too heavily doped p-type region. A short channel may not be practically achievable due to fabrication process limitations or induce the short channel effect. On the other hand, a heavily doped p-GaN layer can cause depletion in the 3DES, leading to channel pinch-off. In this design, the gate-to-source length is set at 1 μm. Further device specifications are listed in Table 1. The Silvaco tool is used to simulate the device's performance.

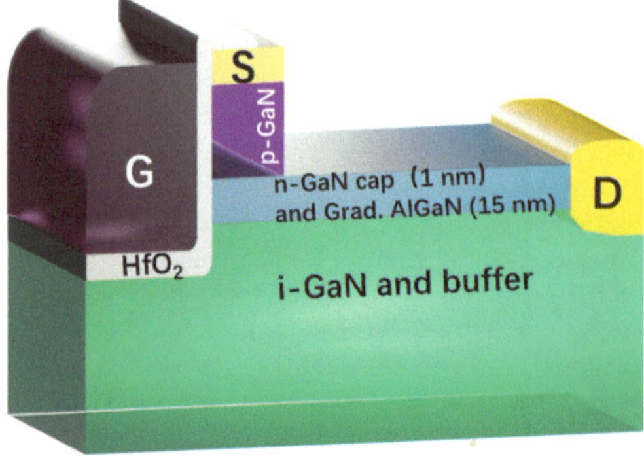

Figure 1. Rendered illustrations of the device structure.

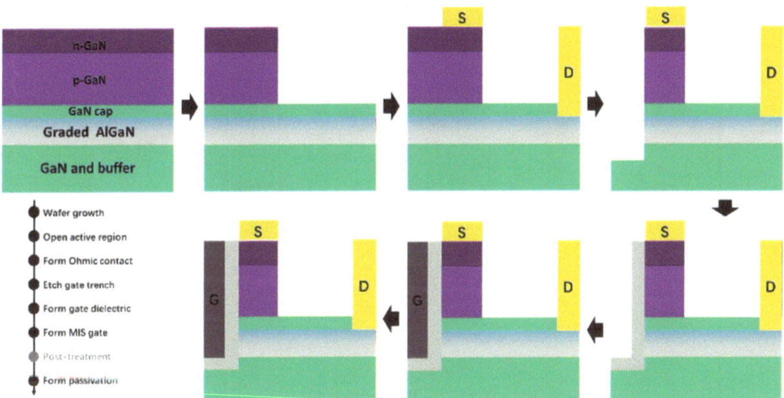

Figure 2. Schematical illustration of the fabrication process of the proposed device.

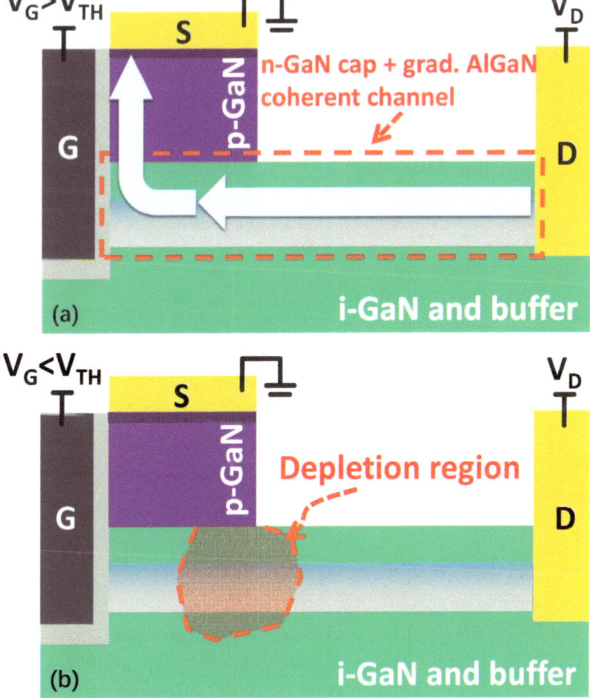

Figure 3. Schematical illustrations of the device's working mechanisms under (**a**) turned-on and (**b**) blocking bias.

Prior to simulation, a calibration of the physical models used in this work is performed. The calibrated device is chosen as a p-GaN gate HEMT due the he similar material composition as the proposed CC-HEMT, such as p-GaN, AlGaN and i-GaN. And the simulated data fits well with the experimental data from the same HEMT device structure with a p-GaN gate [32], as shown in Figure 4a, indicating that the settings of physical models used in this work are reasonable. The detailed settings of the physics models, such as the Shockly–Read–Hall recombination model, Fermi–Dirac static model, electric field dependence model concentration dependent mobility model, and impact ionization model, are the same as our previous publications [16,18,33,34].

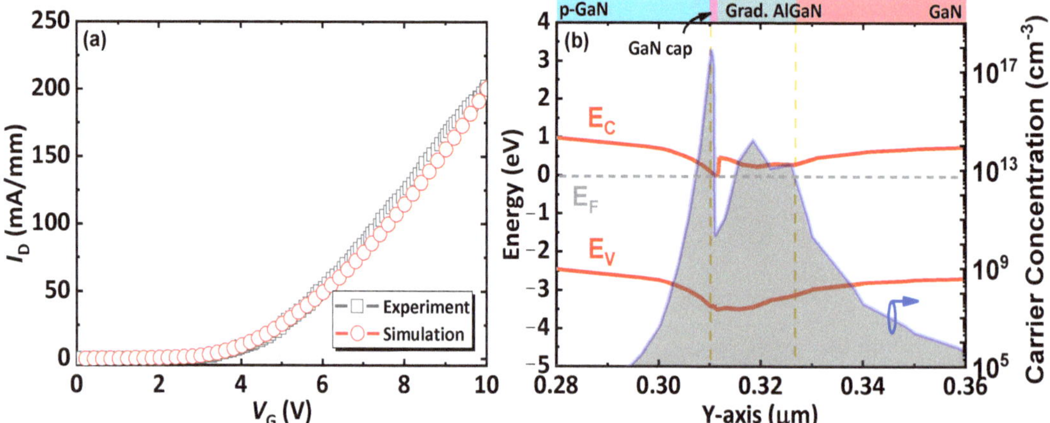

Figure 4. (a) The calibration of the simulation tool; (b) the simulated energy band diagrams and the three-dimensional electron sea distribution at the coherent channel.

Table 1 shows the specifications of the simulated device in this work, including the short name, full name, and the doping concentration and dimensional values.

Table 1. Specifications of the simulated device in this work.

Short Name	Full Name	Value
L_S	Source length	40 nm
H_P	p-GaN height	200 to 400 nm
N_A	p-GaN doping concentration	10^{18} cm^{-3}
/	n-GaN doping concentration	3×10^{20} cm^{-3}
/	GaN cap doping concentration	2×10^{20} cm^{-3}
/	Thickness of n-GaN	10 nm
/	Thickness of GaN cap	1 nm
/	The thickness of graded AlGaN	15 nm

3. Results and Discussion

Figure 4b displays the energy band structure of the proposed CC-HEMT perpendicular to the wafer through the p-GaN pillar. The graded Al fraction lowers the energy band, resulting in the formation of 3DES with high electron density above 10^{13} cm^{-3} in the AlGaN layer, which means the conduction channel is expanded to be a coherent three-dimensional channel instead of the sheet conduction channel of 2DEG in conventional GaN HEMTs. However, since the p-GaN partially depletes the electrons, the Fermi level above the band gap for p-GaN in the p-GaN layer depletion region, and the 3DES region does not span the entire AlGaN layer, as shown in Figure 4b, where the simulated 3DES length is approximately 25 nm. Additionally, the doping and polarization of the thin GaN cap layer contribute to a peak concentration of electrons located near the interface. Although, due to polarization, a valley of the concentration appears at the top of the coherent channel, the lowest concentration of the valley is still higher than 10^{10} cm^{-3}. The formation of 3DES approximately aligns with recent experiments [35] and theoretical calculations [24].

Figure 5a,b represents the transfer performance of the CC-HEMT. A narrow source of 40 nm (L_S) is utilized as an example to improve simulation efficiency. Furthermore, to verify the functionality of the proposed device, the p-GaN doping limit of 10^{18} cm^{-3} is utilized, reflecting current fabrication processes and making for a challenging condition simulation to explore the device performance's boundary. As indicated in the figure, the current transportation capability of the device decreases as the p-GaN height (H_P) increases; this is due to the equivalent increase in the resistance of the channel. Nonetheless, the current flowing through the device is over 3 A/mm in all three samples when the drain voltage is 6 V, which is much higher than the calibrated p-GaN gate HEMT. This high performance is attributable to the high density of 3DES formed in the graded AlGaN layer, which is the key feature of the CC-HEMT. As shown in Figure 5b, threshold voltage (V_{TH}, defined at I_D is 1 mA/mm) is insensitive to H_P, as H_P increases from 200 nm to 400 nm, the varation of V_{TH} is lower than 0.04 V, which means the proposed CC-HEMT has a lager process tolerance. Also, shorter H_P results in higher peak transconductance—This is because of the merit of higher gate controllability through shorter channel length. For the pinch-off region ($V_G < 1V$), H_P does not influence the leakage significantly, mainly because of the stable depletion region of the p-n junction formed by p-GaN and the coherent channel, which does not extend to exceed 200 nm. It should be noted that this paper aims at exploring the use of a coherent channel for realizing a novel AlGaN/GaN HEMT with an enhancement-type functionality with very high-power transmission capability based on p-GaN-on-HEMT architecture and the device performance's boundary, so the comparation between the calibrated device and the proposed device is not performed.

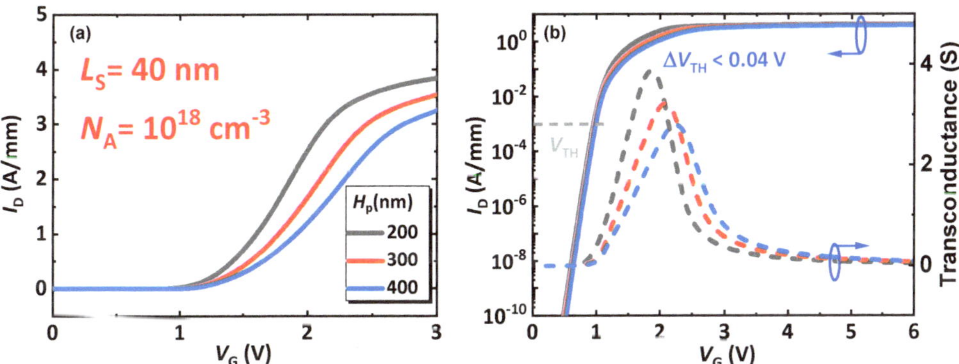

Figure 5. The transfer performance of the device in the (**a**) linear coordinator and (**b**) semi-log coordinator (with transconductance).

However, it should be noted that the maximum Al content and thickness of the AlGaN layer should be traded off while considering the p-GaN doping concentration. High p-type doping may completely deplete the 3DES, which could be prevented by increasing the Al gradation of the AlGaN layer. However, in doing so, the resistance of the layer will also increase, resulting in a reduction of the device's current transportation performance. Consequently, the optimal device configurations should be studied further while considering the aforementioned factors.

Figure 6 gives the output performance of the proposed CC-HEMT with various H_P from 200 nm to 400 nm under V_{GS} of 4 V, where the output curves exhibit good saturation performance, indicating that the channel is resilient to parasitic effects like the short-channel effect. Compared to other recently reported devices, the proposed device boasts a very high current transmission capability [18,22,34,36]. Because of higher series resistance, higher p-GaN thickness turns out to be higher DC R_{ON}, as can be seen in the Figure 6b—the device saturation current drops as a consequence. However, due to the high electron density of 3DES, the lowest saturation current with H_P of 400 nm still stays higher than

5 A/mm, which suggests the proposed device exhibits the desired high potential in power applications.

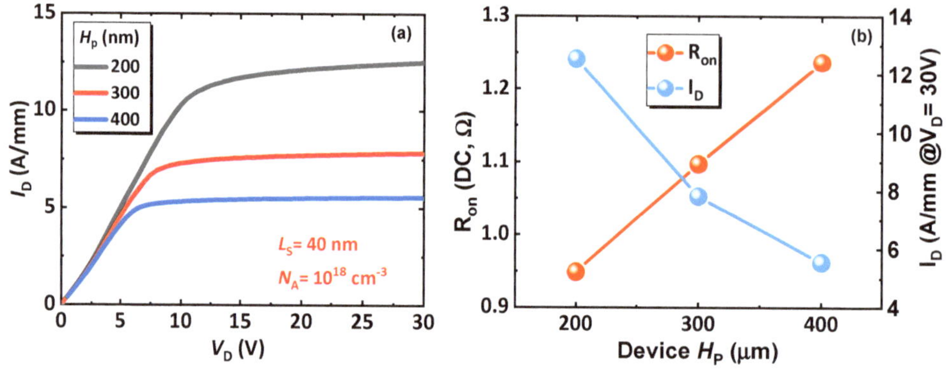

Figure 6. (**a**) Output performance of the proposed device with varying p-GaN heights, and (**b**) the key performance indicators extracted from (**a**).

Figure 7 presents the simulated results of the current density and electric field distribution of the proposed device working in a forwarding or blocking state. It can be seen from Figure 7a that an inversion channel is formed in the p-GaN layer to connect the 3DES and the Ohmic source and the main part of the current flows just through the coherent channel (the coherent channel is highlighted by a zoom out insert) when the device turns on ($V_G = 4V$ and $V_D = 3V$), which indicates that the high current transportation capability is attributable to the high density of 3DES formed in the graded AlGaN layer. Also, the reverse blocking, as in Figure 7b, shows the same behavior as the mechanism of the design—The electric field crowded under the p-GaN pillar ($V_D = 800V$ and other gates are grounded). Meanwhile, the stable depletion region of the p-n junction formed by p-GaN and the coherent channel can reduce the leakage significantly. It is expected that the depletion region of the device can extend towards the drain within the buffer layer, where the material's critical electric field is relatively high so that a higher breakdown voltage can be obtained.

Figure 8 shows Johnson's Figure-Of-Merit (JFOM) of the proposed device with different specifications of the p-GaN pillar. For the JFOM, it is related to the saturation electron velocity of the device V_{sat} and the critical electric field E_C, as below [37,38]:

$$\text{JFOM} = \frac{V_{sat}E_C}{2\pi} \tag{1}$$

The V_{sat} is dependent on the cut-off frequency f_T and the length of the gate L_G (in our case, this should equal the height of the p-GaN, namely the length between the source and the coherent channel):

$$V_{sat} = f_T \cdot 2\pi \cdot L_G \tag{2}$$

We can approximately estimate the E_C by using the breakdown voltage BV and introducing a linear component of fitting, a:

$$E_C = \frac{2BV}{a \cdot L_G} \tag{3}$$

Therefore, if the adjustable parameter is 2, the frequency JFOM can be obtained in a form of:

$$\text{JFOM} = f_T \cdot BV \tag{4}$$

As can be seen in Figure 8, the device features high JFOM values; this can be attributed to the functionality of the coherent channel, where the high-density, large-volume 3DES is formed by the combination of doping (for the n-GaN cap layer) and polarization (for the graded AlGaN layer). Before the best point of doping concentration, higher doping yields higher JFOM; this is because the depletion region between the p-GaN and the coherent channel decreases, and this decreasing trend dominates the JFOM. The concentration exceeding the best point, however, will reduce the mobility of the vertical channel inside the p-GaN, which starts to dominate in lowering the JFOM.

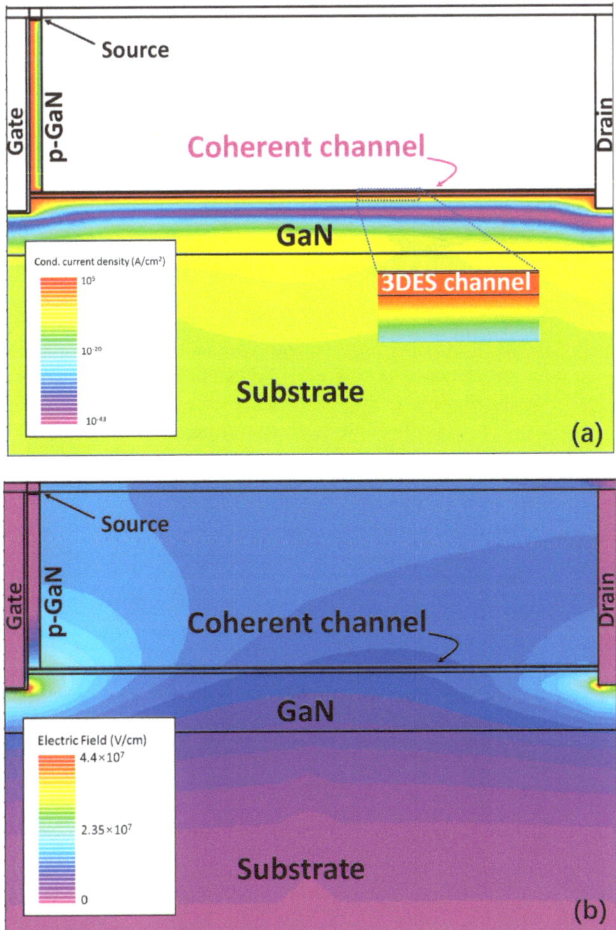

Figure 7. Simulated (**a**) current density in a forward-biased device and (**b**) electric-field distribution in a blocking device. Note that these two figures are from devices with different geometric specifications in order to achieve the best contrast.

Figure 8b can be achieved by fixing the p-GaN doping to 10^{18} cm^{-3}, varying H_P from 200 nm to 400 nm and varing L_S from 20 nm to 40 nm. In this figure, it can be seen that the best height of the p-GaN is around 300 nm for the high JFOM. Lower heights can reduce the BV and, therefore, reduce the JFOM, while higher heights can lower the channel conductivity, as shom in Figure 6b, and result in a lower cut-off frequency. These two factors need to be considered in further studies.

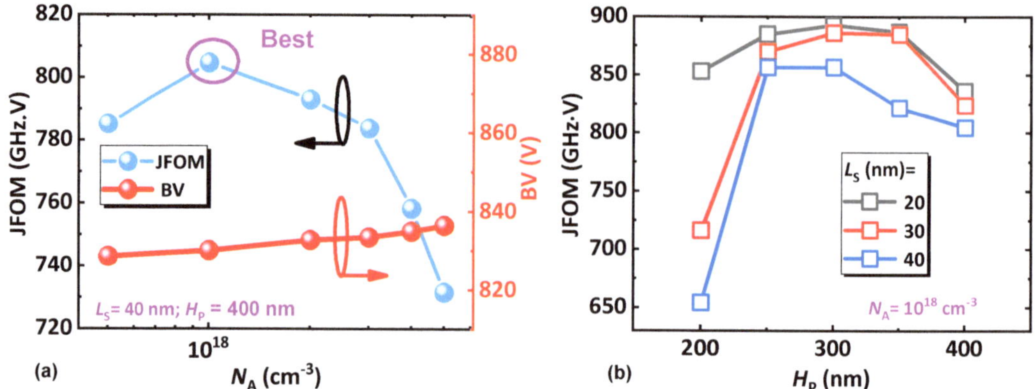

Figure 8. Figure-Of-Merit (FOM) of the proposed device with different (**a**) doping and (**b**) height of p-GaN pillar.

If considering the adjustable factor a in Equation (3), a more comprehensive trend of JFOM vs. H_P can be drawn with various L_S from 20 nm to 40 nm, as in Figure 9. According to the simulation, a THz-level JFOM can be obtained in the best cases. This high performance is the direct consequence of the feat of the coherent channel. In this channel, the polarization layer provides the high-mobility component of the coherent channel, while the doping layer provides extra carriers for current transport. With such a combination, the coherent channel can exhibit high JFOM as well as a high Baliga's FOM (BFOM), as can be seen in Figure 10.

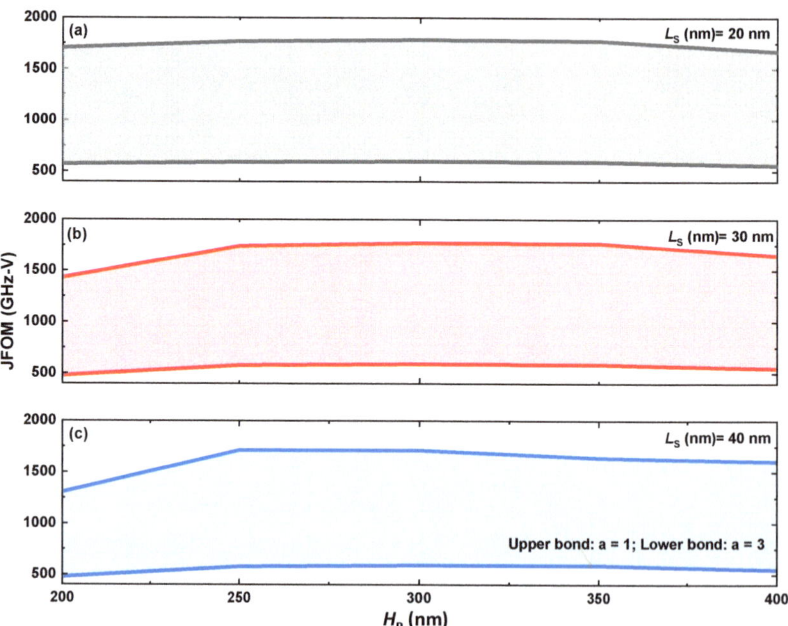

Figure 9. The performance boundary of JFOM when the adjustable parameter varies from 1 to 3. (**a**) L_S of 20 nm. (**b**) L_S of 30 nm. (**c**) L_S of 40 nm.

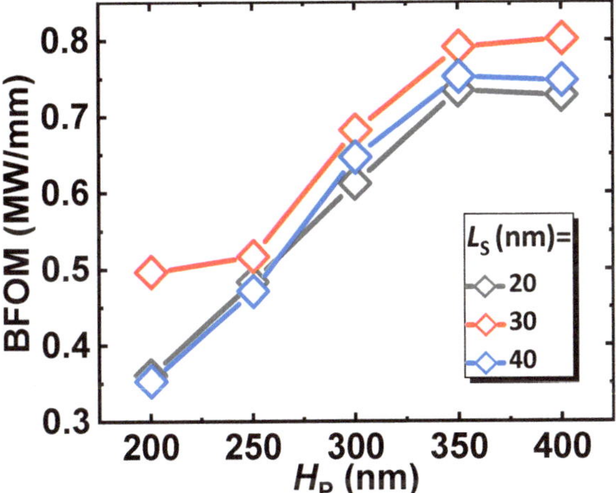

Figure 10. The Baliga's FOM of the proposed device with different specifications.

In particular, the BFOM peaks when the height of the p-GaN reaches around 350 nm; this indicates that the breakdown happens within the p-GaN until it is higher than 350 nm—then the breakdown is the responsibility of the coherent channel. Therefore, we can achieve an even higher BFOM, with high JFOM remaining, by extending the coherent channel length. All these facts suggest that the proposed architecture of CC-HEMT with a coherent channel can be favored in future power applications.

It should also be noted that this research is a proof-of-concept study, and some of the parameters adopted here are ideal. In reality, owing to the limits of the fabrication process, the presence of traps and defects may significantly influence the final performance of the device. Further studies are required to validate the superiority of the proposed device experimentally, which is not the scope of the current study.

4. Conclusions

In conclusion, the proposed CC-HEMT demonstrates outstanding high-power performance, which is attributed to the introduction of the coherent channel by the graded AlGaN layer with the n-GaN cap layer. The graded Al fraction lowers the energy band, resulting in the formation of 3DES with high electron density above 10^{13} cm^{-3} in the AlGaN layer. The device exhibits a remarkable on-state current exceeding 3 A/mm and a high BV of over 800 V, which suggests the proposed device exhibits the desired high potential in power applications. Meanwhile, the proposed CC-HEMT can achieve an even higher BFOM, with high JFOM remaining, by extending the coherent channel length. And the Although further optimization of the device configuration is necessary, the CC-HEMT holds great potential in enhancing the overall performance of future power applications, such as LED power management, wireless power transmission, and charging stations, according to a rigorous numerical analysis presented.

Author Contributions: X.C. completed the simulation in this version with the input of F.W. and Z.W., under the supervision of J.-K.H., Z.W. and J.-K.H. wrote the first draft of the manuscript. X.C. and F.W. revised the manuscript and re-drawn the figures under the supervision of J.-K.H. All authors discussed and analyzed the results. All authors have read and agreed to the published version of the manuscript.

Funding: This research received no external funding.

Data Availability Statement: All data in this study can be accessed upon reasonable request to the corresponding authors.

Acknowledgments: Z.W. acknowledges that there is no project directly funded by CSIRO for this research. This project is a personal collaboration, and no official funding is involved between the affiliations in China and Australia. J.-K.H. thanks the support from the City University of Hong Kong. The authors acknowledge Chengyu Che for his contribution to operating the simulation tool. The author also acknowledges Shengji Wang, Chao Chen, Zirui Wang, and Yuanzhe Yao for their contribution to editing the partial manuscript of the original conference paper.

Conflicts of Interest: The authors declare no conflict of interest.

References

1. Rabkowski, J.; Peftitsis, D.; Nee, H.-P. Silicon carbide power transistors: A new era in power electronics is initiated. *IEEE Ind. Electron. Mag.* **2012**, *6*, 17–26. [CrossRef]
2. Mishra, U.K.; Parikh, P.; Wu, Y.-F. AlGaN/GaN HEMTs-an overview of device operation and applications. *Proc. IEEE* **2002**, *90*, 1022–1031. [CrossRef]
3. Chen, K.J.; Häberlen, O.; Lidow, A.; lin Tsai, C.; Ueda, T.; Uemoto, Y.; Wu, Y. GaN-on-Si power technology: Devices and applications. *IEEE Trans. Electron Devices* **2017**, *64*, 779–795. [CrossRef]
4. Azad, M.T.; Hossain, T.; Sikder, B.; Xie, Q.; Yuan, M.; Yagyu, E.; Teo, K.H.; Palacios, T.; Chowdhury, N. AlGaN/GaN-Based Multimetal Gated High-Electron-Mobility Transistor with Improved Linearity. *IEEE Trans. Electron Devices* **2023**, *70*, 5570–5576. [CrossRef]
5. Chung, J.W.; Hoke, W.E.; Chumbes, E.M.; Palacios, T. AlGaN/GaN HEMT with 300-GHz fmax. *IEEE Electron Device Lett.* **2010**, *31*, 195–197. [CrossRef]
6. Ma, C.-T.; Gu, Z.-H. Review of GaN HEMT Applications in Power Converters over 500 W. *Electronics* **2019**, *8*, 1401. [CrossRef]
7. Hamza, K.H.; Nirmal, D. A review of GaN HEMT broadband power amplifiers. *AEU Int. J. Electron. Commun.* **2020**, *116*, 153040. [CrossRef]
8. Feng, C.; Jiang, Q.; Huang, S.; Wang, X.; Liu, X. Gate-Bias-Accelerated V_{TH} Recovery on Schottky-Type *p*-GaN Gate AlGaN/GaN HEMTs. *IEEE Trans. Electron Devices* **2023**, *70*, 4591–4595. [CrossRef]
9. Uren, M.J.; Karboyan, S.; Chatterjee, I.; Pooth, A.; Moens, P.; Banerjee, A.; Kuball, M. "Leaky Dielectric" Model for the Suppression of Dynamic R_{ON} in Carbon-Doped AlGaN/GaN HEMTs. *IEEE Trans. Electron Devices* **2017**, *64*, 2826–2834. [CrossRef]
10. Wu, N.; Luo, L.; Xing, Z.; Li, S.; Zeng, F.; Cao, B.; Wu, C.; Li, G. Enhanced Performance of Low-Leakage-Current Normally off *p*-GaN Gate HEMTs Using NH$_3$ Plasma Pretreatment. *IEEE Trans. Electron Devices* **2023**, *70*, 4560–4564. [CrossRef]
11. Zhang, W.; Zhang, J.; Xiao, M.; Zhang, L.; Hao, Y. Al$_{0.3}$Ga$_{0.7}$N/GaN (10 nm)/Al$_{0.1}$Ga$_{0.9}$N HEMTs with Low Leakage Current and High Three-Terminal Breakdown Voltage. *IEEE Electron Device Lett.* **2018**, *39*, 1370–1372. [CrossRef]
12. Lee, H.P.; Perozek, J.; Rosario, L.D.; Bayram, C. Investigation of AlGaN/GaN high electron mobility transistor structures on 200-mm silicon (111) substrates employing different buffer layer configurations. *Sci. Rep.* **2016**, *6*, 37588. [CrossRef]
13. Wang, H.; Wei, J.; Xie, R.; Liu, C.; Tang, G.; Chen, K.J. Maximizing the Performance of 650-V p-GaN Gate HEMTs: Dynamic RON Characterization and Circuit Design Considerations. *IEEE Trans. Power Electron.* **2017**, *32*, 5539–5549. [CrossRef]
14. Liu, C.; Chen, X.; Sun, R.; Lai, J.; Chen, W.; Xin, Y.; Wang, F.; Wang, X.; Li, Z.; Zhang, B. On the Abnormal Reduction and Recovery of Dynamic R_{ON} Under UIS Stress in Schottky p-GaN Gate HEMTs. *IEEE Trans. Power Electron.* **2023**, *38*, 9347–9350. [CrossRef]
15. Ye, Y.; Borkar, S.; De, V. A new technique for standby leakage reduction in high-performance circuits. In Proceedings of the 1998 Symposium on VLSI Circuits. Digest of Technical Papers (Cat. No.98CH36215), Honolulu, HI, USA, 11–13 June 1998; pp. 40–41. [CrossRef]
16. Wang, Z.; Wang, Z.; Zhang, Z.; Yang, D.; Yao, Y. On the Baliga's Figure-Of-Merits (BFOM) Enhancement of a Novel GaN Nano-Pillar Vertical Field Effect Transistor (FET) with 2DEG Channel and Patterned Substrate. *Nanoscale Res. Lett.* **2019**, *14*, 128. [CrossRef]
17. Cai, Y.; Zhou, Y.; Chen, K.; Lau, K. High-performance enhancement-mode AlGaN/GaN HEMTs using fluoride-based plasma treatment. *IEEE Electron Device Lett.* **2005**, *26*, 435–437. [CrossRef]
18. Wang, Z. Proposal of a novel recess-free enhancement-mode AlGaN/GaN HEMT with field-assembled structure: A simulation study. *J. Comput. Electron.* **2019**, *18*, 1251–1258. [CrossRef]
19. Wang, F.; Chen, W.; Wang, Z.; Wang, Y.; Lai, J.; Sun, R.; Zhou, Q.; Zhang, B. A low turn-on voltage AlGaN/GaN lateral field-effect rectifier compatible with p-GaN gate HEMT technology. *Semicond. Sci. Technol.* **2021**, *36*, 034004. [CrossRef]
20. Wang, F.; Chen, W.; Sun, R.; Wang, Z.; Zhou, Q.; Zhang, B. An analytical model on the gate control capability in p-GaN Gate AlGaN/GaN high-electron-mobility transistors considering buffer acceptor traps. *J. Phys. D Appl. Phys.* **2021**, *54*, 095107. [CrossRef]
21. Wang, F.; Chen, W.; Li, X.; Sun, R.; Xu, X.; Xin, Y.; Wang, Z.; Shi, Y.; Xia, Y.; Liu, C.; et al. Charge storage impact on input capacitance in p-GaN gate AlGaN/GaN power high-electron-mobility transistors. *J. Phys. D Appl. Phys.* **2020**, *53*, 305106. [CrossRef]
22. Wang, F.; Chen, W.; Xu, X.; Sun, R.; Wang, Z.; Xia, Y.; Xin, Y.; Liu, C.; Zhou, Q.; Zhang, B. Simulation Study of an Ultralow Switching Loss p-GaN Gate HEMT with Dynamic Charge Storage Mechanism. *IEEE Trans. Electron Devices* **2021**, *68*, 175–183. [CrossRef]

23. Efthymiou, L.; Longobardi, G.; Camuso, G.; Chien, T.; Chen, M.; Udrea, F. On the physical operation and optimization of the p-GaN gate in normally-off GaN HEMT devices. *Appl. Phys. Lett.* **2017**, *110*, 123502. [CrossRef]
24. Wang, Z.; Li, L. Two-dimensional polarization doping of GaN heterojunction and its potential for realizing lateral *p–n* junction devices. *Appl. Phys. A* **2022**, *128*, 672. [CrossRef]
25. Wang, Z.; Che, C.; Wang, S.; Chen, C.; Wang, Z.; Yao, Y. A Novel Enhancement-Type GaN HEMT with High Power Transmission Capability Using Extended Quantum Well Channel. In Proceedings of the 2020 IEEE 15th International Conference on Solid-State & Integrated Circuit Technology (ICSICT), Kunming, China, 3–6 November 2020; pp. 1–3. [CrossRef]
26. Anderson, T.J.; Tadjer, M.J.; Hite, J.K.; Greenlee, J.D.; Koehler, A.D.; Hobart, K.D.; Kub, F.J. Effect of Reduced Extended Defect Density in MOCVD Grown AlGaN/GaN HEMTs on Native GaN Substrates. *IEEE Electron Device Lett.* **2016**, *37*, 28–30. [CrossRef]
27. Sokolovskij, R.; Sun, J.; Santagata, F.; Iervolino, E.; Li, S.; Zhang, G.; Sarro, P. Precision Recess of AlGaN/GaN with Controllable Etching Rate Using ICP-RIE Oxidation and Wet Etching. *Procedia Eng.* **2016**, *168*, 1094–1097. [CrossRef]
28. Geng, K.; Chen, D.; Zhou, Q.; Wang, H. AlGaN/GaN MIS-HEMT with PECVD SiN_x, SiON, SiO_2 as Gate Dielectric and Passivation Layer. *Electronics* **2018**, *7*, 416. [CrossRef]
29. Zhang, S.; Liu, X.; Wei, K.; Huang, S.; Chen, X.; Zhang, Y.; Zheng, Y.; Liu, G.; Yuan, T.; Wang, X.; et al. Suppression of Gate Leakage Current in *Ka*-Band AlGaN/GaN HEMT with 5-nm SiN Gate Dielectric Grown by Plasma-Enhanced ALD. *IEEE Trans. Electron Devices* **2021**, *68*, 49–52. [CrossRef]
30. Suria, A.J.; Yalamarthy, A.S.; So, H.; Senesky, D.G. DC characteristics of ALD-grown Al_2O_3/AlGaN/GaN MIS-HEMTs and HEMTs at 600 °C in air. *Semicond. Sci. Technol.* **2016**, *31*, 115017. [CrossRef]
31. Lu, H.; Ma, X.; Hou, B.; Yang, L.; Zhang, M.; Wu, M.; Si, Z.; Zhang, X.; Niu, X.; Hao, Y. Improved RF Power Performance of AlGaN/GaN HEMT Using by Ti/Au/Al/Ni/Au Shallow Trench Etching Ohmic Contact. *IEEE Trans. Electron Devices* **2021**, *68*, 4842–4846. [CrossRef]
32. Hwang, I.; Kim, J.; Choi, H.S.; Choi, H.; Lee, J.; Kim, K.Y.; Park, J.-B.; Lee, J.C.; Ha, J.; Oh, J.; et al. p-GaN Gate HEMTs with Tungsten Gate Metal for High Threshold Voltage and Low Gate Current. *IEEE Electron Device Lett.* **2013**, *34*, 202–204. [CrossRef]
33. Wang, Z.; Yang, D.; Cao, J.; Wang, F.; Yao, Y. A novel technology for turn-on voltage reduction of high-performance lateral heterojunction diode with source-gate shorted anode. *Superlattices Microstruct.* **2019**, *125*, 144–150. [CrossRef]
34. Wang, Z.; Zhang, Z.; Wang, S.; Chen, C.; Wang, Z.; Yao, Y. Design and Optimization on a Novel High-Performance Ultra-Thin Barrier AlGaN/GaN Power HEMT with Local Charge Compensation Trench. *Appl. Sci.* **2019**, *9*, 3054. [CrossRef]
35. Armstrong, A.M.; Klein, B.A.; Baca, A.G.; Allerman, A.A.; Douglas, E.A.; Colon, A.; Abate, V.M.; Fortune, T.R. AlGaN polarization-doped field effect transistor with compositionally graded channel from $Al_{0.6}Ga_{0.4}N$ to AlN. *Appl. Phys. Lett.* **2019**, *114*, 052103. [CrossRef]
36. Hamza, K.H.; Nirmal, D.; Fletcher, A.A.; Arivazhagan, L.; Ajayan, J.; Natarajan, R. Highly scaled graded channel GaN HEMT with peak drain current of 2.48 A/mm. *AEU Int. J. Electron. Commun.* **2021**, *136*, 153774. [CrossRef]
37. Marino, F.A.; Faralli, N.; Ferry, D.K.; Goodnick, S.M.; Saraniti, M.; Varani, L. Figures of merit in high-frequency and high-power GaN HEMTs. In Proceedings of the 16th International Conference on Electron Dynamics in Semiconductors, Optoelectronics and Nanostructures (Edison 16), Montpellier, France, 24–28 August 2009.
38. Fletcher, A.S.A.; Nirmal, D.; Arivazhagan, L.; Ajayan, J.; Varghese, A. Enhancement of Johnson figure of merit in III-V HEMT combined with discrete field plate and AlGaN blocking layer. *Int. J. RF Microw. Comput. Eng.* **2020**, *30*, e22040. [CrossRef]

Article

Performance Improvement of TiO$_2$ Ultraviolet Photodetectors by Using Atomic Layer Deposited Al$_2$O$_3$ Passivation Layer

Yao-Tsung Yang [1], Shih-Chin Lin [2], Ching-Chiun Wang [2], Ying-Rong Ho [3], Jian-Zhi Chen [4] and Jung-Jie Huang [3,*]

[1]   Design and Materials for Medical Equipment and Devices, Da-Yeh University, Changhua 515006, Taiwan; union9900@gmail.com

[2]   Mechanical and Mechatronics System Research Labs, Industrial Technology Research Institute, Hsinchu 310401, Taiwan; shihchin@itri.org.tw (S.-C.L.); juin0306@itri.org.tw (C.-C.W.)

[3]   Department of Electrical Engineering, Da-Yeh University, Changhua 515006, Taiwan; yingrong0702@gmail.com

[4]   Graduate Institute of Photonics, National Changhua University of Education, Changhua 50007, Taiwan; benson900320@yahoo.com.tw

*   Correspondence: jjhuang@mail.dyu.edu.tw; Tel.: +886-4-8511888 (ext. 2227)

Citation: Yang, Y.-T.; Lin, S.-C.; Wang, C.-C.; Ho, Y.-R.; Chen, J.-Z.; Huang, J.-J. Performance Improvement of TiO$_2$ Ultraviolet Photodetectors by Using Atomic Layer Deposited Al$_2$O$_3$ Passivation Layer. *Micromachines* **2024**, *15*, 1402. https://doi.org/10.3390/mi15111402

Academic Editor: Muhammad Ali Butt

Received: 9 October 2024
Revised: 23 October 2024
Accepted: 18 November 2024
Published: 20 November 2024

Abstract: This study employed atomic layer deposition (ALD) to fabricate an Al$_2$O$_3$ passivation layer to optimize the performance of ultraviolet (UV) photodetectors with a TiO$_2$-nanorod-(NR)-containing active layer and a solid–liquid heterojunction (SLHJ). To reduce the processing time and enhance light absorption, a hydrothermal method was used to grow a relatively thick TiO$_2$-NR-containing working electrode. Subsequently, a 5-nm-thick Al$_2$O$_3$ passivation layer was deposited on the TiO$_2$ NRs through ALD, which has excellent step coverage, to reduce the surface defects in the TiO$_2$ NRs and improve the carrier transport efficiency. X-ray photoelectron spectroscopy revealed that the aforementioned layer reduced the defects in the TiO$_2$ NRs. Moreover, high-resolution transmission electron microscopy indicated that following the annealing treatment, Al, Ti, and O atoms diffused across the interface between the Al$_2$O$_3$ passivation layer and TiO$_2$ NRs, resulting in the binding of these atoms to form Al–Ti–O bonds. This process effectively filled the oxygen vacancies in TiO$_2$. Examination of the photodetector device revealed that the photocurrent-to-dark current ratio exhibited a difference of four orders of magnitude (10^{-4} to 10^{-8} A), with the switch-on and switch-off times being 0.46 and 3.84 s, respectively. These results indicate that the Al$_2$O$_3$ passivation layer deposited through ALD can enhance the photodetection performance of SLHJ UV photodetectors with a TiO$_2$ active layer.

Keywords: TiO$_2$ nanorod (NR); Al$_2$O$_3$; ultraviolet (UV) photodetector; atomic layer deposition (ALD)

1. Introduction

Ultraviolet (UV) photodetectors are widely used in various fields, including industry, healthcare, and defense. The materials and structure of the active layer in UV photodetectors considerably influence the sensitivity and stability of these devices. Most currently available UV photodetectors are based on inorganic compounds, and multiple studies have developed various nanostructures to enhance the performance of UV photodetectors. Commonly used two-dimensional (2D) thin-film materials in UV photodetectors include AlN, GaN, Ga$_2$O$_3$, ZnO, and TiO$_2$ [1–6]. In addition, scholars have explored the use of one-dimensional (1D) nanostructures, such as nanowires, nanorods (NRs), and nanotubes, to enhance photodetector performance [7–9]. Given the superior light-harvesting capability and shorter electron transport pathways of 1D structures compared with those of 2D structures, the present study selected TiO$_2$ NRs as the active layer to fabricate UV photodetectors.

To achieve effective light absorption, the active layer of the UV photodetector should comprise a relatively thick film. Common vacuum deposition techniques, such as sputtering [10,11], atomic layer deposition [12,13], and chemical vapor deposition [14,15], provide stable film quality with easily controllable parameters; however, the film deposition rate in these methods is typically low. Alternatively, numerous studies have utilized nonvacuum hydrothermal methods [16,17] to synthesize TiO_2 NRs; these methods are cost-effective and enable rapid film growth. Nevertheless, relevant studies have reported that TiO_2 NRs synthesized through hydrothermal methods often exhibit oxygen vacancy defects [18], which can degrade the performance of UV photodetectors. Research has indicated that TiO_2 NRs can be coated with a thin passivation layer to effectively reduce their oxygen vacancies. For instance, one study used the liquid-phase deposition method to deposit Al_2O_3 passivation layers on TiO_2 NRs in UV photodetectors [18]. This method enabled excellent step coverage to be achieved on 1D TiO_2 NRs and effectively passivated their defects, thereby reducing the switch-off time from 26.5 to 16.5 s during device switching. Moreover, electrochemical deposition has been employed to fabricate TiO_2 NRs/Au/PTTh UV photodetectors [19], which exhibit adequate step coverage and a high detectivity of 1.6×10^{10} Jones. The performance comparison of the UV photodetector studies is shown in Table 1. However, because hydrothermal methods are nonvacuum processes, they result in the generation of a higher number of defects in the passivation layer than vacuum processes. In addition, the sputtering method has been used to fabricate AlN passivation layers [20] for UV photodetectors with TiO_2 NRs. This method offers the advantages of few defects and easy film thickness control, and it reduces the switch-off time of the photodetector to 4.32 s. However, the sputtering process exhibits poor step coverage when the TiO_2 NRs have high aspect ratios. Thus, a more suitable method is required for coating passivation layers on TiO_2 NRs.

Table 1. The performance comparison of UV photodetectors studies.

Active Layer	λ_{cut} (nm)	Sensitivity	Detectivity (Jones)	Ref.
Al_2O_3/TiO_2 NR	365	4380	1.73×10^{10}	this work
TiO_2 NRs/Au/PTTh	365	-	1.67×10^{10}	[19]
AlN/TiO_2 nanorod	350	1360	2.87×10^{9}	[20]
ZnO/TiO_2	365	388	1.10×10^{10}	[21]
AgNW/NiO TiO_2/FTO	365	-	1.60×10^{10}	[22]

A passivation layer should have a low thickness, few defects, the ability to repair film surface defects, and excellent step coverage. Therefore, the present study employed ALD to deposit a high-quality Al_2O_3 passivation layer on the TiO_2-NR-containing active layer of UV photodetectors with a solid–liquid heterojunction (SLHJ). Figure 1 illustrates the energy-level diagram of the SLHJ UV photodetectors in this work. Al_2O_3 can be used to provide surface passivation to reduce the number of oxygen vacancy defects on the TiO_2 NRs surface and can prevent reverse leakage current in SLHJ UV photodetectors, thereby enhancing the performance of photodetectors. Because a sufficient supply of oxygen atoms is maintained in the ALD process, this process can decrease the number of oxygen vacancy defects that are inherently present in the TiO_2 NRs, thereby enhancing the quality of the working electrode of the UV photodetector. This study explored the atomic diffusion between an Al_2O_3 passivation layer and a TiO_2-NR-containing working electrode after thermal treatment. The performance of the SLHJ UV photodetectors with Al_2O_3/TiO_2 NR structures was also examined.

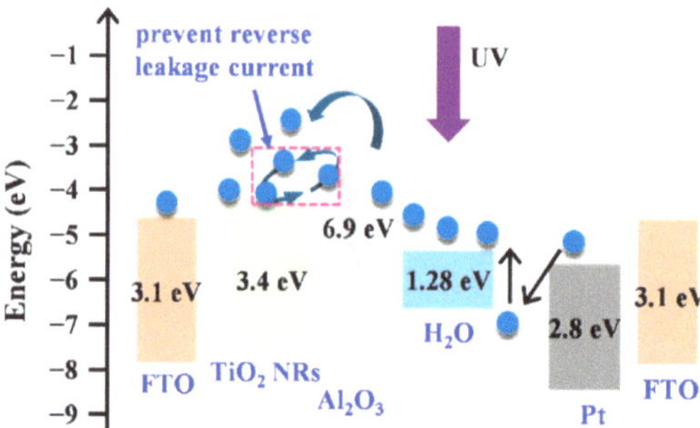

Figure 1. The energy-level diagram of Al_2O_3/TiO_2 NR UV photodetectors.

2. Experiment

The process used in this study to fabricate the Al_2O_3/TiO_2 NR working electrode is illustrated in Figure 2a. Initially, TiO_2 NRs were synthesized using a hydrothermal method. A solution of 6 M HCl and 2 M titanium isopropoxide (TTIP) was prepared. This solution was then combined with deionized water in a Teflon-lined autoclave, in which a fluorine-doped tin oxide (FTO) glass substrate (area: 2 cm × 2 cm, resistivity: 15.5 Ω/m^2) was placed. The autoclave was heated in a circulating oven at 160 °C for 3 h to obtain a working electrode. Next, the 1-μm TiO_2 NRs/FTO was removed, rinsed with deionized water, and dried using nitrogen gas. Subsequently, ALD was employed to deposit an Al_2O_3 passivation layer on the TiO_2 NRs, with trimethylaluminum (TMA) and ultrapure water used as the precursor and oxidizing agent, respectively. During the deposition process, the reaction temperature was set to 90 °C, the process pressure was maintained at 6×10^{-2} Torr, and Ar gas at a flow rate of 100 sccm was used as the purge gas. Each ALD cycle involved the following steps in sequence: TMA injection for 1.5 s, Ar gas purging for 30 s, H_2O injection for 1 s, and Ar purging for 60 s. The ALD growth mechanism is shown in Figure 2b. A total of 50 ALD cycles were performed to grow a 5-nm-thick Al_2O_3 passivation layer on the TiO_2 NRs, which resulted in the formation of an Al_2O_3/TiO_2 NRs/FTO working electrode. Next, the Al_2O_3 passivation layer was subjected to annealing treatments at 300 °C, 400 °C, 500 °C, and 600 °C to enhance its properties, with the heating rate set at 30 °C/min and the temperature hold time set at 1 h. Finally, a Pt/FTO counter electrode was prepared, and a 60-μm-thick spacer (SX1170-60, Solaronix, Aubonne, Switzerland) was placed between the Al_2O_3/TiO_2 NRs/FTO and Pt/FTO electrodes. The spacer was melted by applying pressure on a 100 °C hot plate to bind the working electrode (photoelectrode) and the counter electrode substrate. Deionized water was then injected to create a UV photodetector with SLHJ.

The experimental characterizations conducted in this study were as follows. First, the surface morphologies of the Al_2O_3 and TiO_2 NR films were analyzed through field-emission scanning electron microscopy (JEOL JSM-7000F, Tokyo, Japan) under an accelerating voltage of 15 kV. Second, the microstructures of the TiO_2 NRs coated with Al_2O_3 were examined through high-resolution transmission electron microscopy (HR-TEM, model: JEM-2100 F, Japan). Third, the chemical compositions and bonding of the TiO_2 active layer and Al_2O_3 passivation layer were investigated through X-ray photoelectron spectroscopy (XPS; PHI 5000 VersaProbe, Japan) with Al Kα radiation (photon energy of 1486.6 eV). The energy resolution of the adopted XPS instrument was 0.5 eV full width at half maximum. XPS measurements were conducted at a base pressure of 7.4×10^{-7} Pa in an analyzer chamber. A 2-kV argon ion beam with a current density of 100 A/cm^2 was used to acquire the depth profiles, and the binding energy of each element was calibrated to that of the C1s (284.5 eV) peak. Finally, the current (I)–voltage (V) characteristics of the created photodetector under UV illumination (illumination intensity and wavelength of 15 W and 365 nm, respectively) and dark conditions were explored using the HP 4145B Semiconductor Parameter Analyzer.

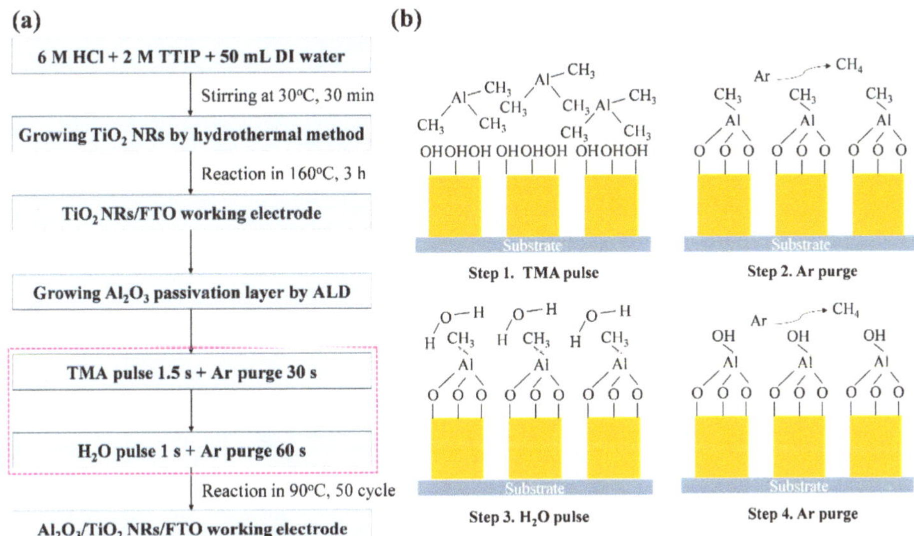

Figure 2. (**a**) Experimental process used to fabricate a working electrode composed of Al_2O_3/TiO_2 NRs and (**b**) growth mechanism of Al_2O_3 passivation layer.

3. Results and Discussion

A 2D thin film of TiO_2 NRs was generated through the reaction of TTIP with water [Reaction (1)]. Subsequently, this layer was subjected to etching with HCl. In this process, the Cl^- ions in HCl molecules reacted with Ti atoms to form $TiCl_3$ [Reaction (2)]. Previous research [18] has indicated that TiO_2 NRs with a preferred (002) orientation can be obtained using 6 M HCl [Figure 3a]. A cross-sectional FE-SEM image of the deposited Al_2O_3 passivation layer is displayed in Figure 3b. The thickness of this layer was approximately 5 nm, and this thickness value was confirmed through HR-TEM [Figure 4c]. Figure 3c shows the EDS elemental analysis results for the ALD-Al_2O_3 film. Mapping analysis results show that the Al_2O_3 film grows uniformly, and the element ratio of Al to O is 40% to 60%. The ALD process involves chemical grafting to grow an Al_2O_3 thin film (Figure 2). A single ALD cycle involves the following steps: First, water vapor was used to attach the OH^- functional groups to the surface of the TiO_2 NRs. Subsequently, TMA gas was introduced to grow the Al_2O_3 thin film, after which CH_4 was removed using Ar gas. In this study, 50 ALD cycles were conducted to grow an Al_2O_3 passivation layer to ensure that this layer was uniformly coated on the TiO_2 NRs, thereby enhancing its passivation effect on the surface defects of these NRs.

$$Ti\{OCH(CH_3)_2\}_4 + 2H_2O \rightarrow TiO_2 + 4(CH_3)_2CHOH \tag{1}$$

$$2Ti + 6HCl \rightarrow 2TiCl_3 + 3H_2\uparrow \tag{2}$$

To explore the effect of the annealing treatment on the microstructure of the Al_2O_3/TiO_2 NRs, HR-TEM was conducted. Figure 4a,d display the cross-sectional HR-TEM image and SAED pattern of the Al_2O_3/TiO_2 NRs produced without annealing treatment. This image reveals that the lattice spacing of the TiO_2 (110) plane was 3.31 Å [Figure 4b]. Moreover, the lattice spacings of the Al_2O_3 (003) and TiO_2 (101) planes were 4.91 Å and 3.47 Å, respectively [Figure 4c]. The analysis results revealed that in the unannealed Al_2O_3/TiO_2 NR structure, the Al_2O_3 thin film was predominantly concentrated on the outer edges of the TiO_2 NRs, and the Al_2O_3 lattice did not form [Figure 4b]. Figure 4e,h display a cross-sectional HR-TEM image and SAED of the Al_2O_3/TiO_2 NR structure after this structure was annealed at 500 °C. The analysis results revealed that after annealing at 500 °C, the lattice spacing of the TiO$_2$ (110) plane was 3.24 Å [Figure 4f]. In the middle section of the Al$_2$O$_3$/NR structure [Figure 4g], the lattice spacings of the TiO$_2$ (110), Al$_2$O$_3$ (003), Al–Ti–O (111), and Al–Ti–O (311) planes were 3.26, 4.86, 3.89, and 2.08, respectively. These results indicate that after annealing at 500 °C, Al$_2$O$_3$ diffused into the TiO$_2$ NRs. As shown in Figure 4g, the thickness of the interface between TiO$_2$ and Al$_2$O$_3$ increased to 10 nm after annealing at 500 °C, which confirmed that Al, Ti, and O atoms diffused across the interface between TiO$_2$ and Al$_2$O$_3$ to form Al–Ti–O bonds. This diffusion process resulted in the effective filling of the oxygen vacancies in TiO$_2$. Thus, annealing can increase the thickness of the Al$_2$O$_3$/TiO$_2$ interface, which reduces the oxygen vacancy defects in the TiO$_2$ NRs, thereby enhancing the photodetector performance.

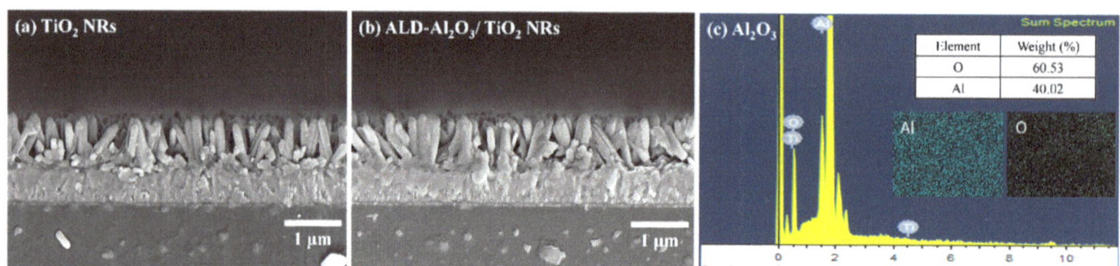

Figure 3. Field-emission scanning electron microscopy images of the surface morphologies of (a) TiO$_2$ NRs, (b) Al$_2$O$_3$/TiO$_2$ NRs, and (c) EDS analysis of ALD-Al$_2$O$_3$.

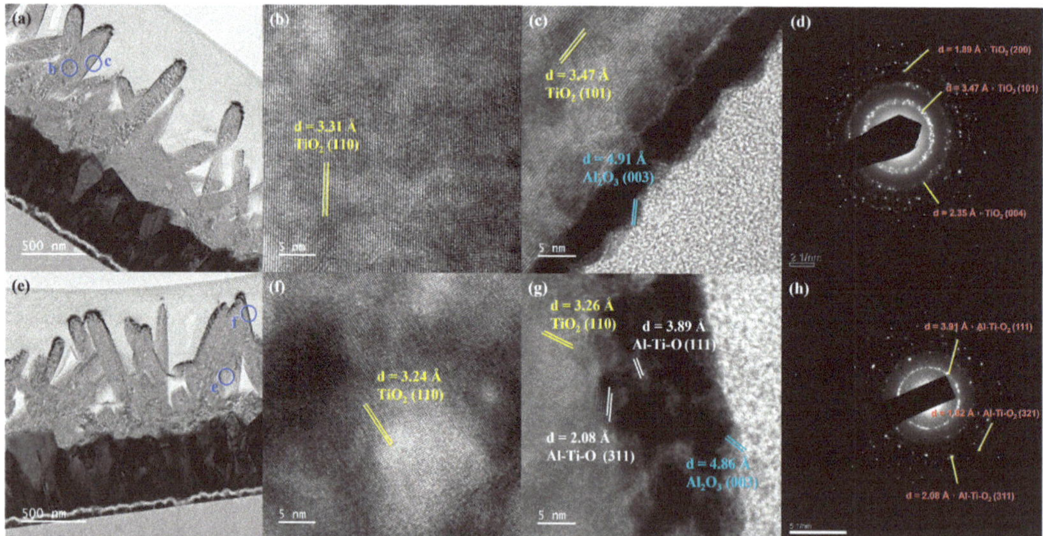

Figure 4. Cross-sectional high-resolution transmission electron microscopy images of (a) Al$_2$O$_3$/TiO$_2$ NRs, (b) the middle portion of Al$_2$O$_3$/TiO$_2$ NRs, (c) the top portion, (d) SAED of Al$_2$O$_3$/TiO$_2$ NRs annealed at 500 °C, (e) Al$_2$O$_3$/TiO$_2$ NRs, (f) the middle portion of Al$_2$O$_3$/TiO$_2$ NRs, and (g) the top portion, (h) SAED of Al$_2$O$_3$/TiO$_2$ NRs.

Figure 5a,b show the Ti $2p$ XPS spectra of the TiO$_2$ NRs before and after they were passivated by the Al$_2$O$_3$ film. The background noise in these spectra was eliminated using the Shirley method in the Spectral Data Processor v4.5 software program, and the spectra were deconvoluted. The binding energy (E_b) was calculated as follows:

$$E_b = hv - E_z - w \tag{3}$$

where E_z is the kinetic energy of the emitted electron, w is the work function, which represents the energy required to remove an electron from the surface of a solid, and hv is the incident photon energy. The binding energy calculations enabled the determination of the proportions of different oxidation states of Ti in TiO_2, namely Ti^{4+}, Ti^{3+}, and Ti^{2+}. Ti^{3+} and Ti^{2+} represent oxygen vacancy states, and Ti^{4+} denotes a stable state. The Ti^{4+} $2p3/2$ peak was located at 459 ± 0.2 eV. As depicted in Figure 5a,b, after the TiO_2 NRs were coated with the Al_2O_3 passivation film, the area proportion of the Ti^{4+} peak in the Ti $2p$ XPS spectrum increased from 52.7% to 64.6%, whereas the combined area proportion of the Ti^{3+} and Ti^{2+} peaks decreased from 47.3% to 35.4%. These results indicated that the oxygen vacancy defects in the TiO_2 NRs were reduced after they were coated with the Al_2O_3 layer. This improvement was attributed to the chemical passivation effect of the Al_2O_3 film, which reduced the surface defect density of the TiO_2 NRs and the number of dangling bonds on their surfaces, thereby enhancing their intrinsic properties. Figure 5c,d show the Al $2p$ and O $1s$ XPS spectra of the Al_2O_3 layer. From Figure 5c, it can be observed that the area percentages of Al-O, Al-OH, and Al-Al were 95.6%, 3.1%, and 1.3%, respectively. This result shows that the bond between Al and oxygen is complete, and the Al_2O_3 film has few defects. However, Figure 5d again proves that the bond between oxygen and aluminum is complete, and the Al-O and Al-OH areas are 95.4% and 4.6%, respectively. In this way, a passivation layer with fewer defects can further improve the characteristics of TiO_2 NRs UV photodetectors.

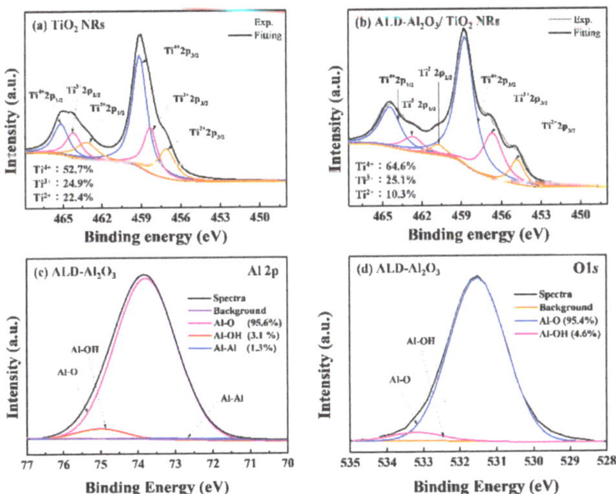

Figure 5. Ti $2p$ spectra of the (**a**) TiO_2 NRs, (**b**) Al_2O_3/TiO_2 NRs annealed at 500 °C and (**c**) Al $2p$, (**d**) O $1s$ of ALD-Al_2O_3 film.

Figure 6 displays the I–V curves measured under UV illumination and dark conditions for SLHJ UV photodetectors with Al_2O_3/TiO_2 NR structures subjected to annealing at different temperatures. The measurements were taken under both ultraviolet illumination and dark conditions. The results revealed that the photodetector containing an Al_2O_3/TiO_2 NR structure annealed at 500 °C exhibited a photocurrent and dark current of 1.22×10^{-5} and 8.96×10^{-9} A, respectively. These results confirm that high photodetection performance can be achieved when the photodetector contains an Al_2O_3 passivation layer that has been deposited through ALD and annealed at an optimal temperature. To evaluate the effect of an Al_2O_3 passivation layer on photocurrent characteristics, the sensitivity, light-dependent resistance (LDR), and detectivity of photodetectors with and without an Al_2O_3 thin-film coating were compared in Table 2. The formulas for calculating these parameters are as follows:

$$Sensitivity = \frac{I_{photo}}{I_{dark}} \tag{4}$$

$$LDR = 20 \ \log \left(\frac{I_{photo}}{I_{dark}} \right) \tag{5}$$

$$Detectivity = \frac{R}{\sqrt{2 \ q \ I_{dark}}} \tag{6}$$

where I_{photo} is the photocurrent, I_{dark} is the dark current, q is the elementary charge of an electron, and R is the responsivity. The calculation results indicated that the sensitivity, LDR, and detectivity of the photodetector with an Al_2O_3/TiO_2 NR structure exhibited a substantial improvement in performance than those of a photodetector with TiO_2 NRs only, with sensitivity increasing from 8 to 4380, LDR increasing from 17.7 dB to 72.8 dB, and detectivity increasing from 3.19×10^7 to 1.73×10^{10} Jones. The chemical passivation effect of the Al_2O_3 film considerably enhanced the sensitivity of the TiO_2 NRs. Therefore, the resulting UV photodetector exhibited an improved ability to detect UV light and a faster response. In general, high sensitivity correlates with high detectivity, which is a quality factor that reflects photodetectors' ability to avoid interference from other light sources during UV detection. Moreover, the LDR is inversely proportional to the light intensity; thus, it can be used to assess the stability of photodetectors under dark conditions. A high LDR value under dark conditions indicates high device stability. Overall, the Al_2O_3 passivation layer improved the stability of the SLHJ UV photodetectors.

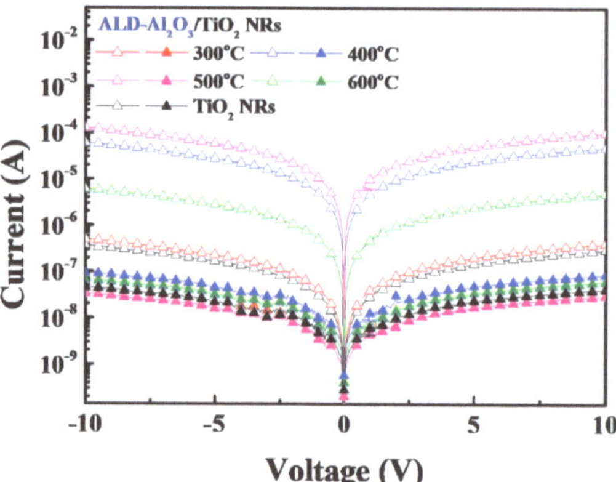

Figure 6. Photocurrent and dark current characteristics of SLHJ UV photodetectors with Al_2O_3/TiO_2 NR structures subjected to annealing under different temperatures.

Figure 7 depicts the on–off switching characteristics of UV photodetectors with and without an Al_2O_3 passivation layer (i.e., with an Al_2O_3/TiO_2 NR structure and a TiO_2 NR structure, respectively). During the measurement process, no external bias was applied (bias = 0), and the switching time and total measurement duration were set to 50 and 275 s, respectively. As displayed in Figure 7a, the Al_2O_3 passivation layer caused the photocurrent density to increase from 1.7 to 7.8 $\mu A/cm^2$. Figure 7b,c displays magnified views of the switch-on and switch-off times of the photodetectors with a TiO_2 NR structure and Al_2O_3/TiO_2 NR structures, respectively. The switch-on time was defined as the time when the current density exceeded 90%, and the switch-off time was defined as the time when the current density dropped below 10%. The switch-on times of the photodetectors with a TiO_2

NR structure and Al_2O_3/TiO_2 NR structure were 0.52 and 0.46 s, respectively. Furthermore, their switch-off times were 7.64 and 3.84 s, respectively. These results indicate that the Al_2O_3 thin film considerably improves the switching characteristics of the photodetector. This improvement was attributed to the chemical passivation effect of this film, which effectively reduced the number of oxygen vacancy defects in the TiO_2 NRs, thereby enhancing the switching characteristics of the UV photodetector.

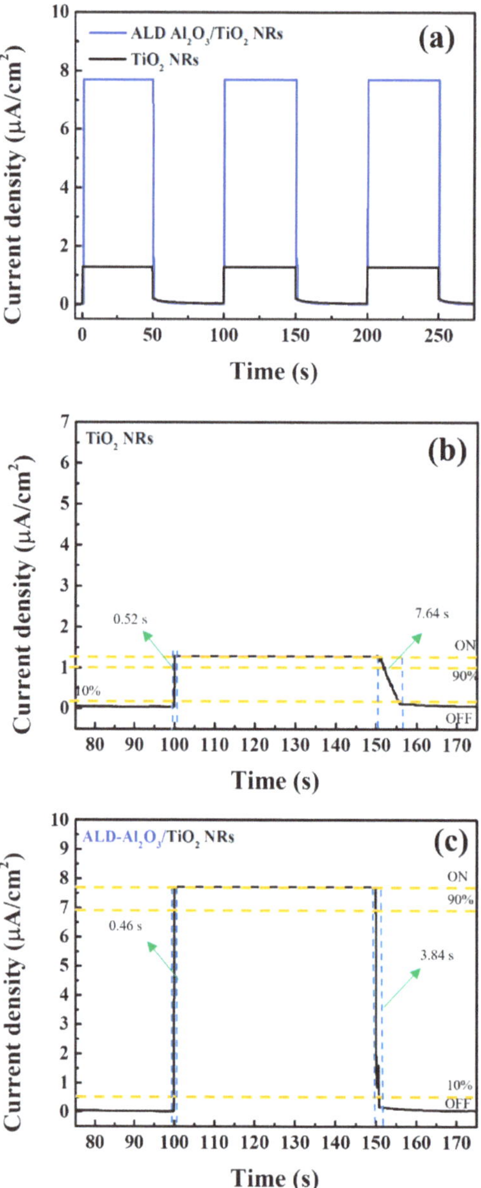

Figure 7. (**a**) Switching characteristics of SLHJ UV photodetectors with a TiO$_2$ NR structure and Al$_2$O$_3$/TiO$_2$ NR structure, (**b**) magnified view of the switch-on and switch-off times of the UV photodetector with the TiO$_2$ NR structure, and (**c**) magnified view of the switch-on and switch-off times of the UV photodetector with the Al$_2$O$_3$/TiO$_2$ NR structure.

Table 2. Photocurrent, dark current, sensitivity, light-dependent resistance (LDR), and detectivity of ultraviolet (UV) photodetectors with solid–liquid heterojunction (SLHJ) and Al_2O_3/TiO_2 nanorod (NR) structures annealed at various temperatures.

Annealing Temperature	I_{photo} (A)	I_{dark} (A)	Sensitivity	LDR (dB)	Detectivity (Jones)
TiO_2 NRs	3.66×10^{-7}	4.38×10^{-8}	8	17.7	3.19×10^7
300 °C	4.81×10^{-7}	4.65×10^{-8}	10	20.3	4.49×10^7
400 °C	6.12×10^{-5}	8.96×10^{-8}	683	56.7	4.55×10^9
500 °C	1.37×10^{-4}	3.13×10^{-8}	4380	72.8	1.73×10^{10}
600 °C	5.81×10^{-6}	7.19×10^{-8}	81	38.1	4.77×10^8

4. Conclusions

This study improved the photodetection performance of UV photodetectors with SLHJ- and TiO_2-NR-containing active layers by depositing an Al_2O_3 passivation layer on this active layer through ALD. The results of HR-TEM results confirmed that after the Al_2O_3/TiO_2 NR structure was annealed, an Al–Ti–O interface was formed between TiO_2 and Al_2O_3. This interface was formed because Al diffused into the TiO_2 NRs during thermal annealing, thereby passivating their surface defects. XPS analyses indicated that the Al_2O_3 passivation layer reduced the Ti^{3+} and Ti^{2+} defects in the TiO_2 NRs, with the combined area proportion of these two defects in the Ti $2p$ XPS spectrum decreasing from 47.3% to 35.4%. This reduction in oxygen vacancy defects was attributed to the filling effect provided by the ALD-Al_2O_3 process. Accordingly, the sensitivity of the ALD-Al_2O_3/TiO_2 NR SLHJ ultraviolet photodetector increased from 8 to 4380, and its detectivity improved from 3.19×10^7 to 1.73×10^{10} Jones. The photocurrent-to-dark-current ratio exhibited a difference of four orders of magnitude. Additionally, the turn-on and turn-off times of the photodetector decreased by 0.06 and 3.8 s, respectively. The aforementioned results indicate that the ALD-Al_2O_3 passivation layer is highly effective and has the potential for future applications in enhancing the performance of photodetector devices.

Author Contributions: Conceptualization, J.-J.H.; methodology, J.-J.H. and Y.-R.H.; validation, S.-C.L., C.-C.W. and Y.-R.H.; formal analysis, Y.-T.Y. and Y.-R.H.; investigation, Y.-T.Y., Y.-R.H. and J.-Z.C.; data curation, J.-Z.C.; writing—original draft preparation, J.-J.H. and Y.-R.H.; writing—review and editing, J.-J.H.; project administration, J.-J.H. All authors have read and agreed to the published version of the manuscript.

Funding: The authors thank the National Science and Technology Council of Taiwan for financially supporting this research under Contract No. 112-2221-E-212-003.

Data Availability Statement: The raw and processed data supporting the findings are available upon reasonable request.

Conflicts of Interest: The authors declare that they have no competing interest.

References

1. Chatterjee, A.; Agnihotri, V.; Porwal, S.; Khan, S.; Baraik, K.; Ganguli, T.; Bose, A.; Raghavendra, S.; Dixit, V.K.; Sharma, T.K. Impact of oxygen plasma power on the performance of Ga_2O_3 passivated GaN ultraviolet photodetectors. *Appl. Surf. Sci.* **2024**, *672*, 160861. [CrossRef]
2. Mohammad, F.K.; Ramizy, A.; Ahmed, N.M.; Yam, F.K.; Hassan, Z.; Beh, K.P. Fabrication and photoresponsive characteristics of ultraviolet GaN p-i-n photodetector based AlN:Al_2O_3 passive layer. *Opt. Mater.* **2024**, *149*, 1115055. [CrossRef]
3. Razeen, A.S.; Kotekar-Patil, D.; Tang, X.S.; Yuan, G.; Ong, J.; Radhakrishnan, K.; Tripathy, S. Enhanced near-UV responsivity of AlGaN/GaN HEMT based photodetectors by nanohole etching of barrier surface. *Mat. Sci. Semicon. Proc.* **2024**, *173*, 108115. [CrossRef]
4. Rajamanickam, S.; Mohammad, S.M.; Razak, I.A.; Muhammad, A.; Abed, S.M. Enhanced sensitivity from Ag micro-flakes encapsulated Ag-doped ZnO nanorods-based UV photodetector. *Mater. Res. Bull.* **2023**, *161*, 112148. [CrossRef]
5. Mohammed, A.A.A.; Lim, W.F. High photosensitivity performance vertical structured metal-semiconductor based ultraviolet photodetector using Ga_2O_3 thin film sputtered on n-type Si(100). *Mater. Sci. Eng. B* **2024**, *308*, 117613. [CrossRef]

6.   Huang, J.J.; Lin, C.H.; Ho, Y.R.; Chang, Y.U. Aluminium oxide passivation films by liquid phase deposition for TiO_2 ultraviolet solid–liquid heterojunction photodetectors. *Surf. Coat. Technol.* **2020**, *391*, 125684. [CrossRef]
7.   Zhou, J.W.; Qiao, Q.; Tan, Y.F.; Wu, C.; Hu, J.W.; Qiu, X.F.; Wu, S.H.; Zheng, J.; Wang, R.; Zhang, C.X.; et al. The improvement of polymer photodetector based on 1D-ZnO nanorod arrays/0D-ZnO quantum dots composite film. *Opt. Mater.* **2023**, *142*, 114086. [CrossRef]
8.   Karagoz, E.; Altaf, C.T.; Yaman, E.; Yildirim, I.D.; Erdem, E.; Celebi, C.; Fidan, M.; Sankir, M.; Sankir, N.D. Flexible metal/semiconductor/metal type photodetectors based on manganese doped ZnO nanorods. *J. Alloys Compd.* **2023**, *959*, 170474. [CrossRef]
9.   Abdulrahman, A.F.; Abd-Alghafour, N.M.; Almessiere, M.A. A high responsivity, fast response time of ZnO nanorods UV photodetector with annealing time process. *Opt. Mater.* **2023**, *141*, 113869. [CrossRef]
10.   Nithya, G.; Kumar, K.N.; Shaik, H.; Reddy, S.; Sen, P.; Prakash, N.G.; Ansari, M.A. Simulation and Deposition of Tungsten Oxide (WO_3) films using DC sputtering Towards UV Photodetector for High Responsivity. *Physica B* **2024**, *695*, 416555.
11.   Kumar, M.; Dhar, J.C. Enhanced UV photodetector performance using sputtered Mg-doped ZnO thin film. *Opt. Mater.* **2024**, *157*, 116059. [CrossRef]
12.   Chen, Y.C.; Chen, D.B.; Zeng, G.; Li, X.X.; Li, Y.C.; Zhao, X.F.; Chen, N.; Wang, T.Y.; Yang, Y.G.; Zhang, D.W.; et al. High performance solar-blind photodetectors based on plasma-enhanced atomic layer deposition of thin Ga_2O_3 films annealed under different atmosphere. *J. Alloys Compd.* **2023**, *936*, 168127. [CrossRef]
13.   Chu, S.Y.; Yeh, T.H.; Lee, C.T.; Lee, H.Y. Mg-doped beta-Ga_2O_3 films deposited by plasma-enhanced atomic layer deposition system for metal-semiconductor-metal ultraviolet C photodetectors. *Mat. Sci. Semicon. Proc.* **2022**, *142*, 106471. [CrossRef]
14.   Jehad, A.K.; Fidan, M.; Ünverdi, Ö.; Çelebi, C. CVD graphene/SiC UV photodetector with enhanced spectral responsivity and response speed. *Sens. Actuators A Phys.* **2023**, *355*, 114309. [CrossRef]
15.   Almaviva, S.; Marinelli, M.; Milani, E.; Prestopino, G.; Tucciarone, A.; Verona, C.; Verona-Rinati, G.; Angelone, M.L.; Pillon, M. Extreme UV photodetectors based on CVD single crystal diamond in a p-type/intrinsic/metal configuration. *Diam. Relat. Mater.* **2009**, *18*, 101–105. [CrossRef]
16.   Samriti; Prateek; Joshi, M.C.; Gupta, R.K.; Prakash, J. Hydrothermal synthesis and Ta doping of TiO_2 nanorods: Effect of soaking time and doping on optical and charge transfer properties for enhanced SERS activity. *Mater. Chem. Phys.* **2022**, *278*, 125642. [CrossRef]
17.   Naceur, J.B.; Jrad, F.; Souiwa, K.; Rhouma, F.B.; Chtourou, R. Hydrothermal reaction time effect in wettability and photoelectro-chemical properties of TiO_2 nanorods arrays films. *Optik* **2021**, *239*, 166794. [CrossRef]
18.   Huang, J.J.; Ho, Y.R.; Chang, Y.H.; Lin, C.H.; Ou, S.L. Al_2O_3-passivated TiO_2 nanorods for solid-liquid heterojunction ultraviolet photodetectors. *J. Mater. Sci.* **2021**, *56*, 6052–6063.
19.   Zhang, H.J.; Abdiryim, T.; Jamal, R.; Liu, X.; Niyaz, M.; Xie, S.Y.; Liu, H.L.; Kadir, A.; Serkjan, N. Coal-based carbon quantum dots-sensitized TiO_2 NRs/PTTh heterostructure for self-powered UV detection. *Appl. Surf. Sci.* **2022**, *605*, 154797. [CrossRef]
20.   Huang, J.J.; Ho, Y.R. Performance improvement of TiO_2 nanorods ultraviolet photodetector by AlN thin film passivation. *Mat. Sci. Semicon. Proc.* **2023**, *166*, 107756. [CrossRef]
21.   Zhou, M.; Wu, B.; Zhang, X.; Cao, S.; Ma, P.; Wang, K.; Fan, Z.; Su, M. Preparation and UV photoelectric properties of aligned ZnO-TiO_2 and TiO_2-ZnO core-shell structured heterojunction nanotubes. *ACS Appl. Mater. Interfaces* **2020**, *12*, 38490–38498. [CrossRef] [PubMed]
22.   Nguyen, T.T.; Patel, M.; Kim, S.; Mir, R.A.; Yi, J.; Dao, V.; Kim, J. Transparent photovoltaic cells and self-powered photodetectors by TiO_2/NiO heterojunction. *J. Power Sources* **2021**, *481*, 228865. [CrossRef]

 *micromachines*

Article

Internally Harmonic Matched Compact GaN Power Amplifier with 78.5% PAE for 2.45 GHz Wireless Power Transfer Systems

Caoyu Li [1], Ziliang Zhang [1,2], Yi Pei [3], Changchang Chen [3], Gang Feng [1,4] and Yuehang Xu [1,2,*]

1   Yangtze Delta Region Institute (Huzhou), University of Electronic Science and Technology of China, Huzhou 313001, China; licaoyu@csj.uestc.edu.cn (C.L.); 202221020901@std.uestc.edu.cn (Z.Z.); fenggang@uestc.edu.cn (G.F.)
2   School of Electronic Science and Engineering, University of Electronic Science and Technology of China, No. 2006, Xiyuan Ave., West Hi-Tech Zone, Chengdu 611731, China
3   Dynax Semiconductor Inc., Suzhou 215300, China; yi.pei@dynax-semi.com (Y.P.); changchang.chen@dynax-semi.com (C.C.)
4   National Key Laboratory of Wireless Communications, University of Electronic Science and Technology of China, No. 2006, Xiyuan Ave., West Hi-Tech Zone, Chengdu 611731, China
*   Correspondence: yuehangxu@uestc.edu.cn

Abstract: In this paper, a high-efficiency compact power amplifier is designed and fabricated with a 0.25 µm GaN high electron mobility transistor (HEMT) to meet the demands of a high integration level and high efficiency for microwave wireless power transfer (WPT) systems. The proposed power amplifier (PA) is implemented using an internally matched method to achieve a compact circuit size. The output second and third harmonic impedances can be optimized through output matching circuits, eliminating the need for additional harmonic matching networks. This approach simplifies the design of matching circuits and reduces the circuit size. Furthermore, the input third harmonic has been controled for improving the efficiency of DC-to-RF conversion. The total size of the proposed PA is 13.4×13.5 mm^2. The test results obtained from the continuous wave (CW) testing indicate that the output power of the power amplifier at 2.45 GHz reaches 43.75 dBm. Additionally, the large-signal gain is measured at 15.75 dB, and the power-added efficiency (PAE) achieves a value of 78.5%.

Keywords: GaN HEMT; power amplifiers; high efficiency; 2.45 GHz

Citation: Li, C.; Zhang, Z.; Pei, Y.; Chen, C.; Feng, G.; Xu, Y. Internally Harmonic Matched Compact GaN Power Amplifier with 78.5% PAE for 2.45 GHz Wireless Power Transfer Systems. *Micromachines* **2024**, *15*, 1354. https://doi.org/10.3390/mi15111354

Academic Editors: Zeheng Wang and Jing-Kai Huang

Received: 17 October 2024
Revised: 31 October 2024
Accepted: 1 November 2024
Published: 6 November 2024

1. Introduction

Wireless power transfer (WPT) systems represent a promising trend in the future development of electronic power delivery methods [1]. The dimensions of the circuit are also critical, as the high integration level required for beam-forming structures necessitates a compact design for power amplifiers. Consequently, compact, high-efficiency, and high-power amplifiers play a vital role in WPT systems. The most crucial metric for evaluating a WPT system is its transfer efficiency. Gallium nitride (GaN) high electron mobility transistors (HEMTs) are particularly well-suited for applications demanding high power and efficiency due to their exceptional electrical characteristics.

The GaN power amplifier can be realized through three primary methods: a Monolithic Microwave Integrated Circuit (MMIC), an external matching circuit (EMC), and an internally matched circuit. In comparison to an MMIC and EMC, the internally matched power amplifier demonstrates significant advantages in terms of device performance, size, and production cost [2]. To satisfy the design requirements for miniaturization, high efficiency, and high output power, the internally matched power amplifier has increasingly become a focal point of research and has found widespread applications in satellite navigation and radar detection fields [3–5].

The theoretical drain efficiency (DE) of class E and class F power amplifiers is 100% [6–8]. Both design methodologies can satisfy the requirement for high efficiency at a single frequency. However, a significant limitation of class E power amplifiers is that the output capacitance of the transistor has a pronounced effect on the maximum operating frequency, particularly in the microwave range [9]. The theoretical DE of class F power amplifiers can also reach 100%, but this is contingent upon achieving impedance matching across all harmonics. Typically, impedance matching for the second and third harmonics necessitates additional circuit design; however, it should be noted that any insertion loss introduced by higher harmonic matching circuits may adversely affect power-added efficiency (PAE) more than the efficiency gains achieved through such matching [10].

In the microwave frequency range, a power amplifier (PA) designed with class F architecture is recognized as one of the most effective methodologies for enhancing the power-added efficiency (PAE) of narrow-band power amplifiers. Recent years have witnessed extensive research focused on harmonic termination techniques aimed at developing high-efficiency power amplifiers [11]. To meet the harmonic impedance requirements of a class F power amplifier, it is essential to manage not only the output harmonic impedance but also the input harmonic impedance. Various techniques have been proposed for harmonic tuning, including open-ended stubs integrated into external matching circuits [2,12,13], on-chip LC tuning circuits [14–16], and off-chip LC tuning circuits [17–19].

In 2009, F. M. Ghannouchi proposed a 2.45 GHz inverse class F power amplifier (PA) with a power-added efficiency (PAE) of 71.5% [20]. A. A. Ismail reported a 2.45 GHz class F PA characterized by high linearity; however, its PAE is only 62% [21]. Both of these studies employed externally matched circuits, which tend to occupy larger physical sizes compared to internally matched PAs. In 2018, Takumi Sugitani et al., from Mitsubishi Electric Corporation [14], presented a high-power and high-efficiency power amplifier designed for microwave heating applications, achieving an output power exceeding 57 dBm and demonstrating a drain efficiency (DE) greater than 70% at the frequency of 2.45 GHz. This internally matched power amplifier incorporates an on-chip LC tuning circuit situated near the gate terminals of the HEMT to regulate the input second harmonic impedance while utilizing an open-ended stub to manage the output second harmonic impedance. The on-chip LC tuning circuit offers minimal area occupation along with superior consistency in both amplitude and phase; however, it lacks tunability.

To address the growing demand for compact and high-efficiency power amplifiers in microwave wireless power transfer systems, we have designed an internally-matched power amplifier that operates with high efficiency at a frequency of 2.45 GHz, utilizing 0.25 μm GaN HEMT technology. The implementation of LC components is achieved through bonding wire and thin-film circuit techniques to ensure a compact design. Additionally, harmonic tuning methods are employed to enhance overall efficiency.

2. Overall Design

The GaN HEMT under consideration is fabricated utilizing an advanced GaN chip production line provided by Dynax Semiconductor Inc. This device achieves a typical power density of 6.0 W/mm, a drain efficiency (DE) of 73% at 6 GHz, and a typical gain of 19.0 dB. To meet the power requirement of 20 W while ensuring high power-added efficiency (PAE) with a compact internally matched power amplifier (PA), a single-stage topology has been employed. The overall gate size of the GaN HEMT die measures 655×2725 μm. The schematic diagram illustrating the internally matched power amplifier circuit topology is presented in Figure 1.

In order to meet the output power requirements and maximize the power-added efficiency (PAE) of the power amplifier, the transistor operates in deep class AB mode. The gate voltage (Vgs) is set at -2.7 V, while the drain voltage is maintained at 28 V, resulting in a quiescent current of 50 mA for the transistor. Input and output matching are achieved through an LC tuning circuit that utilizes a parallel plate capacitor along with bonding wire as inductance elements. Both input harmonic matching and output harmonic matching are considered and analyzed as follows.

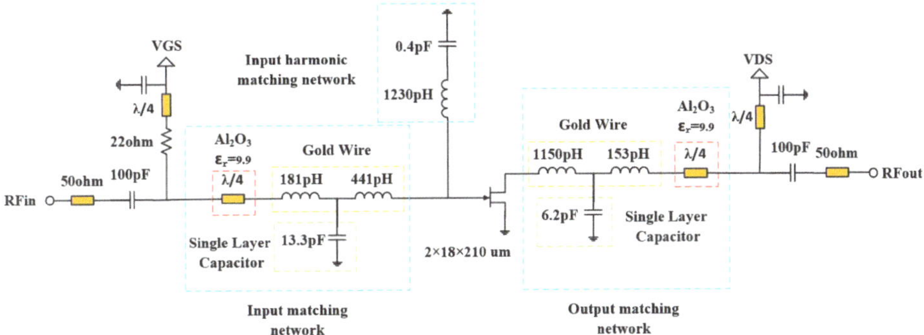

Figure 1. Schematic diagram of the internally matched power amplifier circuit topology.

3. High-Efficiency Matching Circuit Design

Firstly, a simulation of load pull and source pull for the fundamental frequency is conducted at 2.45 GHz for the GaN HEMT die by employing an accurate model [22]. This large-signal model provides precise impedance values while taking into account the influence of the knee voltage characteristic of GaN HEMTs. The fundamental output impedance of the GaN HEMT is measured to be 5.7 + j*10.2 Ω, whereas its input impedance is determined to be 1.1 + j*3 Ω. Regarding amplifier bandwidth issues, according to the Bode-Fano criterion, there exists a constraint between the bandwidth and reflection coefficient as follows [23]:

$$BW = \frac{\pi \omega_0}{R \ln\left(\frac{1}{\tau_{min}}\right)} \tag{1}$$

$$n = \left(\frac{R_L}{R_S}\right)^{\frac{1}{K}} \tag{2}$$

In which, R_L denotes the load impedance, K stands for the number of nodes in the T-type network, and n represents the impedance conversion ratio. From (1) and (2), increasing the LC network's order will expand the bandwidth, but it will also result in greater physical space requirements and increased insertion loss. Since the power amplifier required to be designed is a high-efficiency power amplifier operating at 2.45 GHz, it is necessary to minimize the influence of the working bandwidth on the output power and PAE. So, the output matching circuit only needs to use an L-C-L network to implement the matching of the virtual part, while increasing the impedance of the real part to 10 Ω, and then transforming the real part to 50 Ω through the quarter-wavelength transformation line. The input matching circuit also uses an L-C-L network and quarter-wavelength transformation line to achieve the best match with the source impedance between 1.1 + j*3 Ω and 50 Ω.

Second, the GaN HEMT undergoes 2nd and 3rd load/source pull simulations using precise modeling techniques. Figure 2 shows the 2nd and 3rd load pull simulation results. In Figure 2, the impedance point of the optimal PAE is offset compared with the ideal class F power amplifier due to the influence of the parasitic parameters of the transistor.

The class F power amplifier requires that the input 2nd and 3rd harmonic impedance is in the short state. The 2nd source pull simulation results show that the 2nd harmonic optimal source impedance is 1.4 + j*0.595 Ω. The 3rd source pull simulation results show that the 3rd harmonic optimal source impedance is 1.41 + j*1.79 Ω. The impedance point of the optimal PAE is slightly offset compared with the ideal class F power amplifier due to the influence of the parasitic parameters of the transistor. In order to achieve optimal harmonic source impedance, we need to increase the off-chip harmonic tuning circuit to achieve this requirement. Due to the accuracy of L and C values, the harmonic tuning circuit may cause a frequency mismatch of fundamental frequency. To avoid power and gain loss at operation frequency and to simplify the input matching network, only the 3rd harmonic tuning circuit is applied in the input matching network. The input harmonic matching network is shown in Figure 1. The 3rd harmonic tuning circuits are LC series resonant circuits, which are realized in the same way as the fundamental frequency matching circuit. The resonant frequency of a series L-C resonant circuit can be calculated by (3).

$$f = \frac{1}{2\pi\sqrt{LC}} \tag{3}$$

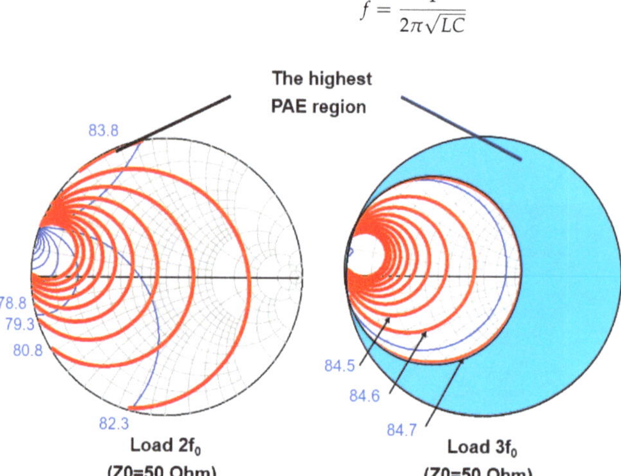

Figure 2. Simulated PAE dependence with the output harmonic impedance.

By setting the values of L and C reasonably, the resonant frequency of the LC resonant circuit is the third harmonic frequency, respectively, so as to realize the short circuit state of the input third harmonic. The simulation curve of input matching network impedance with and without harmonic tuning is given by Figures 3 and 4.

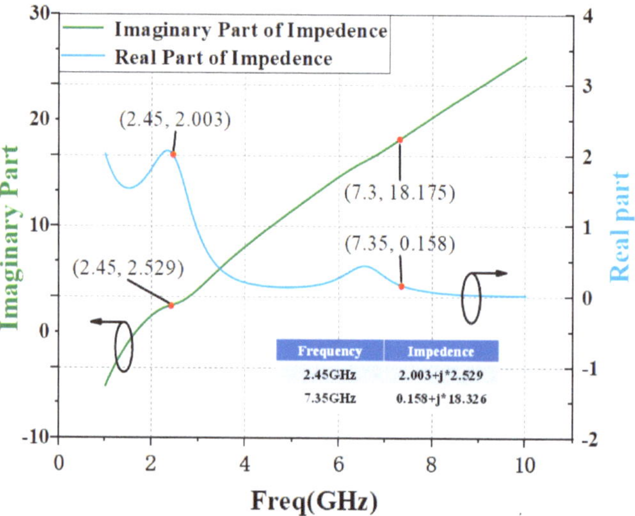

Figure 3. Simulation results of impedance curve of input matching circuit without harmonic tuning circuit.

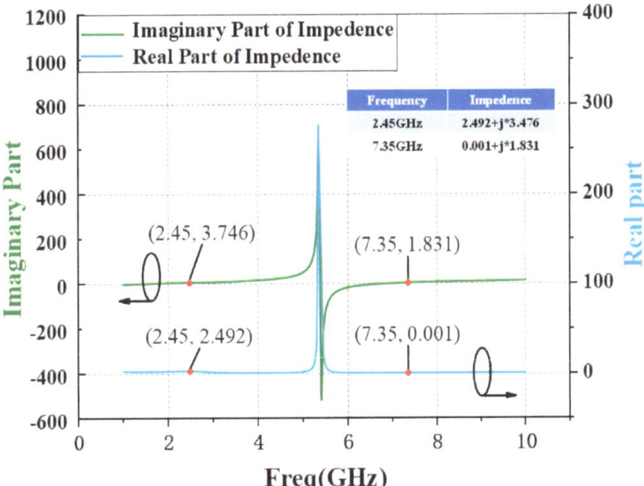

Figure 4. Simulation results of impedance curve of input matching circuit with harmonic tuning circuit.

Figure 5 gives the schematic diagram of the output match network. The simulation results of the impedance curve of the output matching circuit are given by Figure 6. It can be seen that the 2nd harmonic impedance achieved by the output matching circuit without harmonic tuning is $0.511 + j*31.466\ \Omega$. If harmonic tuning is added, the 2nd harmonic impedance achieved by the output matching circuit is $0.002 + j*2.003\ \Omega$. Combined with the 2nd load pull simulation results in Figure 2, it can be seen that the output matching circuit without the output 2nd harmonic tuning will transform the 2nd harmonic impedance to the region with the highest efficiency, while the output matching circuit with the output 2nd harmonic tuning will transform the 2nd harmonic impedance into the "trap" with low efficiency. The case of the 3rd harmonic is the same as the case of the 2nd harmonic. Therefore, for output matching, we adopt the method of combining fundamental frequency matching with the 2nd and 3rd harmonic matching, rather than the harmonic tuning method.

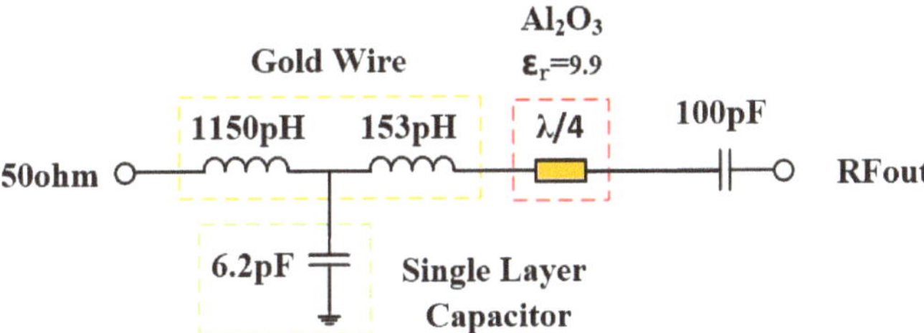

Figure 5. Schematic diagram of output matching circuit.

Figure 7 shows the simulated result of large-signal performance with and without a harmonic tuning circuit. The output power and gain at saturation in the two cases have almost no difference. The biggest difference in PAE between the with and without harmonic tuning cases is 4% when the PA saturates.

The impedance transformation for fundamental frequency matching is achieved using an alumina substrate which has a permittivity of 9.9 and the thickness of the substrate is 0.254 mm, with a characteristic impedance of $24.1\ \Omega$. These matching circuits are designed to realize the optimal output harmonic impedance.

The components in the matching circuit are realized by different substrate materials, and then all the components are glued to the inside of the tube shell by conducting resin, and finally connected by gold wire. The inductance L is achieved using a bonding wire with a diameter of 25 µm. Capacitor C is implemented using thin-film technology, and the substrate material is made of alumina ceramics with a relative dielectric constant of 9.9. The capacitance value of the parallel plate capacitor can be calculated by (4).

$$C = \varepsilon_0 \varepsilon_{rd} \frac{A}{d} = \varepsilon_0 \varepsilon_{rd} \frac{W \cdot l}{d} \tag{4}$$

Here, W denotes the width of the capacitor, and l denotes its length. The ε_0 represents the permittivity of the medium of the capacitor, and ε_{rd} represents the permittivity of the vacuum.

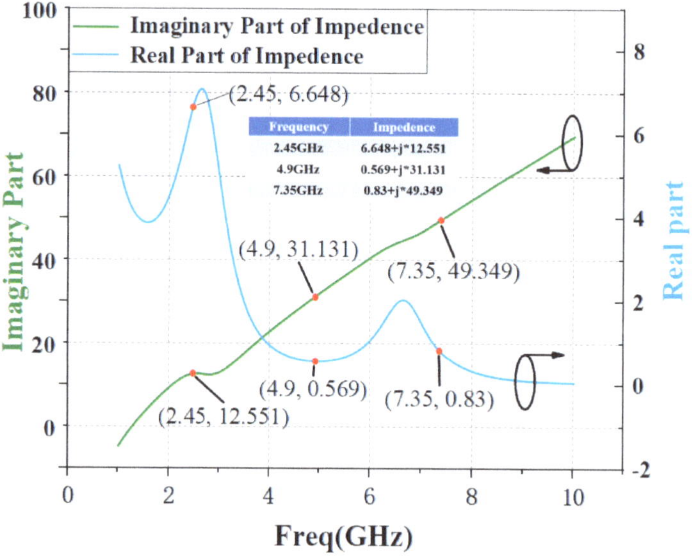

Figure 6. Impedance curve simulated result of output matching circuit.

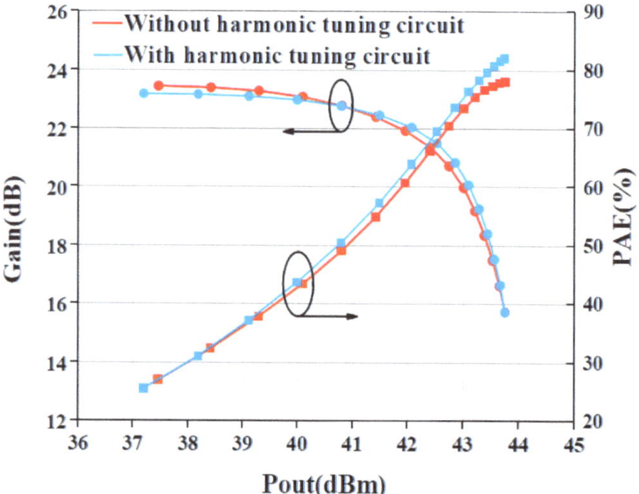

Figure 7. Simulated result of large-signal performance with and without harmonic tuning circuit.

4. Results and Discussion

We evaluated the RF performance of an internally-matched power amplifier with the measurement fixture. Figure 8 is the top-view picture of the GaN power amplifier using a harmonic tuning circuit and measurement fixture. The detail of the proposed PA is given with an enlarged photo shown in Figure 9. The metal cavity underneath the measurement fixture is used to improve the heat dissipation of the internally-matched power amplifier. The width of a single gate finger is 210 μm. The package size is 13.4 × 13.5 mm^2. The internally-matched power amplifier is tested under the condition of a CW signal. The drain bias voltage is 28 V, and the quiescent drain current is 50 mA.

We assessed the RF performance of an internally matched power amplifier using a measurement fixture. Figure 8 presents a top-view image of the GaN power amplifier, which incorporates a harmonic tuning circuit and measurement fixture. Detailed information regarding the proposed power amplifier is provided in an enlarged photograph shown in Figure 9. The metal cavity located beneath the measurement fixture serves to enhance heat dissipation for the internally matched power amplifier. Each gate finger has a width of 210 μm, and the package dimensions are 13.4 × 13.5 mm^2. The internally matched power amplifier was evaluated under continuous wave (CW) signal conditions, with a drain bias voltage set at 28 V and a quiescent drain current of 50 mA.

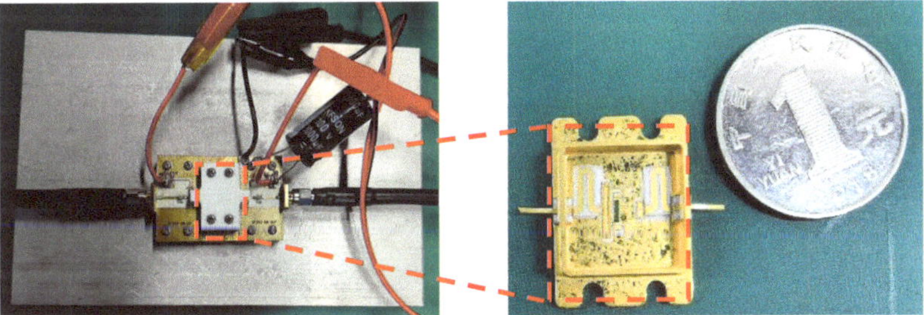

Figure 8. Picture of the GaN power amplifier using the harmonic tuning circuit and measurement fixture.

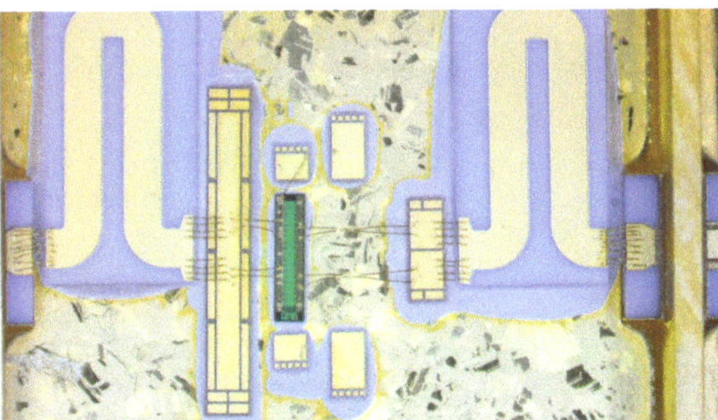

Figure 9. Picture of the detail of the proposed internally matched power amplifier.

4.1. Small-Signal Measurement

Figure 10 is the picture of the small-signal parameter measurement setup. Vector Network Analyzer ZVB 8 from Rohde & Schwarz is applied to measure the small-signal parameters. A 51 dB attenuator is applied to prevent the overpowering of the Vector Network Analyzer. Figure 11 presents a comparison between simulated and measured small signal results from 1 to 3 GHz. It can be seen that the measured input reflection coefficient is less than −8 dB, which is 3 dB lower than the simulated one. The measured small-signal gain is better than 19.8 dB, which is 2 dB lower than the simulated one. The discrepancy between the measured and simulated S-parameters is mainly caused by the

SMA connector at the input and output of the measurement fixture and the inaccuracy of the GaN active device model.

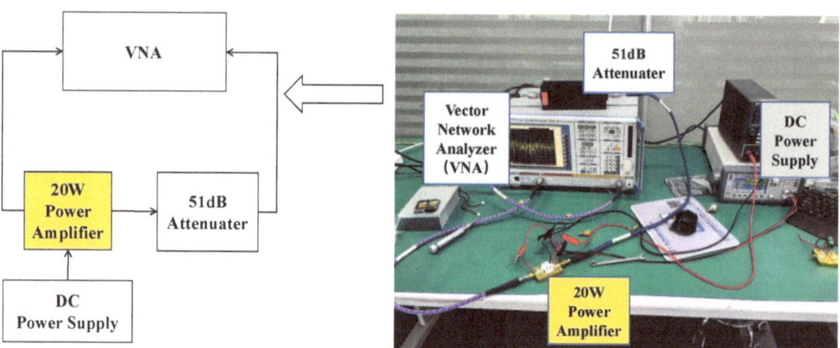

Figure 10. Small-signal parameter measurement setup.

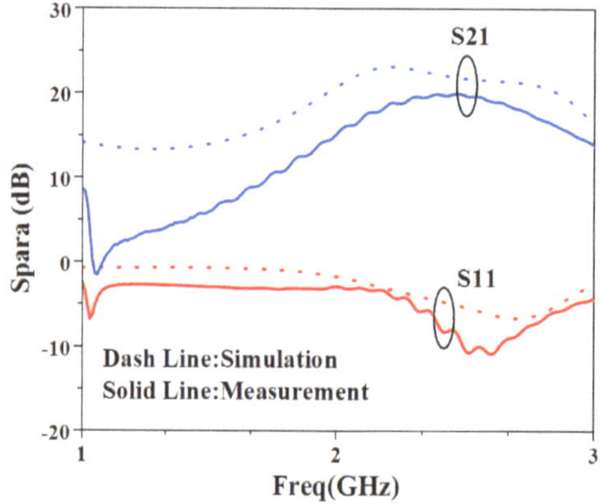

Figure 11. Measured and simulated S-parameter results.

4.2. Large-Signal Measurement

Figure 12 is the picture of the large-signal parameter measurement setup. Vector Signal Generator SMCV100B from Rohde & Schwarz (Munich, Germany) is applied to generate the RF signal. A Driver amplifier is applied to provide enough input power for DUT. Power Meter N1912A from Keysight (Santa Rosa, CA, USA) is applied to measure the accurate output power of DUT. The DC current is measured by the DC power supply. A continuous wave (CW) is implemented to test the performance of the proposed PA. The input power range is from 14 dBm to 28 dBm. The operation frequency is 2.45 GHz.

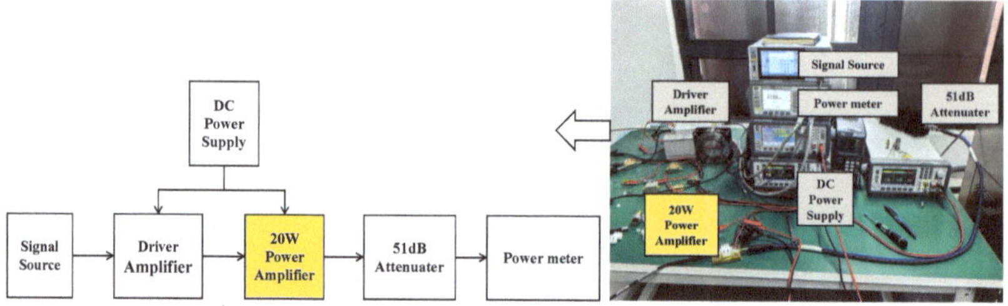

Figure 12. Large-signal parameter measurement setup.

Figure 13 gives the comparison result of measured and simulated data of the proposed PA. The gain and PAE curves are shown in Figure 13. The saturated power of the internally-matched PA is 43.75 dBm (23.71 W), while the PAE is 78.5% when input with 28 dBm power. The gain is 21.2 dB in the linear range and 15.75 dB at saturation power. The P_{1dB} is 42.13 dBm. In addition, we also measured the output power and PAE of the power amplifier without an input harmonic tuning circuit, as shown in Figure 14. The measurement of PA without a harmonic tuning circuit is done after removing the bonding wire of the input harmonic tuning circuit. When the input power is 28 dBm, the output power of the power amplifier is 43.73 dBm (23.6 W) while the PAE is 76.0%. This indicates that the PA is still in a high-efficiency mode, which means the output harmonic load is much more important than the input harmonic load. In our design, the output matching has already satisfied the output harmonic load condition so that the PA can perform well without input harmonic tuning. It can be seen from Figure 15 that the PAE of the harmonic tuning circuit is 2.5% higher than that of the non-harmonic tuning circuit under the same output power, which is mainly due to the reduction in the dynamic drain current as shown in Figure 16. In addition, it can be seen that there is a certain gap between the measured result and the simulated result. From Figures 13 and 14, the simulation results with harmonic tuning have a larger gap with measured results than the simulation without harmonic tuning. And, all the measured results show lower output power when the input power is low. This indicates that the harmonic load of the GaN model is not correct in the low-power region. It can be noticed that the simulated PAE is lower than the measured results; the inaccuracies in the Pdc representation within the GaN model contribute to these issues. Additionally, discrepancies between the modeling test environment and the application test environment lead to variations in simulated Pdc. This has a significant influence on PAE and gain simulation in low-power regions.

$$Q_{di} = P_{out}(W) \cdot \text{PAE} / \text{Dimension}\left(\text{mm}^2\right) \tag{5}$$

Table 1 displays the comparison of the latest advancements in correlation design performance. We define Q_{di} which is shown in (5) to compare the performance of each power amplifier. It can be seen from Table 1 that most of the externally-matched PA has good PAE performance at 2.45 GHz. However, the circuit size of externally matched PAs is much larger than this work. Although MMIC PAs have a smaller circuit size, our work performs with better power and PAE. Our work can achieve 23-watt output power with a circuit size of 13.4 × 13.5 mm², which makes it suitable for wireless power transfer systems.

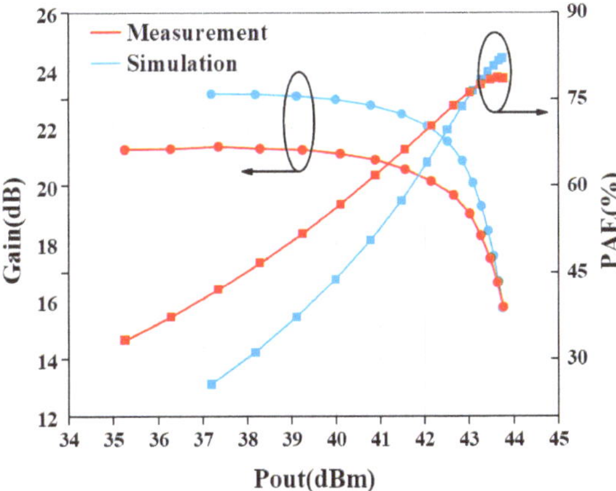

Figure 13. Simulated and measured large-signal performance with harmonic tuning circuit.

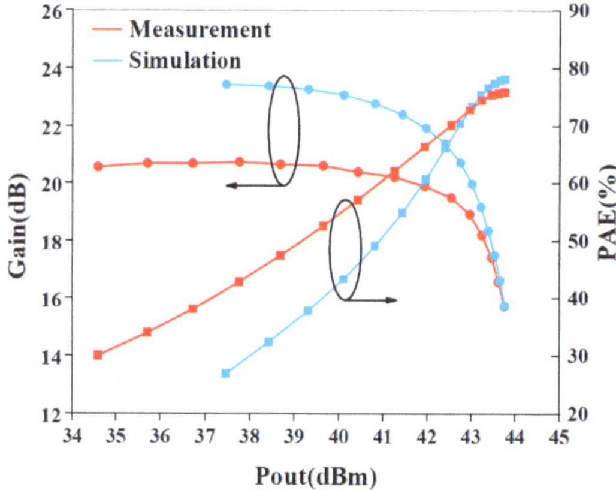

Figure 14. Simulated and measured large-signal performance without harmonic tuning circuit.

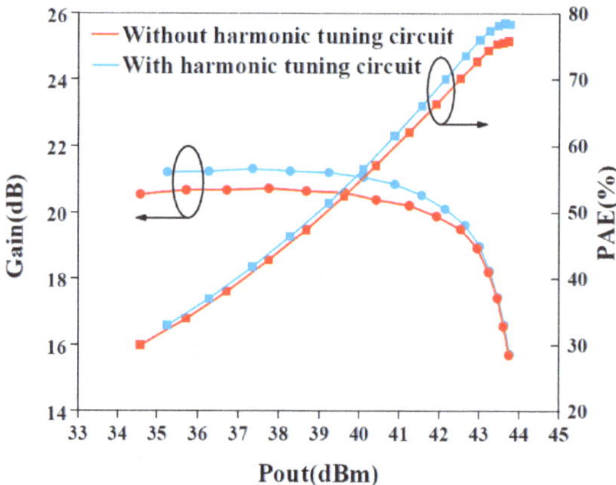

Figure 15. Measured large-signal performance with and without harmonic tuning circuit.

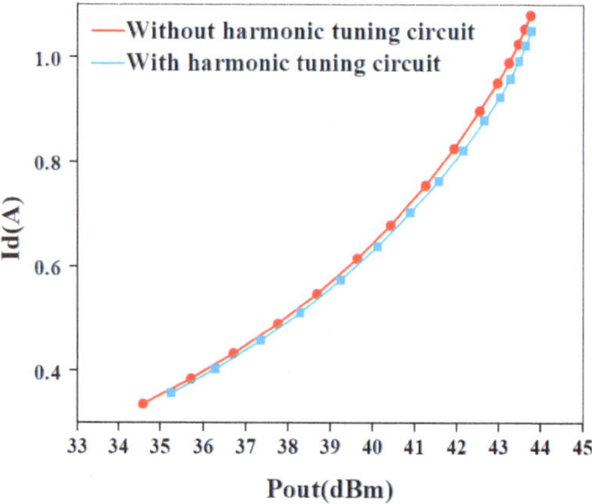

Figure 16. Comparison of measured dynamic drain current in two cases.

Table 1. Comparison of state-of-the-art PA under CW operation.

Reference	Freq (GHz)	Pout (W)	Gain (dB)	PAE (%)	Type	Q_{di}	Size (mm^2)
2018 [14]	2.45	450	13	67	Internally-matched	NA	NA
2020 [24]	2.4	33	13.6	78.8	Externally-matched	1.27	31 × 66
2013 [25]	3.1	10	15	82	Externally-matched	0.46	40 × 45
2009 [26]	3.5	11	12	78	Externally-matched	0.1	80 × 110
2021 [27]	2	7	10	74	Externally-matched	0.17	51 × 59
2023 [28]	4.25	11.8	11.5	55.3	Externally-matched	0.14	62 × 77
2022 [10]	2.21	11.5	15.6	82.6	Externally-matched	0.51	29 × 64
2022 [29]	1.97	10.4	10.8	79.3	Externally-matched	0.69	26 × 46
2024 [30]	2.6–3.6	12.0	8.5	50.8	MMIC	45.8	3.5 × 3.8
2021 [31]	2.6–3.8	5.25	10.2	55	MMIC	29.6	6.5 × 1.5
This work	2.45	23.7	15.75	78.5	Internally-matched	10.28	13.4 × 13.5

5. Conclusions

In order to meet the high-power, high-efficiency, and miniaturization requirements of a microwave wireless power transfer system, a high-power and high-efficiency internally matched power amplifier based on an off-chip LC tuning circuit at 2.45 GHz is designed in this paper. The second and third harmonic matching impedance can be reached without an extra harmonic circuit, which simplifies the output matching design and reduces the circuit size. Further more, the input harmonic tuning circuit has the advantages of a moderate circuit area, excellent harmonic control ability, and low loss. Based on the proposed method, an internally-matched power amplifier at 2.45 GHz is fabricated by using a 0.25 μm GaN HEMT. The results show that output power is 43.75 dBm, large-signal gain is over 15.75 dBm, and PAE is 78.5% at 2.45 GHz. The proposed PA is implemented with a size of 13.4 × 13.5 mm^2. The compact size, high efficiency, and high power can allow the PA in this work to potentially be usefed in wireless power transfer systems.

Author Contributions: Conceptualization, C.L.; methodology, C.L. and Z.Z.; validation, C.L., Y.P., C.C. and Z.Z.; writing—original draft preparation, Z.Z.; writing—review and editing, C.L. and G.F.; supervision, Y.X.; project administration, Y.X.; funding acquisition, Y.X. All authors have read and agreed to the published version of the manuscript.

Funding: This work is Supported by the Sichuan Science and Technology Program under Grant No.2024NSFJQ0022.

Data Availability Statement: The data presented in this study are available in this article.

Acknowledgments: Thanks to Chang Wu and Jielong Liu from HuBei Jiufengshan laboratory (Wuhan, Hubei, China) for the technical support of this work.

Conflicts of Interest: Authors Yi Pei and Changchang Chen were employed by the Dynax Semiconductor Inc. The remaining authors declare that the research was conducted in the absence of any commercial or financial relationships that could be construed as a potential conflict of interest.

References

1. Zhang, Z.; Pang, H.; Georgiadis, A.; Cecati, C. Wireless Power Transfer—An Overview. *IEEE Trans. Ind. Electron.* **2019**, *66*, 1044–1058. [CrossRef]
2. Yamasaki, T.; Kittaka, Y.; Minamide, H.; Yamauchi, K.; Miwa, S.; Goto, S.; Nakayama, M.; Kohno, M.; Yoshida, N. A 68% efficiency, C-band 100W GaN power amplifier for space applications. In Proceedings of the 2010 IEEE MTT-S International Microwave Symposium, Anaheim, CA, USA, 23–28 May 2010; pp. 1384–1387.
3. Li, R.; Zhong, S. Design of Internal-matching High Power Class-F Doherty Power Amplifier. In Proceedings of the 2023 International Conference on Microwave and Millimeter Wave Technology (ICMMT), Qingdao, China, 14–17 May 2023; pp. 1–3.
4. Gao, H.; Liu, D.; Zhou, Z. S-band Broadband Power Amplifier Based on Internal Matching and Micro-assembly Technology. In Proceedings of the 2020 International Conference on Microwave and Millimeter Wave Technology (ICMMT), Shanghai, China, 17–20 May 2020; pp. 1–3.
5. Pengelly, R.S.; Wood, S.M.; Milligan, J.W.; Sheppard, S.T.; Pribble, W.L. A Review of GaN on SiC High Electron-Mobility Power Transistors and MMICs. *IEEE Trans. Microw. Theory Tech.* **2013**, *60*, 1764–1783. [CrossRef]
6. Liu, C. Analysis of class-F power amplifiers with a second-harmonic input voltage manipulation. *IEEE Trans. Circuits Syst. II Exp. Briefs* **2020**, *67*, 225–229. [CrossRef]
7. Dhar, S.K.; Sharma, T.; Zhu, N.; Darraji, R.; Mclaren, R.; Holmes, D.G.; Mallette, V.; Ghannouchi, F.M. Input-harmonic-controlled broadband continuous class-F power amplifiers for sub-6-GHz 5G applications. *IEEE Trans. Microw. Theory Techn.* **2020**, *68*, 3120–3133. [CrossRef]
8. Liu, C.; Cheng, Q.-F. A Novel Compensation Circuit of High-Efficiency Concurrent Dual-Band Class-E Power Amplifiers. *IEEE Microw. Wirel. Components Lett.* **2018**, *28*, 720–722. [CrossRef]
9. Cumana, J.; Grebennikov, A.; Sun, G.; Kumar, N.; Jansen, R.H.T. An extended topology of parallel-circuit class-E power amplifier to account for larger output capacitances. *IEEE Trans. Microw. Theory Techn.* **2011**, *59*, 3174–3183. [CrossRef]
10. Sheikhi, A.; Hemesi, H. Analysis and Design of the Novel Class-F/E Power Amplifier with Series Output Filter. *IEEE Trans. Circuits Syst. II Express Briefs* **2022**, *69*, 779–783. [CrossRef]
11. Stameroff, A.; Pham, A.-V. Wide bandwidth inverse class F power amplifier with novel balun harmonic matching network. In Proceedings of the 2012 IEEE/MTT-S International Microwave Symposium, Digest, Montreal, QC, Canada, 17–22 June 2012; pp. 1–3.
12. Yamanaka, K.; Morimoto, T.; Chaki, S.; Nakayama, M.; Hirano, Y. X-band internally harmonic controlled GaN HEMT amplifier with 57% power added efficiency. In Proceedings of the 2011 6th European Microwave Integrated Circuit Conference, Manchester, UK, 10–11 October 2011; pp. 61–64.
13. Yoshioka, T.; Kosaka, N.; Hangai, M.; Yamanaka, K. An S-band 240 W output 54% PAE GaN power amplifier with broadband output matching network for both fundamental and 2nd harmonic frequencies. In Proceedings of the 2016 IEEE MTT-S International Microwave Symposium (IMS), San Francisco, CA, USA, 22–27 May 2016; pp. 1–4.
14. Sugitani, T.; Iyomasa, K.; Hangai, M.; Kawamura, Y.; Nishihara, J.; Shinjo, S. 2.45 GHz ISM-Band 450W High Efficiency GaN Pallet Amplifier for Microwave Heating. In Proceedings of the 2018 Asia-Pacific Microwave Conference (APMC), Kyoto, Japan, 6–9 November 2018; pp. 1621–1623.
15. Gu, L.; Feng, W.; Yang, B.; Che, W. A high efficiency X-band internally-matched GaN power amplifier using on-chip harmonic tuning Technology. In Proceedings of the 2022 IEEE MTT-S International Microwave Workshop Series on Advanced Materials and Processes for RF and THz Applications (IMWS-AMP), Guangzhou, China, 13–15 November 2022; pp. 1–3.
16. Miwa, S.; Kamo, Y.; Kittaka, Y.; Yamasaki, T.; Tsukahara, Y.; Tanii, T.; Kohno, M.; Goto, S.; Shima, A. A 67% PAE, 100 W GaN power amplifier with on-chip harmonic tuning circuits for C-band space applications. In Proceedings of the 2011 IEEE MTT-S International Microwave Symposium, Baltimore, MD, USA, 5–10 June 2011; pp. 1–4.
17. Chéron, J.; Campovecchio, M.; Barataud, D.; Reveyrand, T.; Stanislawiak, M.; Eudeline, P.; Floriot, D. Wideband 50W packaged GaN HEMT with over 60% PAE through internal harmonic control in S-band. In Proceedings of the 2012 IEEE MTT-S International Microwave Symposium Digest, Montreal, QC, Canada, 17–22 June 2012; pp. 1–3.
18. Lu, Y.; Cao, M.; Wei, J.; Zhao, B.; Ma, X.; Hao, Y. 71% PAE C-band GaN power amplifier using harmonic tuning technology. *Electron. Lett.* **2014**, *50*, 1207–1209. [CrossRef]
19. Motoi, K.; Matsunaga, K.; Yamanouchi, S.; Kunihiro, K.; Fukaishi, M. A 72% PAE, 95-W, single-chip GaN FET S-band inverse class-F power amplifier with a harmonic resonant circuit. In Proceedings of the 2012 IEEE/MTT-S International Microwave Symposium Digest, Montreal, QC, Canada, 17–22 June 2012; pp. 1–3.
20. Ghannouchi, F.M.; Ebrahimi, M.M.; Helaoui, M. Inverse Class F Power Amplifier for WiMAX Applications with 74% Efficiency at 2.45 GHz. In Proceedings of the 2009 IEEE International Conference on Communications Workshops, Dresden, Germany, 14–18 June 2009; pp. 1–5.
21. Ismail, A.A.; Younis, A.T.; Abduljabbar, N.A.; Mohammed, B.A.; Abd-Alhameed, R.A. A 2.45-GHz class-F power amplifier for CDMA systems. In Proceedings of the 2015 Internet Technologies and Applications (ITA), Wrexham, UK, 8–11 September 2015; pp. 428–433.
22. Xu, Y.; Wang, C.; Sun, H.; Wen, Z.; Wu, Y.; Xu, R.; Yu, X.; Ren, C.; Wang, Z.; Zhang, B.; et al. A Scalable Large-signal Multi-harmonic Model of Al-GaN/GaN HEMTs and Its Application in C-band High Power Amplifier MMIC. *IEEE Trans. Micro Wave Theory Tech.* **2017**, *65*, 2836–2846. [CrossRef]

Micromachines **2024**, *15*, 1354

23. Smith, M.C.; Dixit, R. Future trends in filter technology for military multifunction systems. In Proceedings of the 2012 IEEE International Conference on Wireless Information Technology and Systems (ICWITS), Maui, HI, USA, 11–16 November 2012; pp. 1–4.
24. Suzuki, A.; Hara, S. 2.4GHz High Efficiency GaN Power Amplifier using Matching Circuit Less Design. In Proceedings of the 2020 4th Australian Microwave Symposium (AMS), Sydney, NSW, Australia, 13–14 February 2020; pp. 1–2.
25. Chen, K.; Peroulis, D. A 3.1-GHz Class-F Power Amplifier with 82% Power-Added-Efficiency. *IEEE Microw. Wirel. Components Lett.* **2013**, *23*, 436–438. [CrossRef]
26. Saad, P.; Nemati, H.M.; Thorsell, M.; Andersson, K.; Fager, C. An inverse class-F GaN HEMT power amplifier with 78% PAE at 3.5 GHz. In Proceedings of the 2009 European Microwave Conference (EuMC), Rome, Italy, 29 September–1 October 2009; pp. 496–499.
27. Tanaka, S.; Asami, H. 2-GHz Class-E Power Amplifier Using a Compact Redundancy-Free Harmonic Tuning Circuit. In Proceedings of the 2020 50th European Microwave Conference (EuMC), Utrecht, The Netherlands, 12–14 January 2021; pp. 13–16.
28. Kondo, T.; Fujiwara, K.; Yaginuma, K.; Mizojiri, S.; Hara, S.; Tanba, N. Development and environmental evaluation of 10- W class-F power amplifier for microwave-discharge ion thruster on satellites. In Proceedings of the 2023 Asia-Pacific Microwave Conference (APMC), Taipei, Taiwan, 5–8 December 2023; pp. 387–389.
29. Tanaka, S.; Iisaka, N. A 2-GHz 79%-PAE Power Amplifier with a Novel Harmonic Tuning Circuit Using Only CRLH TLs. In Proceedings of the 2021 51st European Microwave Conference (EuMC), London, UK, 4–6 April 2022; pp. 358–361.
30. Jiang, X.; Bo, C.; Wu, X.; Dong, Q.; Yang, S.; Wei, K.; Liu, X.; Luo, W. A Simple Design Method for Harmonic-Tuned GaN MMIC Power Amplifier Using Real-to-Real LPF Matching Network. *IEEE Microw. Wirel. Technol. Lett.* **2024**, *34*, 528–531. [CrossRef]
31. Zhao, Y.; Li, X.; Gai, C.; Liu, C.; Qi, T.; Hu, B.; Hu, X.; Chen, W.; Helaoui, M.; Ghannouchi, F.M. Theory and Design Methodology for Reverse-Modulated Dual-Branch Power Amplifiers Applied to a 4G/5G Broadband GaN MMIC PA Design. *IEEE Trans. Microw. Theory Tech.* **2021**, *69*, 3120–3131. [CrossRef]

 *micromachines*

Article

A Sub-1 ppm/°C Reference Voltage Source with a Wide Input Range

Yuchi Xiao [1], Chunlai Wang [1,2], Hongyang Hou [1] and Weihua Han [1,3,*]

1　School of Physical Science and Technology, Lanzhou University, Lanzhou 730000, China;
　　220220939211@lzu.edu.cn (Y.X.); allan.w@bstsemi.com (C.W.); 220220938921@lzu.edu.cn (H.H.)
2　BASALT Semiconductor Co., Ltd., Wuhan 430000, China
3　Guangzhou Institute of Blue Energy, Guangzhou 510555, China
*　Correspondence: hanwh@lzu.edu.cn

Abstract: With the continuous advancement of electronic technology, the application of high-voltage integrated circuits is becoming increasingly prevalent in fields such as power systems, medical devices, and industrial automation. The reference circuit within high-voltage integrated circuits must not only exhibit insensitivity to temperature variations but also maintain stability across a broad voltage supply. This paper presents a bandgap reference (BGR) source capable of operating over a wide input range. This BGR employs a high-order curvature compensation method to eliminate nonlinear voltage terms, resulting in minimal temperature drift. The circuit achieves an impressive temperature coefficient (TC) of 0.88 ppm/°C over a temperature range from −40 °C to 130 °C. To ensure stable operation within a 4–40 V range, the design incorporates a pre-regulation circuit that stabilizes the supply voltage of the BGR core at a fixed value, thereby enhancing the ability to withstand variations in power supply voltage.

Keywords: bandgap reference; wide input range; high-order curvature compensation; temperature coefficient

Citation: Xiao, Y.; Wang, C.; Hou, H.; Han, W. A Sub-1 ppm/°C Reference Voltage Source with a Wide Input Range. *Micromachines* **2024**, *15*, 1273. https://doi.org/10.3390/mi15101273

Academic Editors: Zeheng Wang and Jing-Kai Huang

Received: 29 August 2024
Revised: 17 October 2024
Accepted: 17 October 2024
Published: 21 October 2024

1. Introduction

Due to its low TC and high power supply rejection ratio (PSRR), the BGR voltage source, serving as a fundamental component in integrated circuits, has been extensively utilized in DC–DC converters, analog-to-digital converters, low-dropout regulators, and other mixed-signal applications [1–3]. However, the large variation in the supply voltage range can lead to fluctuations in the reference voltage provided by the BGR, consequently impacting circuit operation. High performance BGRs with a wide input range (e.g., 4–40 V) are extensively applied in high-voltage integrated circuits, including smart grids, energy storage systems, and electric vehicles [4].

In addition to supply independence, an ideal BGR should exhibit minimal sensitivity to temperature variations. A conventional BGR achieves a temperature-independent output by combining two voltages with opposite TCs [5]. Specifically, a complementary-to-absolute-temperature (CTAT) voltage comes from the base-emitter voltage V_{BE} of bipolar transistors, while the difference between the base-emitter voltages of two bipolar transistors operating at unequal current densities yields a directly proportional-to-absolute-temperature (PTAT) voltage. However, the presence of higher-order temperature term $T \ln T$ in the base-emitter voltage causes the output reference voltage to still fluctuate with temperature. As a result, the linear combination of positive and negative TC voltages often leads to a TC exceeding 10 ppm/°C. Previous studies have proposed various methods to mitigate the effects of the higher-order term $T \ln T$. Lavrentiadis et al. employed MOSFET transistors operating in the sub-threshold region instead of bipolar transistors to generate reference voltage [6], achieving a TC of 24 ppm/°C. Du et al. [7] referenced the curvature compensation method from [8] to compensate for the $T \ln T$ nonlinearity; the author reported a typical TC of 1.5 ppm/°C but reached a current consumption of 267 μA. Various methodologies have been proposed by designers to generate a compensatory current that resembles the shape of the $T \ln T$ curve [9–11]. For example, Malcovati et al. utilized a T^2 current to compensate for nonlinear terms [9], while Cao and Shen et al. employed the emitter current and the base current of the transistor for compensation [10,11]. However, an analysis of the provided expressions for compensation current reveals that the compensation term does not adequately fit the nonlinear term $T \ln T$, resulting in a temperature drift exceeding 1 ppm/°C.

This paper presents a BGR circuit designed to operate effectively over a broad voltage range. The proposed BGR incorporates advanced compensation techniques aimed at minimizing temperature drift. Experimental findings demonstrate that utilizing a 0.18 μm Bipolar-CMOS-DMOS (BCD) process, the reference voltage achieves a TC of less than 1 ppm/°C across a power supply range from 4 V to 40 V. To accommodate the wide-input voltage range required for the power supply and to enhance the precision of the BGR circuit, a customary approach involves first generating a low-voltage power supply using a pre-regulator circuit. This pre-regulated supply stabilizes voltage around 3.3 V, which subsequently powers the core BGR circuit. Within this wide-input range, the BGR core circuit produces a stable 1.2 V reference voltage with minimal TC. Figure 1 illustrates the block diagram of designed BGR.

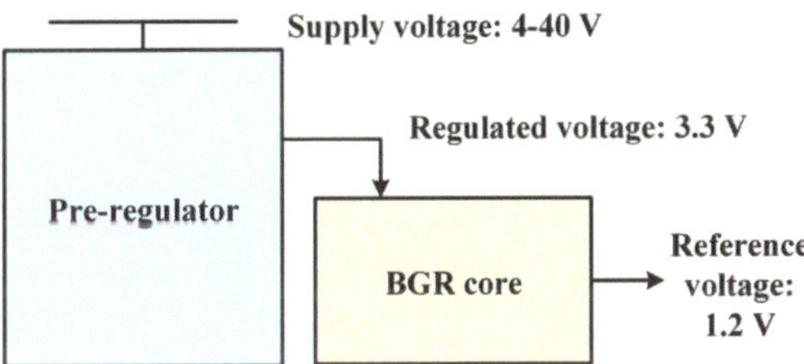

Figure 1. Block diagram of designed BGR.

2. Materials and Methods

2.1. Pre-Regulator Circuit

The pre-regulator circuit discussed in this article is illustrated in Figure 2a. Transistors M_{P1}, M_{P2}, M_{P3} along with the resistor R_S collectively form the start-up circuit, ensuring stable and reliable operation of the pre-regulator module. A self-biased current source is implemented using a cascode current mirror, providing a current that remains independent of the supply voltage. This current flows through the diode-connected MOS transistors to generate the regulated output voltage. High-voltage devices are utilized in the pre-regulator circuit to efficiently convert a wide range of input voltages into a lower, more stable output voltage. Incorporating a feedback loop minimizes the impact of power supply fluctuations on the output voltage stability. In contrast to the pre-regulator circuit described in [11], this design utilizes a PMOS transistor for driving rather than an NMOS transistor. This choice allows for a reduction in the minimum operating voltage of the pre-regulator circuit. As illustrated in Figure 2b, the PMOS-driven circuit achieves a lower minimum operating voltage by one threshold voltage compared to its NMOS-driven counterpart. During operation, the regulator maintains a tiny drop-out voltage relative to the supply, ensuring normal operation even under low-power conditions and reducing energy loss from the power source.

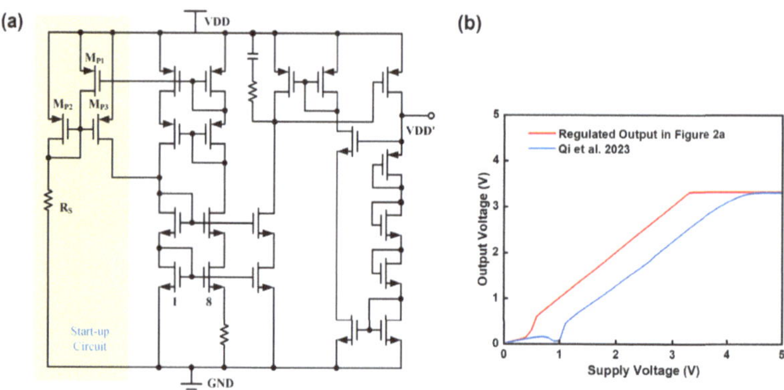

Figure 2. (**a**) Schematic of proposed pre-regulator circuit. (**b**) Regulated output voltages of two structures [11].

2.2. Nonlinearity in BGR

Figure 3 presents a current-mode BGR circuit [12], where the operational amplifiers OP1 and OP2 adjust the gate voltage of the PMOS devices to equalize V_X, V_Y, and V_Z. Transistor Q_2 consists of n unit transistors connected in parallel, and Q_1 is a single unit transistor. This configuration generates PTAT and CTAT currents, with their expressions given as follows.

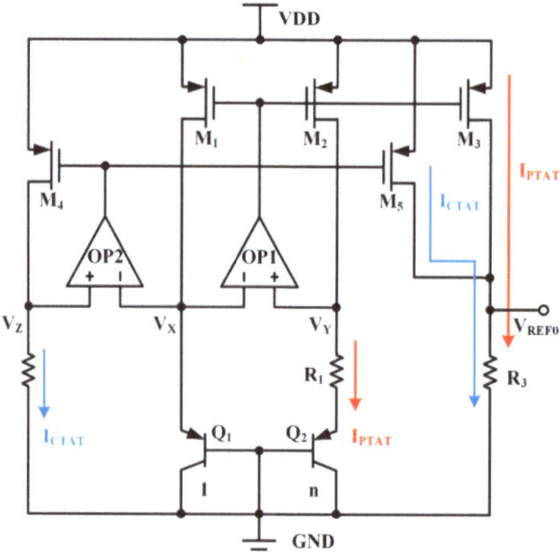

Figure 3. Schematic of a traditional current mode BGR.

$$I_{\text{PTAT}} = \frac{V_{EB1} - V_{EB2}}{R_1} = \frac{V_T \ln n}{R_1} \tag{1}$$

$$I_{\text{CTAT}} = \frac{V_{EB1}}{R_2} \tag{2}$$

Here, V_{EB1} and V_{EB2} are emitter–base voltages of the bipolar junction transistor Q_1 and Q_2. $V_T = kT/q$ is the thermal voltage, where k is the Boltzmann constant, T is the absolute temperature, q is the electron charge. The PTAT and CTAT currents are copied and passed through a resistor R_3 to generate a zero-TC voltage.

$$
\begin{aligned}
V_{REF0} &= (I_{PTAT} + I_{CTAT}) \cdot R_3 \\
&= \left(\frac{V_T \ln n}{R_1} + \frac{V_{EB1}}{R_2} \right) \cdot R_3
\end{aligned}
\tag{3}
$$

However, the base–emitter voltage of a BJT is not a purely linear function of temperature; it can be expressed as [13]

$$
\begin{aligned}
V_{BE}(T) = {}& V_G(T) - \left(\frac{T}{T_r} \right) V_G(T_r) + \left(\frac{T}{T_r} \right) V_{BE}(T_r) \\
& (\eta \quad \delta) \left(\frac{kT}{q} \right) \ln \left(\frac{T}{T_r} \right)
\end{aligned}
\tag{4}
$$

where $V_G(T)$ is the bandgap voltage at temperature T, $\eta = 4 - n$, n represents the order of temperature dependence of carrier mobility, δ is the order of temperature dependence of collector current. T_r refers to reference temperature.

Equation (4) shows that besides a CTAT term, $V_{BE}(T)$ includes a $T \ln(T/T_r)$ term, which introduces nonlinearity. The presence of this higher-order term causes the reference voltage to exhibit curvature as temperature varies. To achieve temperature independence of the reference voltage, it is necessary to eliminate this nonlinear term.

2.3. Implementation of High-Order Compensation

When high-precision reference voltages are required, the nonlinear terms in V_{BE} cannot be neglected. Exponential current generators have been employed in curvature compensation [14]. The compensation current is generated by sub-threshold MOS transistors, which exhibit significant deviations across different fabrication processes. The efficacy of this approach using a single compensation current is suboptimal, often necessitating multi-segment compensation for the circuit. Therefore, it is essential to generate a term that more accurately matches the nonlinear component.

Figure 4 shows the complete circuit of the proposed BGR core, based on the current-mode BGR circuit discussed in Section 2.2. In the circuit, transistors M_7 and M_8 have the same aspect ratio W_n/L_n. M_1, M_2, and M_6 have identical device dimensions. Bipolar junction transistors Q_3 and Q_4 consist of the same units in parallel. Currents for both PTAT and CTAT have been designed to be 100 nA. Since $V_{GS7} + V_{BE3} = V_{GS8} + V_{BE4}$, we have

$$
\begin{aligned}
V_{GS7} - V_{GS8} &= V_{BE4} - V_{BE3} = V_T \ln \frac{I_{C4}}{I_{C3}} = V_T \ln \frac{I_9}{I_7} \\
&= \sqrt{\frac{2}{\mu_n C_{ox} \frac{W_n}{L_n}}} \left(\sqrt{I_7} - \sqrt{I_8} \right)
\end{aligned}
\tag{5}
$$

where V_{GS7} and V_{GS8} are gate-source voltages of MOS transistors M_7 and M_8, μ_n represents the mobility of electrons, and C_{ox} is the gate-oxide capacitance per unit area. The current flowing through M_8 can be expressed as

$$
\begin{aligned}
I_8 &= I_R + I_{B4} \\
&= \frac{V_{BE4}}{R_5} + \frac{1}{\beta} I_9
\end{aligned}
\tag{6}
$$

where V_{BE4} is the base–emitter voltage of bipolar junction transistor Q_4, β represents the common-emitter current gain of Q_4. Thus, we obtain the output current as follows:

$$I_9 = I_7 \cdot \exp\left(\frac{\sqrt{\frac{2}{\mu_n C_{ox} \frac{W_n}{L_n}}}\left(\sqrt{I_7} - \sqrt{I_8}\right)}{V_T}\right)$$

$$= \frac{V_T \ln n}{R_1} \cdot \exp\left(\frac{\sqrt{\frac{2}{\mu_n C_{ox} \frac{W_n}{L_n}}}}{V_T}\left(\sqrt{\frac{V_T \ln n}{R_1}} - \sqrt{\frac{V_{BE4}}{R_5} + \frac{1}{\beta}I_9}\right)\right) \tag{7}$$

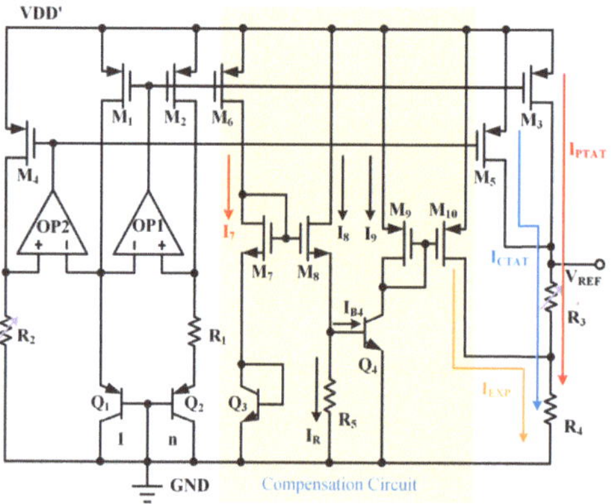

Figure 4. Proposed BGR core circuit with compensation.

The resistor values of R_1 and R_5 can be determined through numerical simulations using specialized software (such as MATLAB R2023b). Nonlinear terms can be eliminated by applying compensating current. Figure 5 illustrates that compared with former research, the generated current effectively matches the $T\ln(T/T_r)$ term across a wide temperature range. The error between these curves remains below 10% across temperatures ranging from 10 °C to 130 °C.

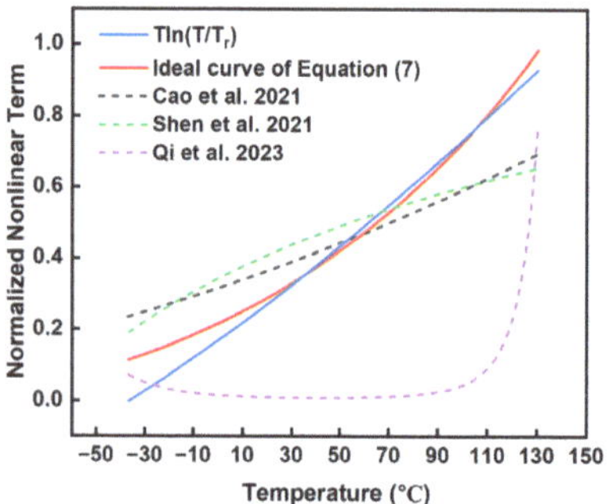

Figure 5. Normalized curve of $T\ln(T/T_r)$, I_9, and compensation terms used in [9–11].

Transistor M_{10} copies the current of M_9 to obtain the compensating current I_{EXP} and injects it into resistor R_4, thereby eliminating the nonlinear term and generating reference voltage.

$$V_{REF} = (I_{PTAT} + I_{CTAT}) \cdot (R_3 + R_4) + I_{EXP} \cdot R_4 \tag{8}$$

Due to the errors and mismatches inherent in chip manufacturing, it is essential to adjust the resistance values in the circuit to minimize temperature drift caused by process variations. Notably, R_2 governs the linear trend of the reference voltage, while R_3 controls the magnitude of the output voltage. Resistor trimming networks are employed at R_2 and R_3 to modify these resistances.

Fuse trimming is a relatively conventional method for adjusting resistance values, where each resistor in a series resistor network is connected in parallel with a fuse. Each fuse is associated with two trimming pads, and prior to trimming, the fuse short-circuits the resistor. During the trimming process, a voltage pulse applied across the terminals of the fuse burns it out, gradually increasing the resistance value of the resistor network.

Figure 6 illustrates the resistance trimming network utilizing the fuse trimming technique employed in this work. The fixed resistance R_F of R_2 and R_3 are 6.66 MΩ and 6.44 MΩ, respectively, and the unit trimming resistance R_T connected in parallel with the fuse are 40 kΩ and 60 kΩ.

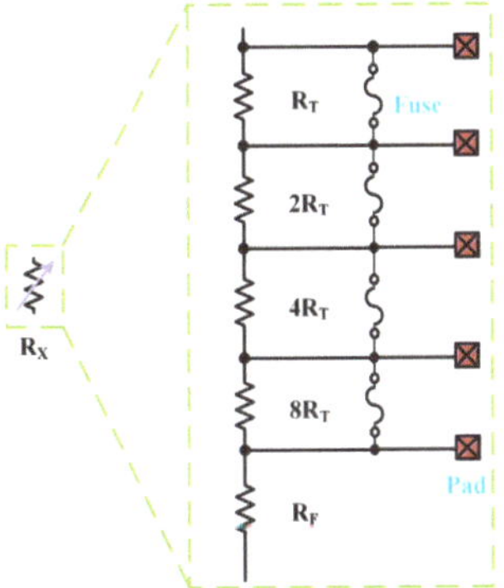

Figure 6. Resistance trimming network.

3. Results

The BGR is designed using a 0.18 μm BCD process. The schematic depiction of the overall circuit is illustrated in Figure 7. In the BGR core, we use the same start-up circuit as in the pre-regulator circuit. The mirrored current from the current source is injected into the BJT branch to help the circuit generate the reference voltage more quickly. The output of the BGR circuit is shown in Figure 8 across a temperature range of −40 to 130 °C. The output voltage from the structure depicted in Figure 3 exhibits a TC of 7.74 ppm/°C, as illustrated in Figure 8a. After implementing the advanced compensation methods proposed in this paper, the TC of the reference voltage was improved to 0.88 ppm/°C.

The untrimmed output of the BGR under different process corners is illustrated in Figure 9a (T, S, F are abbreviations for typical, slow, and fast, representing carrier mobilities. The first letter refers to NFETs, while the last letter refers to PFETs.). The variation between corners and temperatures reaches 6.5%. Through fuse trimming, the vertical discrepancy in output voltage across various corners is maintained within a margin of 1 mV. Table 1 presents the TCs of the proposed BGR before and after trimming. The results indicate that the circuit is capable of delivering a stable reference voltage despite significant temperature variations.

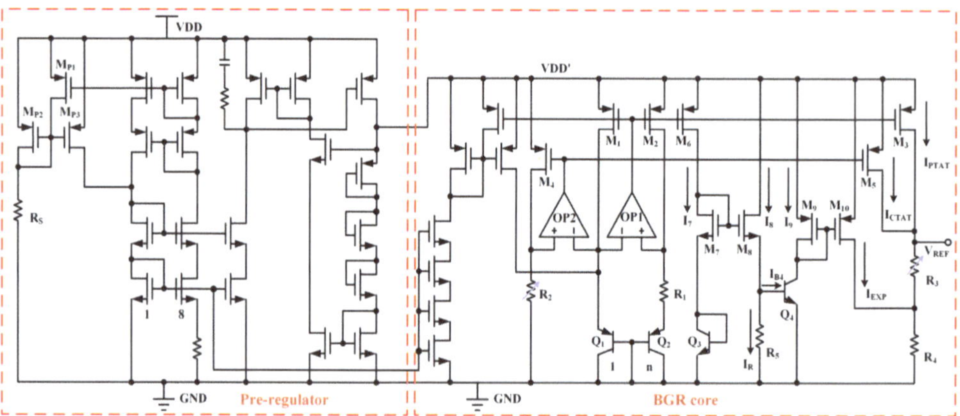

Figure 7. Proposed BGR structure.

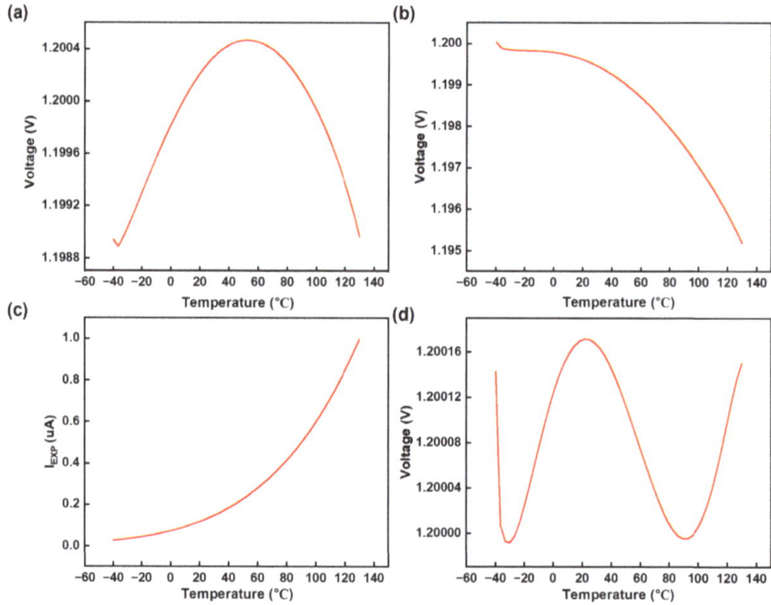

Figure 8. BGR output under temperature variation. (**a**) BGR output using structure in Figure 3. (**b**) BGR output without I_{EXP}. (**c**) Compensation current I_{EXP}. (**d**) BGR output after compensation.

Table 1. Simulation TC under different corners.

Simulation Corner	TC Without Trimming	TC With Trimming
TT	0.88 ppm/°C	0.88 ppm/°C
FF	11.57 ppm/°C	1.84 ppm/°C
SS	32.42 ppm/°C	1.37 ppm/°C

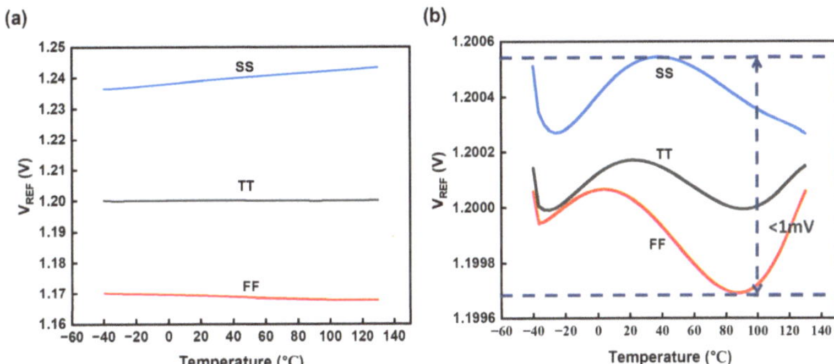

Figure 9. Output voltage with temperature across corners. (**a**) Reference without trimming. (**b**) Reference after trimming.

Figure 10a shows the output curves of the circuit at different supply voltages across varying temperatures. It can be observed that the changes in supply voltage result in an impact of less than 0.001% on the circuit output, which can be neglected. This observation is also supported by the PSRR curve depicted in Figure 10b. At a supply voltage of 4 V, the circuit achieves a PSRR of −167.38 dB at 10 Hz, −114.62 dB at 1 kHz, and −51.8 dB at 100 kHz. The results depicted in Figure 10 demonstrate that the circuit exhibits a robust resilience to power supply fluctuations.

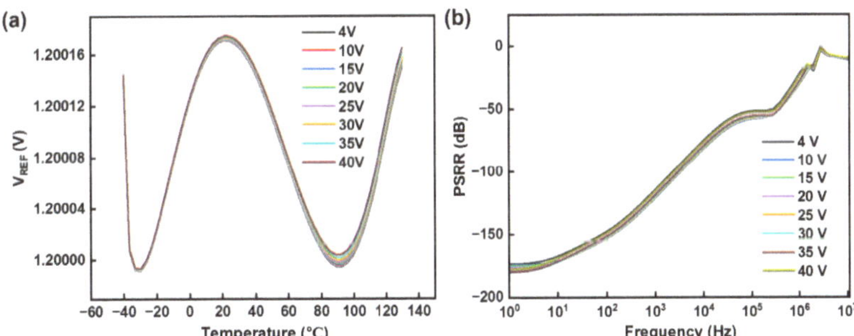

Figure 10. (**a**) Output voltage with temperature across various supply voltages. (**b**) PSRR across various supply voltages.

Table 2 presents the performance of the BGR and compares it with other published results. Compared with other works, our design achieves the lowest TC. In addition to excellent temperature independence, our circuit also exhibits remarkable rejection of voltage variations. In [6], the use of MOSFETs operating in the sub-threshold region can reduce the power consumption of the circuit to some extent; however, the TC of the reference circuit will greatly increase.

Table 2. Performance summary and comparison.

Parameter \ Design	[6]	[7]	[9]	[10]	[11]	[This Work]
Process (μm)	0.18	0.18 BCD	0.18 BCD	0.18 BCD	0.18 BCD	0.18 BCD
Supply voltage (V)	9–15	4.5–35	2.47–10	3–18	4.4–35	4–40
Reference (V)	2.507	2.5	1.218	2.048	1.205	1.2
Temperature (°C)	−20–80	−40–125	−50–125	−40–125	−40–125	−40–130
TC (ppm/°C)	24	1.5	9	1.2	3.83	0.88
Current consumption (μA)	8.47	267	-	279.6	23	9
PSRR (dB)	−60@10 Hz −53@1 kHz −81@1 MHz	−102.82@10 Hz	−86@120 Hz	-	−96@120 Hz	−167.38@10 Hz −114.62@1 kHz −51.8@100 kHz

4. Discussion

The proposed circuit not only achieves superior performance metrics but also demonstrates feasibility for practical applications requiring high precision and stability, such as power management systems and precision analog circuits. The advanced compensation techniques employed effectively address common challenges associated with temperature-induced variations, making this design a robust and reliable solution for modern integrated circuits.

However, compared to existing achievements, the minimum start-up voltage of our design remains relatively high, primarily constrained by the operating voltage of the BGR core. To further reduce the minimum operating voltage of the circuit, we could explore redesigning the BGR core module using processes characterized by lower threshold voltages and reduced static currents. This approach would also contribute to a decrease in the circuit's power consumption. Future work could investigate further optimization and integration strategies to enhance performance and adaptability across different semiconductor processes.

5. Conclusions

This article presents a low-TC BGR circuit that employs an advanced high-order temperature compensation technique and can operate over a wide range of temperatures and supply voltages through the implementation of a pre-regulation module. With a minimum TC of 0.88 ppm/°C and a PSRR of −167.38 dB at 10 Hz, this circuit demonstrates good resilience against temperature and voltage variations.

Author Contributions: Conceptualization, Y.X.; methodology, Y.X. and H.H.; software, Y.X.; validation, C.W.; formal analysis, Y.X. and H.H.; investigation, Y.X.; resources, C.W.; data curation, Y.X.; writing—original draft preparation, Y.X.; writing—review and editing, C.W.; visualization, Y.X.; supervision, W.H.; project administration, W.H.; funding acquisition, W.H. All authors have read and agreed to the published version of the manuscript.

Funding: This research received no external funding.

Data Availability Statement: The original contributions presented in the study are included in the article. Further inquiries can be directed to the corresponding author.

Acknowledgments: This project received experimental facilities and technical support from Wuhan BASALT Semiconductor Co., Ltd.

Conflicts of Interest: Author Chunlai Wang was employed by the company BASALT Semiconductor Co., Ltd. The remaining authors declare that the research was conducted in the absence of any commercial or financial relationships that could be construed as a potential conflict of interest.

References

1. Park, J.H.; Kim, J.K. A non-isolated dual-input DC-DC converter with wide input voltage range for renewable energy sources. In Proceedings of the 2017 IEEE 3rd International Future Energy Electronics Conference and ECCE Asia (IFEEC 2017—ECCE Asia), Kaohsiung, Taiwan, 3–7 June 2017; pp. 654–658. [CrossRef]
2. Yang, Y.; Li, Z.; Chen, C.; Wang, C. A High Voltage Bootstrapped Switch Based SAR ADC for Battery Management System. In Proceedings of the 2023 IEEE International Conference on Integrated Circuits, Technologies and Applications (ICTA), Hefei, China, 27–29 October 2023; pp. 110–111. [CrossRef]
3. Wei, X.; Liu, W.; Zhang, Z.; Yu, Z. A Wide Input Voltage Range LDO with Fast Transient Response. In Proceedings of the 2023 6th International Conference on Electronics Technology (ICET), Chengdu, China, 12–15 May 2023; pp. 114–119. [CrossRef]
4. Huang, W.; Yang, X.; Ling, C. A bandgap voltage reference design for high power supply. In Proceedings of the 2011 IEEE International Conference on Anti-Counterfeiting, Security and Identification, Xiamen, China, 24–26 June 2011; pp. 184–187. [CrossRef]
5. Widlar, R. New developments in IC voltage regulators. *IEEE J. Solid-State Circuits* **1971**, *6*, 2–7. [CrossRef]
6. Lavrentiadis, C.; Gogolou, V.; Siskos, S. Nano-Watt Voltage References for High Supply Voltages. In Proceedings of the 2022 Panhellenic Conference on Electronics & Telecommunications (PACET), Tripolis, Greece, 2–3 December 2022; pp. 1–6. [CrossRef]
7. Du, Y.; Wang, X.; Wu, W.; Han, Z.; Qi, Z.; Wang, Z.; Hu, S. A 14-ppm/V 1.5-ppm/°C Reference Voltage Source with a Wide Input Range of 4.5–35 V. In Proceedings of the 2023 8th International Conference on Integrated Circuits and Microsystems (ICICM), Nanjing, China, 20–23 October 2023; pp. 92–97. [CrossRef]
8. Malcovati, P.; Maloberti, F.; Fiocchi, C.; Pruzzi, M. Curvature-compensated BiCMOS bandgap with 1-V supply voltage. *IEEE J. Solid-State Circuits* **2001**, *36*, 1076–1081. [CrossRef]
9. Cao, Y.; Wen, Z. A 2.47-10V Curvature-corrected Bandgap Reference for LDO. In Proceedings of the 2021 5th International Conference on Communication and Information Systems (ICCIS), Chongqing, China, 15–17 October 2021; pp. 23–28. [CrossRef]
10. Shen, Y.; Lv, C.; Dong, L.; Xin, Y.; Zhang, B.; Guo, Z.; Xue, Z.; Geng, L. A 1.2 ppm/°C TC, 0.094% Inaccuracy 3–18 V Supply-Range Bandgap Reference with Self-Compensation. In Proceedings of the 2021 IEEE International Conference on Integrated Circuits, Technologies and Applications (ICTA), Zhuhai, China, 24–26 November 2021; pp. 139–140. [CrossRef]
11. Qi, E.; Fang, C.; Zhang, Y.; Cheng, Y.; Wang, N. A wide input low quiescent current without operational amplifier bandgap reference circuit. In Proceedings of the 2023 8th International Conference on Integrated Circuits and Microsystems (ICICM), Nanjing, China, 20–23 October 2023; pp. 326–330. [CrossRef]
12. Sheng, J.; Chen, Z.; Shi, B. A 1V supply area effective CMOS Bandgap reference. In Proceedings of the 2003 5th International Conference on ASIC Proceedings, Beijing, China, 21–24 October 2003; Volume 1, pp. 619–622. [CrossRef]
13. Tsividis, Y. Accurate analysis of temperature effects in $I_c - V_{BE}$ characteristics with application to bandgap reference sources. *IEEE J. Solid-State Circuits* **1980**, *15*, 1076–1084. [CrossRef]
14. Liao, X.; Zhang, Y.; Zhang, S.; Liu, L. A 3.0 μ Vrms, 2.4 ppm/°C BGR With Feedback Coefficient Enhancement and Bowl-Shaped Curvature Compensation. *IEEE Trans. Circuits Syst. I Regul. Pap.* **2024**, *71*, 2424–2433. [CrossRef]

micromachines

Article

A Study on the Frequency-Domain Black-Box Modeling Method for the Nonlinear Behavioral Level Conduction Immunity of Integrated Circuits Based on X-Parameter Theory

Xi Chen *, Shuguo Xie , Mengyuan Wei and Yan Yang

School of Electronic and Information Engineering, Beihang University, Beijing 100191, China
* Correspondence: chenxi@buaa.edu.cn

Abstract: During circuit conduction immunity simulation assessments, the existing black-box modeling methods for chips generally involve the use of time-domain-based modeling methods or ICIM-CI binary decision models, which can provide approximate immunity assessments but require a high number of tests to be performed when carrying out broadband immunity assessments, as well as having a long modeling time and demonstrating poor reproducibility and insufficient accuracy in capturing the complex electromagnetic response in the frequency domain. To address these issues, in this paper, we propose a novel frequency-domain broadband model (Sensi-Freq-Model) of IC conduction susceptibility that accurately quantifies the conduction immunity of components in the frequency domain and builds a model of the IC based on the quantized data. The method provides high fitting accuracy in the frequency domain, which significantly improves the accuracy of circuit broadband design. The generated model retains as much information within the frequency-domain broadband as possible and reduces the need to rebuild the model under changing electromagnetic environments, thereby enhancing the portability and repeatability of the model. The ability to reduce the modeling time of the chip greatly improves modeling efficiency and circuit design. The results of this study show that the "Sensi-Freq-Model" reduces the broadband modeling time by about 90% compared to the traditional ICIM-CI method and improves the normalized mean square error (NMSE) by 18.5 dB.

Keywords: integrated circuit (IC); models of integrated circuits for RF immunity behavioral simulation-conducted immunity modeling (ICIM-CI); immunity modeling; electromagnetic compatibility (EMC) modeling; direct power injection (DPI); X-parameters

Citation: Chen, X.; Xie, S.; Wei, M.; Yang, Y. A Study on the Frequency-Domain Black-Box Modeling Method for the Nonlinear Behavioral Level Conduction Immunity of Integrated Circuits Based on X-Parameter Theory. *Micromachines* **2024**, *15*, 658. https://doi.org/10.3390/mi15050658

Academic Editors: Zeheng Wang and Jing-Kai Huang

Received: 1 April 2024
Revised: 14 May 2024
Accepted: 15 May 2024
Published: 17 May 2024

1. Introduction

The immunity of integrated circuits is related to the electromagnetic safety of electronic devices (Figure 1) [1]. Researchers J. Loeckx and G. Gielen [2] found that small circuit topology changes can increase the immunity of integrated circuits by several orders of magnitude during DPI testing [3]. In light of this finding, being able to predict whether a device will pass sensitivity testing before it is manufactured is of great importance in terms of cost reduction [4,5]. However, chip manufacturers in general do not provide immunity models for their chips; as such, circuit designers must obtain a model of the chip's behavior by utilizing specific testing methods. This behavioral model is called the "black-box model", and involves extraction without knowing the internal physical details of the chips; instead, through the mapping of input and output signals, an abstract mathematical expression is obtained to describe the relationship between the recorded input signal x(t) and the output signal y(t). When the black-box model and the DUT are inspired by the same input signal, the output of the black-box model should effectively be as close as possible to the actual response of the DUT. The use of behavioral models protects the intellectual property (IP) rights of the device manufacturer and reduces the difficulty involved in modeling [6]. The authors of articles [7] and [8] used artificial neural networks (ANNs), Volterra levels [9], time-domain behavioral models such as envelope domain models [10], nonlinear impulse response models [11], and two-path memory models [12] in their studies, and constructed time-domain-sensitive behavioral models of devices to describe the output behavioral effects of the devices. However, time-domain models usually require a large number of tests to be conducted and for characterizations in the time domain to be made in order to generate a model that is accurate over a specific frequency range [13]. In the aforementioned techniques, the models can only be used to focus on IC faults (detection efficiency, jitter, etc.) in the time domain, and it is difficult to analyze mismatches between ports and high harmonics, which in turn prevents them from meeting the needs of broadband EMC applications. Moreover, one model only supports the simulation of a single frequency point under a single interference, and a large number of simulation models need to be established if the simulation is performed in the broadband frequency range. Therefore, frequency-domain models are more suitable for simulating distributed components over a bandwidth. In recent years, the ICIM-CI [14] model has been shown to predict chip immunity. The ICIM-CI [15] model is able to approximate and replace accurate chip or circuit simulations by using a look-up table and power distribution network (PDN)-based approach and enables the use of the model in the frequency domain [16].

Figure 1. External interference causes sensitization of the chip on the board.

However, the model still needs to depend on time-domain measurements when obtaining the IB (input behavior) netlist, which leads to the need for a large number of tests in broadband multi-frequency applications as well as redundant designs, which makes the process cumbersome and time-consuming. This then leads to a lack of model adaptability and the need to rebuild the model frequently when the criterion is changed, which in turn increases the amount of work required and overall time consumption. In addition, the ICIM-CI model has limitations in handling the nonlinear characteristics of the chip when it is under interference, and linear assumptions often need to be made [17], which leads to discrepancies between the model and the actual measurement results. In addition, since the model is based on the go–no-go decision criterion, it is not capable of parametric simulation, which limits its ability to accurately predict and meticulously analyze the performance of integrated circuits in complex electromagnetic environments and also obstructs circuit designers from integrating the chip into the cascade simulation of the whole circuit board. Due to this issue, the model is unable to comprehensively assess the electromagnetic characteristics of the circuit, resulting in a negative impact on overall design efficiency. Therefore, although the ICIM-CI model provides a framework for frequency-domain analysis, its limitations in terms of efficiency and nonlinear characteristic handling [18], as well as its inability to support overall quantitative simulation at the board level, are important challenges that need to be overcome.

In this study, we present the "Sensi-Freq-Model", a novel immunity black-box model for the rapid frequency-domain characterization of chips based on X-parameter theory, which aims to address the limitations of existing models. The model is accurate in characterizing the output characteristics of chips in both time and frequency domains and supports the parametric simulation of chips in circuits. The main advantage of the model is that it can adapt to changes in the sensitivity criteria so that circuit designers can quickly adapt to changes in design specifications or complex electromagnetic environments without having to dedicate long periods of time to reconstructing the model, which greatly improves the model's versatility and modeling efficiency. This flexibility and efficiency are especially important for modern circuit design, in particular in the pursuit of high-precision and fast iterative engineering.

The present paper is organized as follows: Section 2 describes the structure and theoretical basis of the model. Section 3 illustrates the extraction process and simulation results through two simulation examples. Lastly, in Section 4, we verify the modeling results and analyze them through the use of a test modeling example.

2. Sensi-Freq-Model Structure

In general, chips exhibit nonlinearity when sensitized by conducted interference [19], and this nonlinear effect is the main cause of EMC failures on chips. This is why the assumption of linearity often leads to biased final predictions [20,21]. When a small signal is injected, at this moment, the system exhibits linear characteristics and the harmonic frequencies at the output are negligible. The behavior of the chip can be sufficiently characterized using the scattering parameters in this case; however, with an increasing number of injected signals, the system will exhibit nonlinearity, and the range available for the scattering parameter will continue to decrease. The harmonic response cannot be ignored (Figure 2), at which point the output signal will produce multi-harmonic spectral mapping on the chip. Therefore, based on the characteristics of the chip-conducted interference response, the Sensi-Freq-Model modeling theory is proposed in combination with the X-parameter theory [22,23].

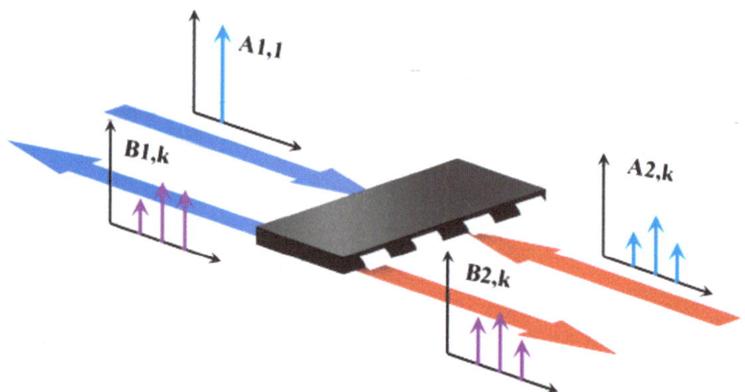

Figure 2. Schematic diagram of multi-harmonic spectrum mapping for a two-port network. Blue color represents inputs and outputs of port 1; red color represents inputs and outputs of port 2.

We introduce a set of multivariate complex functions $F_{pk}(.)$ into the frequency domain that relate all relevant input spectral components A_{qn} to the output spectral components B_{pk} (where q and p range from 1 to the number of signal ports and m and n range from zero to the highest harmonic index). The mathematical expression is as follows:

$$B_{pk} = F_{pk}\left(A_{11}, A_{12}\ldots, A_{1n}, A_{21}, A_{22}, \ldots, A_{2n}, \ldots A_{q1}, A_{q2}, \ldots A_{qn}\right) \tag{1}$$

where A_{11} denotes the input interference fundamental frequency.

While the chip is under the interference condition, it usually comprises a large-signal interference input component A_{11}, a DC excitation component, a regular drive signal input component, and other input components (harmonic frequency components). At this time, the relatively small regular drive signal input component satisfies the superposition principle, and all of the large-signal excitation and large-signal response in the excitation can be expressed by Equation (2):

$$LSOP = \begin{cases} DCS^{(LOSP)} = \{DCS_q\} \\ RFS^{(LOSP)} = A_{11} \\ DCR_p^{(LOSP)} = X_p^{(FDCR)}(\{DCS_q\}, |A_{11}|) \\ B_{p,k}^{(LOSP)} = X_{p,k}^{(F)}(\{DCS_q\}, |A_{11}|)P^k \end{cases} \tag{2}$$

where $DCS^{(LSOP)}$ denotes the DC excitation present in the chip related to the large-signal DC bias excitation DCS_q at port q.

$RFS^{(LSOP)}$ denotes the RF interference excitation present in the chip equal to the large-signal interference input component A_{11}.

$DCR_p^{(LSOP)}$ denotes the DC response at the large-signal operating point of port p related to the large-signal DC bias excitation DCS_q at port q, and the large-signal interference input component A_{11}, with the parameter $X_p^{(FDCR)}$ used to denote the X-parameter element of the excitation portion of the DC bias voltage.

$B_{p,k}^{(LSOP)}$ denotes the system response at the large-signal operating point related to the large-signal DC bias excitation DCS_q at port q and the large-signal interference input component A_{11}, with the parameter $X_{p,k}^{(F)}$ used to denote the X-parameter element of the large-signal operating point's influence [24].

Therefore, combining the effects of the large-signal nonlinear mapping and the linear non-analytic mapping that describe the co- and cross-frequency disturbances caused by the small-signal incidence, the scattering wave at the response port can then be described by Equation (3):

$$\begin{aligned} B_{pk} \cong{}& X_{pk}^F(refLOSP_{in})P^k + \sum_{\substack{q=1 \\ l=1 \\ (q,l)\neq(1,1)}}^{\substack{q=N \\ l=K}} X_{pk,ql}^S(refLOSP_{in})P^{k-l}A_{ql} \\ &+ \sum_{\substack{q=1 \\ l=1 \\ (q,l)\neq(1,1)}}^{\substack{q=N \\ l=K}} X_{pk,ql}^T(refLOSP_{in})P^{k+l}A_{ql}^* \end{aligned} \tag{3}$$

where $P = e^{j\varphi(A_{11})}$ is a unit-length phase quantity with the same phase as A_{11}. X_{pk}^F, $X_{pk,ql}^S$, and $X_{pk,ql}^S$ represent frequency-domain X-parameter elements that describe co-frequency disturbances caused by a small signal incident on the port of the device during testing. $X_{pk,ql}^T$ represents a frequency-domain X-parameter element that describes the cross-frequency disturbances caused by a small signal incident on the port of the device during testing. LSOPin represents the excitation portion of the LSOP, and refLSOPS represents the corresponding reference excitation.

The A-wave of incidence and the B-wave of scattering have two sets of indexes: p and q refer to the port number of the chip, while k and l refer to the number of harmonics.

3. Results

3.1. Immunity Modeling Based on Simulated Circuits

In this section, a method for extracting immunity models using chip-simulated circuits is presented and the accuracy of the Sensi-Freq-Model is verified. An analog circuit model of an operational amplifier built based on white-box theory is used. The circuit structure and specific parameters are shown in Figure 3. This white-box model was built to extract the Sensi-Freq-Model and verify the difference between its output in the simulation software and the white-box model. The op-amp is a forward amplifier circuit, and the parameters set for normal operation are Vin^+: f = 10 kHz, V = 100 mV sine wave signal, and the DC bias voltage set to $\text{V}^+ = +15$ V and $\text{V}^- = -10$ V. When the chip is in normal operation, the output of the circuit is as seen in Figure 4. Based on the DPI test method of IEC62132-4 [3], interference signals of different frequencies and powers are applied to the op-amp's input ports and power supply ports. The interference signals are selected for continuous waveforms according to Section 5 of IEC 62132-4. Here, the interference noise signal is simulated using the signal source module (power source-N Frequencies and Power Levels) in the ADS (advanced design system). The output of the op-amp is shown in Figures 4–7. Interference signals of different levels, when observing the output response sensitivity characteristics of the monitoring port, can be mainly divided into four phenomena (Table 1), which can be used to describe the faults as four types of situations according to the IC performance level specified in IEC62132-1 [25]. The specific description is shown in Table 1, and its output waveform schematic is shown in Figures 4–7.

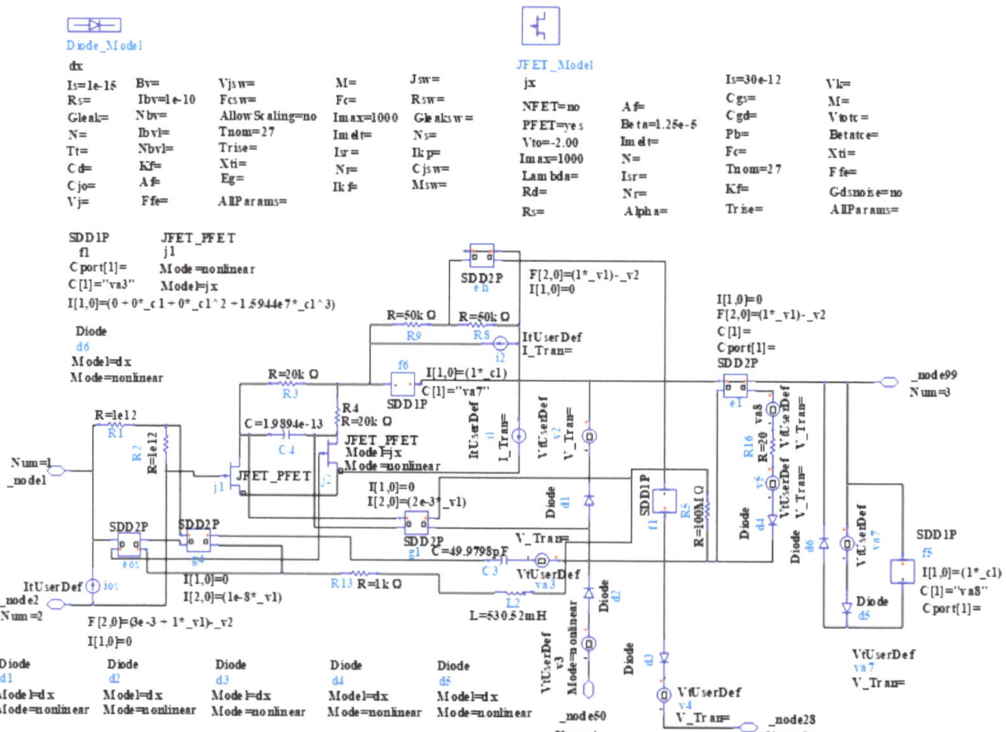

Figure 3. The constructed op-amp white-box model is used to extract the Sensi-Freq-Model and verify its accuracy. Component labeling in blue, component parameters in black.

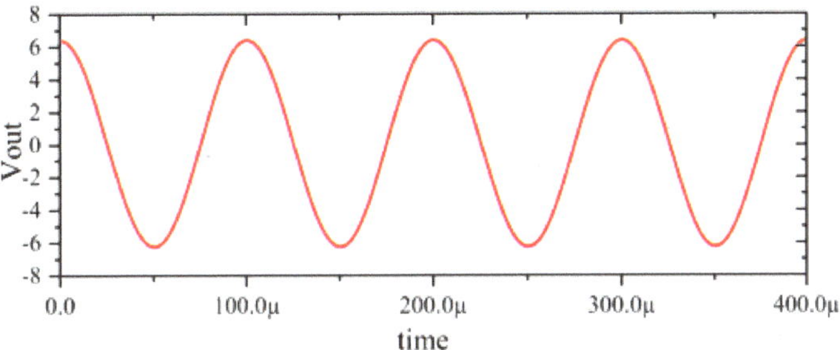

Figure 4. Performance without interference.

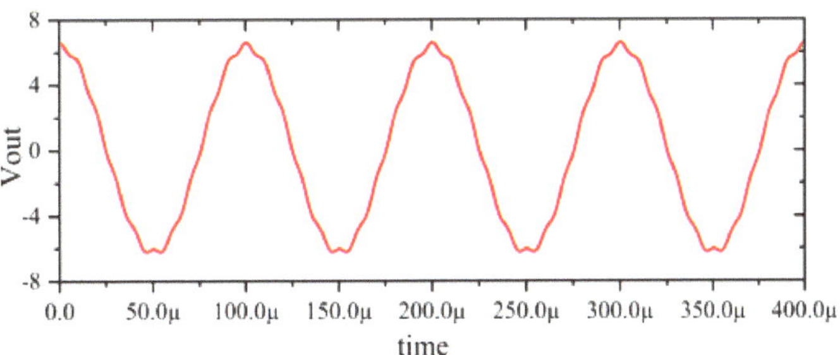

Figure 5. Slight jitter in the output waveform when subjected to −25 dBm interference, not exceeding the tolerance requirements.

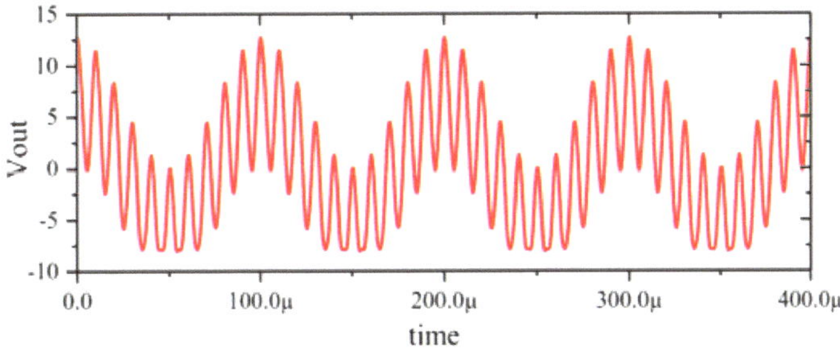

Figure 6. Jittery output waveforms exceeding tolerance requirements when subjected to 0 dBm interference.

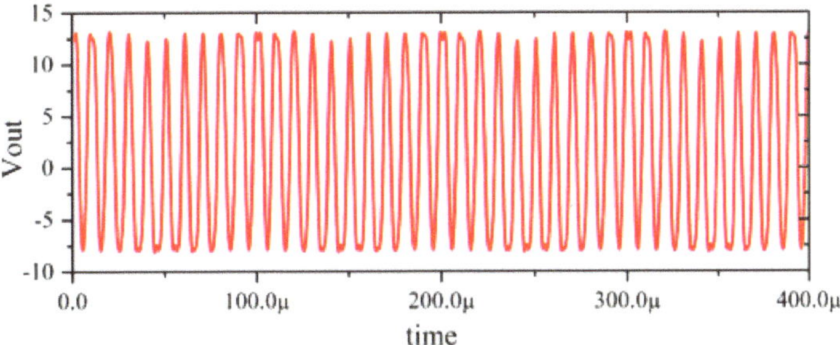

Figure 7. Severe jitter in the output waveform exceeding the tolerance requirement when subjected to 10 dBm interference.

Table 1. Output type and description of the operational amplifier after interference.

Case	Criteria	IC Performance Level	Description
Figure 4		Class A_{IC}	Normal output
Figure 5		Class A_{IC}	All monitored functions of the IC perform within the defined tolerances during and after exposure to disturbance.
Figure 6	$\Delta \text{Vout}_{p\text{-}p} \leq 13.2\,\text{mV}$	Class C_{IC}	The output waveform experiences distortion or jitter. The IC does not perform within the defined tolerances during exposure and does not return to normal operation. It returns to normal operation via manual intervention.
Figure 7		Class C_{IC}	The output waveform experiences serious distortion or jitter. The IC does not perform within the defined tolerances during exposure and does not return to normal operation by itself. It returns to normal operation via manual intervention.

Through the use of simulation, the Sensi-Freq-Model is extracted, a simulation model is built, the accuracy of the model is tested and it is verified as to whether the model can accurately reflect the output response of the chip under different disturbed situations to compare the built Sensi-Freq-Model with the traditional ICIM-CI model in the frequency-domain immunity prediction curves.

3.1.1. Model Extraction

During normal operation of the chip, the interference injection signal is injected from V− and Vin+ (Figures 8 and 9) separately, and the power of the interference signal is in the range from −40 dBm to 20 dBm. Each parameter in Equation (2) is solved according to the proposed method outlined in Section 2 to complete the operational amplifier immunity model.

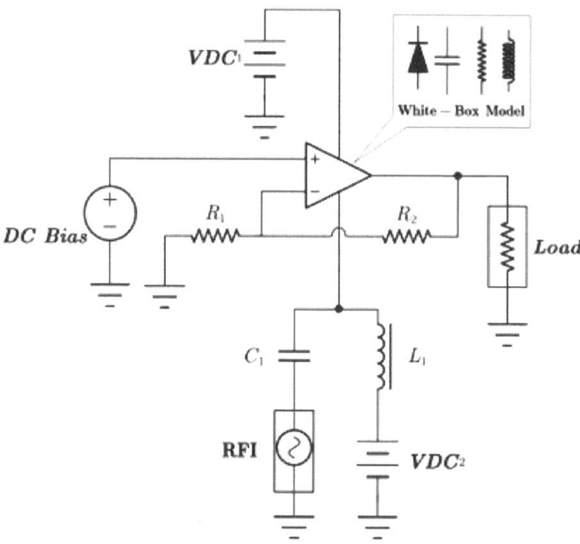

Figure 8. Extraction of injection immunity models for the power terminal of V−. *DC* Bias is a *DC* source; R_1 and R_2 are resistors; C_1 is the capacitor; L_1 is the inductor; Load is the matching load; RFI is a source of radio frequency interference; *VDC* is the *DC* voltage source for the amplifier.

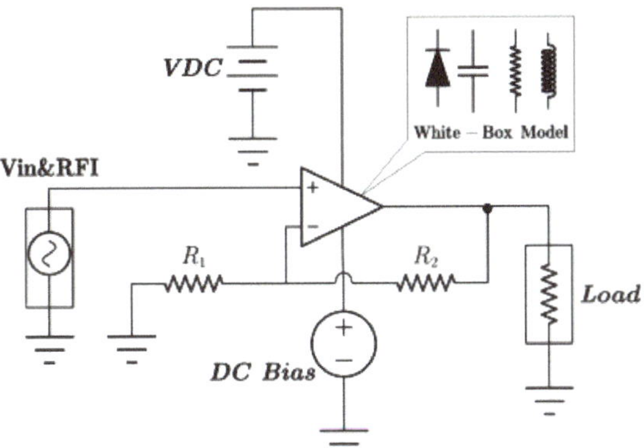

Figure 9. Extraction of the injected immunity model for the input terminal of Vin+. *DC* Bias is a *DC* source; R_1 and R_2 are resistors; Load is the matching load; Vin and RFI are the amplifier's function signal and interference signal injection source; *VDC* is the *DC* voltage source for the amplifier.

3.1.2. Model Verification

The accuracy of the extracted immunity model is examined by first verifying whether the model can output an accurate time-domain response waveform at a single frequency. As an example, the frequency of interference injection to the input is set at 100 kHz to verify the accuracy of the model. It can be seen that the model is able to accurately simulate the response waveforms of the device under various disturbed/unperturbed states, such as normal operation (Figure 10), distorted output waveform (Figure 11), distorted and jittered output waveform (Figure 12), and severely distorted output signal (Figure 13). These results show that the model is able to accurately predict the disturbed behavior and provide quantitative waveforms.

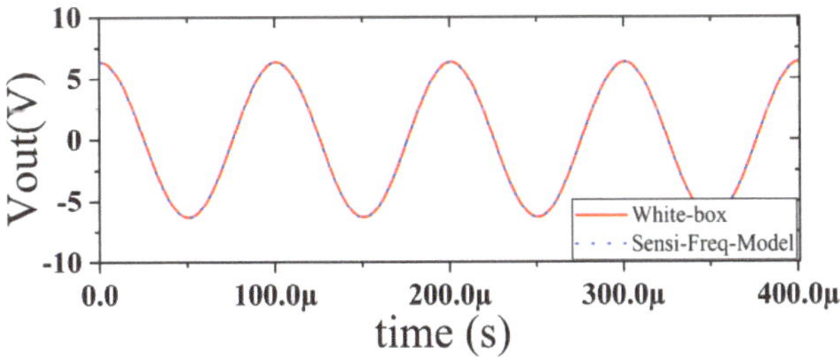

Figure 10. Comparison of simulation model output and actual output results during normal operation of the chip.

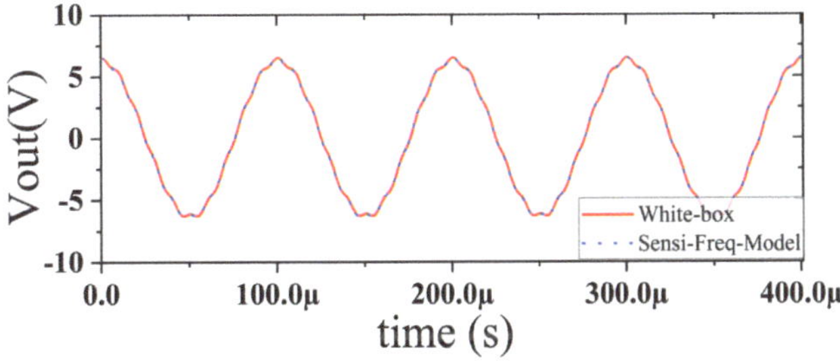

Figure 11. Comparison of simulation model output and actual output when slight jitter occurs at the chip output (interference injection −25 dBm).

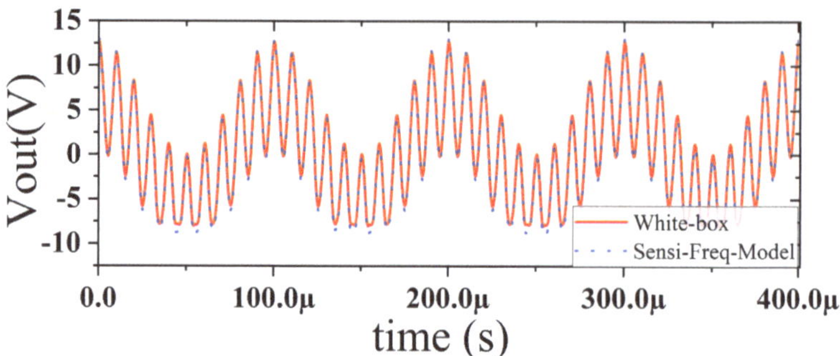

Figure 12. Comparison of simulation model output and actual output results when severe jitter occurs at the chip output (when interference is injected at 0 dBm).

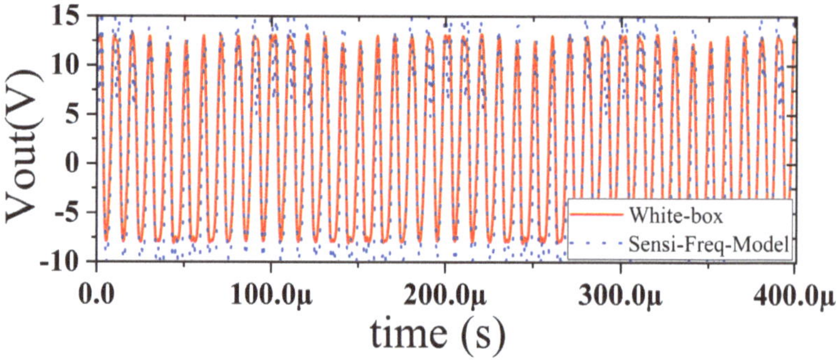

Figure 13. Comparison of simulation model output and actual output results when the chip output is severely disturbed (interference injection at 10 dBm).

3.1.3. Discussion

As can be seen from the results displayed in the above section, the Sensi-Freq-Model can directly output the response waveform of the chip after being perturbed; therefore, it is only necessary to build broadband immunity prediction curves for different immunization standards after one test. Compared to the traditional ICIM-CI model, which needs to determine the immunity criteria before establishing the immunity prediction curves, the Sensi-Freq-Model does not need to be re-modeled due to the change in the test criteria, which will result in significant savings in overall modeling time. In this section, the immunity criterion is set to the condition that the allowable change in peak-to-peak output voltage $\Delta \mathrm{Vout}_{p\text{-}p}$ is $\leq 5\%$ for comparison purposes, and Figure 14 shows the comparison of the results of using the method proposed in this paper and the traditional ICIM-CI modeling method with white-box simulation when interference is injected from the power supply side through the V- port. It can be seen that, under this immunity criterion, both modeling methods predict the sensitivity better because the chip has higher linearity.

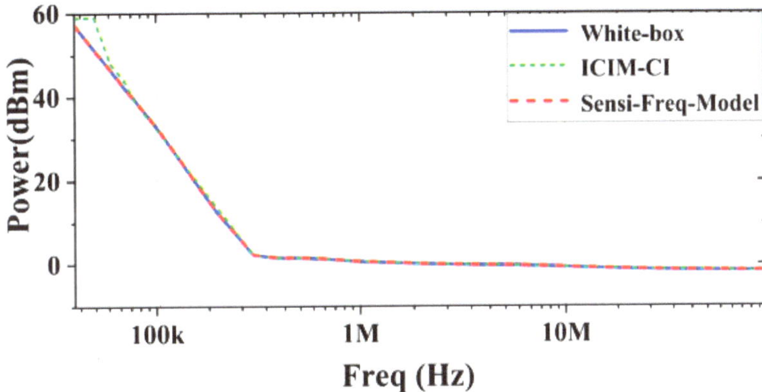

Figure 14. Comparison of the sensitivity prediction of the two modeling methods with the actual white-box output results (at the power supply side's V− with the immunity criterion at $\Delta \text{Vout}_{\text{p-p}} \leq 5\%$).

When injecting interference to Vin+ and setting $\Delta \text{Vout}_{\text{p-p}} \leq 5\%$, it can be seen that the prediction accuracy of the ICIM-CI model deteriorates when the system has a nonlinear response due to interference; in contrast, the Sensi-Freq-Model's prediction accuracy is still relatively satisfactory (Figure 15).

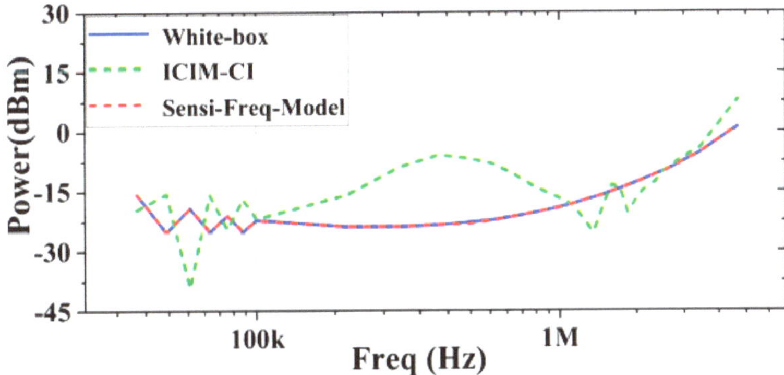

Figure 15. Comparison of the sensitivity prediction of the two modeling methods with the actual output results (at the Vin+ port with the immunity criterion at $\Delta \text{Vout}_{\text{p-p}} \leq 5\%$).

Changing the immunization criterion to output voltage peak-to-peak $\Delta \text{Vout}_{\text{p-p}} \leq 10\%$, the ICIM-CI requires that the model be rebuilt; however, the Sensi-Freq-Model can provide the immunity curve directly based on the output waveforms, and it can also be seen that the prediction accuracy of ICIM-CI is still lower than the prediction accuracy of the Sensi-Freq-Model (Figure 16).

It can be seen that the Sensi-Freq-Model can provide a very accurate output response in both time and frequency domains. Compared to ICIM-CI, even when the immunity criterion is changed, the model can still accurately predict the sensitivity phenomena of the device under examination without the need for re-measurement and modeling, which will greatly reduce the time required for modeling and testing.

3.2. Immunity Modeling Based on Measurements

In this section, we will verify the accuracy of the methodology by obtaining a Sensi-Freq-Model immunity model of the device using actual instrumentation and performing

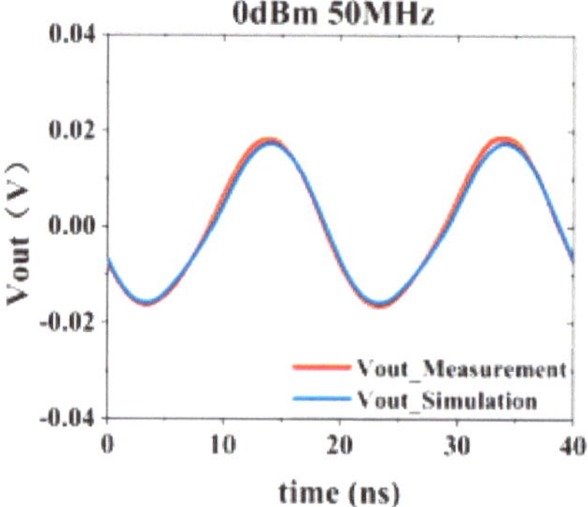

Figure 20. Injection of 0 dBm interference.

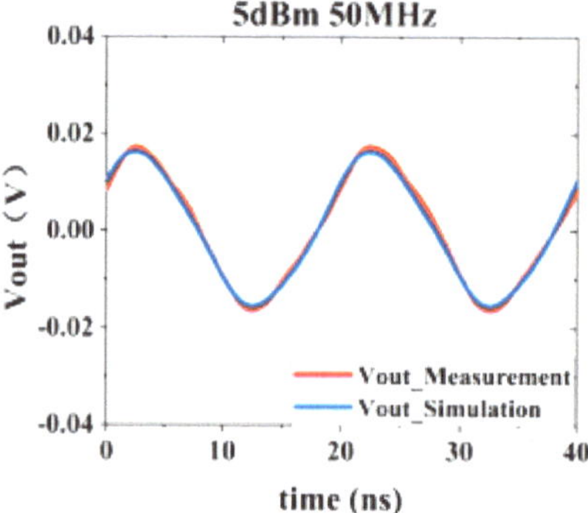

Figure 21. Injection of 5 dBm interference.

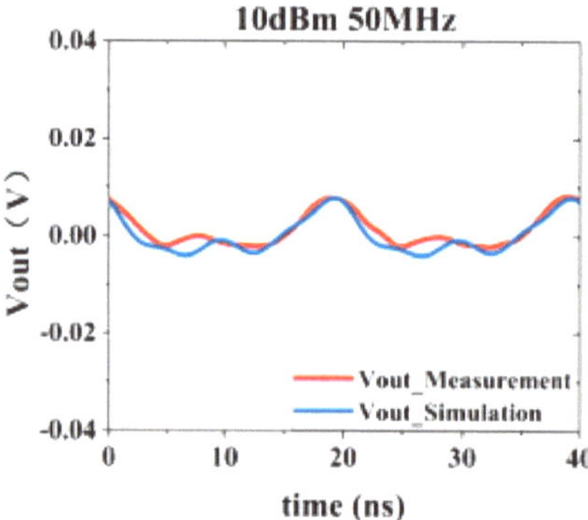

Figure 22. Injection of 10 dBm interference.

Figure 23 shows the comparison between the curves of the sensitivity level using the Sensi-Freq-Model method and the conventional ICIM-CI with the measured results when the condition of the immunity criterion $\Delta \text{Vout}_{p\text{-}p} \leq 30$ mV is introduced. It can be seen that the op-amps increase their immunity to disturbances as the disturbance frequency increases, and this trend can be predicted using both modeling methods; it is obvious, however, that using the method proposed in this paper (Sensi-Freq-Model) is more accurate than the traditional ICIM-CI modeling method in terms of prediction accuracy.

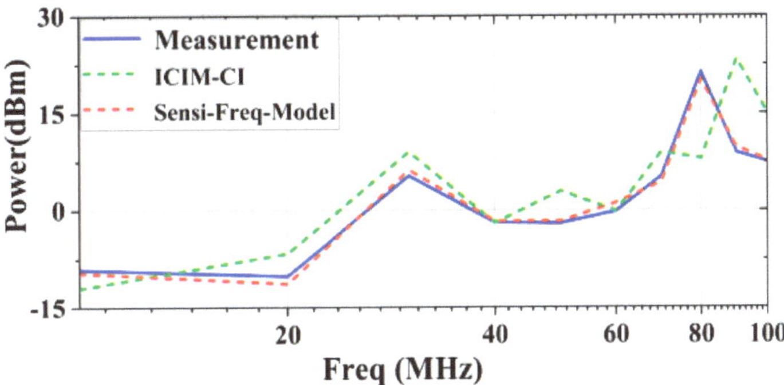

Figure 23. Comparative comparison of immunity prediction and DPI measurements for the two modeling approaches.

3.2.3. Discussion

Table 2 comprehensively demonstrates the comparison of the two modeling methods in the time and frequency domains in terms of modeling accuracy, modeling time, and whether cascade quantization simulation can be carried out. It can be seen that the Sensi-Freq-Model has obvious advantages in terms of modeling accuracy, modeling time, and support for quantization output and cascade simulation. Compared with ICIM-CI, using the method outlined in this paper (Sensi-Freq-Model) not only improves the modeling accuracy in the frequency domain by around 18.5 dB but also has significant advantages in terms of being able to output quantized waveforms at a single point and support cascade simulation, as well as improving the overall modeling time.

Table 2. Comparison of the NMSE and modeling times of different models.

Signal Type	Modeling Method	NMSE (dB)	Modeling Time	Supports Cascade Quantization Simulation
Time domain	Sensi-Freq-Model	−30.92	21 s	Yes
	ICIM-CI	No waveform output		No
Frequency domain	Sensi-Freq-Model	−31.3352	0.58 h	Yes
	ICIM-CI	−12.7982	8.3 h	No

In contrast, the Sensi-Freq-Model method provides circuit designers with more flexibility in designing circuit boards by providing quantized waveform output from monitoring ports. Specifically, based on the specific waveforms output by the Sensi-Freq-Model, designers are able to not only work with different degrees of redundancy to meet diverse design needs but also adjust the immunity standard-setting guidelines for the chips on the board without having to rebuild the chip immunity model. In addition, designers can further optimize the board layout using the actual immunity waveform output data provided by the Sensi-Freq-Model. This form of layout adjustment based on real-world data is difficult to achieve in traditional behavioral-level black-box models of frequency-domain conduction immunity. With this approach, immunization problems in circuit design can be more accurately addressed and solved, improving the reliability and efficiency of the design.

4. Conclusions

In this study, we validate the proposed X-parameter-based IC frequency-domain conduction sensitivity modeling method, the Sensi-Freq-Model, by comparing simulation and real cases, and the results prove its effectiveness and accuracy in describing and predicting the conduction sensitivity of ICs. Compared with the traditional ICIM-

CI modeling method, the Sensi-Freq-Model significantly reduces the time required for modeling and achieves a reduction of 18.5 dB in normalized mean square error (NMSE) in the frequency domain, which demonstrates its advantages in terms of efficiency and accuracy. In addition, the method provides quantifiable simulation results in the time domain, supporting the need for quantitative simulation of circuit board cascades and enhancing its application scope and utility. The Sensi-Freq-Model's modeling process relies on only unclassified information and quickly obtains highly accurate conduction immunity predictions from measurements alone over a wide range of frequencies, even in the absence of a full-impedance model of the integrated circuit and the surrounding PCB. Even in situations where a full impedance model of the IC and its surrounding PCB is not present, interference information can be accurately captured in the time-frequency domain over a wide range of frequencies, without being limited by the criteria for determining chip susceptibility. In light of the above, the Sensi-Freq-Model not only meets the needs of most IC terminal users to predict potential EMI in electronic devices but also provides an efficient and accurate modeling tool for circuit design and EMC analysis.

Author Contributions: Conceptualization, X.C. and S.X.; methodology, X.C. and S.X.; software, X.C.; validation, X.C. and M.W.; formal analysis, X.C.; investigation, X.C.; resources, S.X.; data curation, X.C.; writing—original draft preparation, X.C.; writing—review and editing, X.C.; visualization, Y.Y.; supervision, M.W.; project administration, S.X.; funding acquisition, S.X. All authors have read and agreed to the published version of the manuscript.

Funding: This research received no external funding.

Data Availability Statement: The original contributions presented in the study are included in the article.

Conflicts of Interest: The authors declare no conflict of interest.

References

1. Masetti, G.; Graffi, S.; Golzio, D.; Kovács-V, Z.M. Failures Induced on Analog Integrated Circuits by Conveyed Electromagnetic Interferences: A Review. *Microelectron. Reliab.* **1996**, *36*, 955–972. [CrossRef]
2. Loeckx, J.; Gielen, G. Assessment of the DPI Standard for Immunity Simulation of Integrated Circuits. In Proceedings of the 2007 IEEE International Symposium on Electromagnetic Compatibility, Honolulu, HI, USA, 9–13 July 2007.
3. *IEC 62132-4*; Integrated Circuits—Measurement of Electromagnetic Immunity—Part 4: Direct RF Power Injection Method. International Electrotechnical Commission: Geneva, Switzerland, 2006.
4. Sicard, E.; Jianfei, W.; Shen, R.J.; Li, E.-P.; Liu, E.-X.; Kim, J.; Cho, J.; Swaminathan, M. Recent Advances in Electromagnetic Compatibility of 3D-ICs—Part I. *IEEE Electromagn. Compat. Mag.* **2015**, *4*, 79–89. [CrossRef]
5. Sicard, E.; Jianfei, W.; Shen, R.; Li, E.-P.; Liu, E.-X.; Kim, J.; Cho, J.; Swaminathan, M. Recent Advances in Electromagnetic Compatibility of 3D-ICs—Part II. *IEEE Electromagn. Compat. Mag.* **2016**, *5*, 65–74. [CrossRef]
6. Gazda, C.; Ginste, D.V.; Rogier, H.; Couckuyt, I. An Immunity Modeling Technique to Predict the Influence of Continuous Wave and Amplitude Modulated Noise on Nonlinear Analog Circuits. In Proceedings of the 2013 International Symposium on Electromagnetic Compatibility, Brugge, Belgium, 2–6 September 2013.
7. Chahine, I.; Kadi, M.; Gaboriaud, E.; Louis, A.; Mazari, B. Characterization and Modeling of the Susceptibility of Integrated Circuits to Conducted Electromagnetic Disturbances up to 1 GHz. *IEEE Trans. Electromagn. Compat.* **2008**, *50*, 285–293. [CrossRef]
8. Čeperić, V.; Barić, A. Modelling of Electromagnetic Immunity of Integrated Circuits by Artificial Neural Networks. In Proceedings of the Electromagnetic Compatibility, 2009 20th International Zurich Symposium, Zurich, Switzerland, 12–16 January 2009.
9. Schroter, M.; Pehlke, D.R.; Lee, T.-Y. Compact Modeling of High-Frequency Distortion in Silicon Integrated Bipolar Transistors. *IEEE Trans. Electron Devices* **2000**, *47*, 1529–1535. [CrossRef]
10. Wang, H.; Bao, J.; Wu, Z. Multislice Behavioral Modeling Based on Envelope Domain for Power Amplifiers. *J. Syst. Eng. Electron.* **2009**, *20*, 274–277. [CrossRef]
11. Dghais, W.; Bellamine, F.H. Discrete Controlled Pre-Driver FIR Model for Hybrid IBIS Model AMS Simulation. In Proceedings of the 2016 IEEE 20th Workshop on Signal and Power Integrity (SPI), Turin, Italy, 8–11 May 2016.
12. Ngoya, E.; Quindroit, C.; Nebus, J.-M. On the Continuous-Time Model for Nonlinear-Memory Modeling of RF Power Amplifiers. *IEEE Trans. Microw. Theory Tech.* **2009**, *57*, 3278–3292. [CrossRef]
13. Lafon, F.; Ramdani, M.; Perdriau, R.; Drissi, M.; de Daran, F. An Industry-Compliant Immunity Modeling Technique for Integrated Circuits. In Proceedings of the International Symposium on Electromagnetic Compatibility (EMC 09), Kyoto, Japan, 11 October 2010.
14. *IEC 62433-4*; Integrated Circuit EMC IC Modeling Part 4: ICIM-CI, Integrated Circuit Immunity Model, Conducted Immunity. International Electrotechnical Commission Std.: Geneva, Switzerland, 2009.

15. Boyer, A.; Sicard, E. A Case Study to Apprehend RF Susceptibility of Operational Amplifiers. In Proceedings of the 2019 12th International Workshop on the Electromagnetic Compatibility of Integrated Circuits (EMC Compo), Hangzhou, China, 21–23 October 2019.

16. Khan, Q.M.; Koohestani, M.; Levant, J.-L.; Ramdani, M.; Perdriau, R. Validation of IC Conducted Emission and Immunity Models Including Aging and Thermal Stress. *IEEE Trans. Electromagn. Compat.* **2023**, *65*, 780–793. [CrossRef]

17. Lafon, F.; de Daran, F.; Ramdani, M.; Perdriau, R.; Drissi, M. Immunity Modeling of Integrated Circuits: An Industrial Case. *IEICE Trans. Commun.* **2010**, *E93.B*, 1723–1730. [CrossRef]

18. Wang, Z.; Zhou, C.; Liu, T.; Zhao, S.; Liang, Z. Nonlinear Behavior Immunity Modeling of an LDO Voltage Regulator under Conducted EMI. *IEEE Trans. Electromagn. Compat.* **2016**, *58*, 1016–1024. [CrossRef]

19. Magerl, M.; Stockreiter, C.; Baric, A. Analysis of Re-Radiated RF Harmonic Disturbance Caused by Integrated Circuit Input Pin Nonlinearity. In Proceedings of the 2019 International Symposium on Electromagnetic Compatibility—EMC EUROPE, Barcelona, Spain, 2–6 September 2019.

20. Duipmans, L.; Milosevic, D.; van der Wel, A.; Baltus, P. Distortion Contribution Analysis for Identifying EM Immunity Failures. *IEEE Trans. Circuits Syst. I Regul. Pap.* **2020**, *67*, 2767–2779. [CrossRef]

21. Redouté, J.; Steyaert, M. *EMC of Analog Integrated Circuits*; Springer: Dordrecht, The Netherlands, 2010; p. 243.

22. Verspecht, J.; Root, D.E.; Wood, J.; Cognata, A. Broad-Band, Multi-Harmonic Frequency Domain Behavioral Models from Automated Large-Signal Vectorial Network Measurements. In Proceedings of the IEEE MTT-S International Microwave Symposium Digest, Long Beach, CA, USA, 17 June 2005. [CrossRef]

23. Root, D.E.; Verspecht, J.; Sharrit, D.D.; Wood, J.N.; Cognata, A. Broad-Band Poly-Harmonic Distortion (PHD) Behavioral Models from Fast Automated Simulations and Large-Signal Vectorial Network Measurements. *IEEE Trans. Microw. Theory Tech.* **2005**, *53*, 3656–3664. [CrossRef]

24. Verspecht; Root, D.E. Polyharmonic Distortion Modeling. *IEEE Microw. Mag.* **2006**, *7*, 44–57. [CrossRef]

25. *IEC 62132-1*; Integrated Circuits, Measurement of Electromagnetic Immunity, 150 KHz–1 GHz: General Conditions and Definitions—Part 1. International Electrotechnical Commission Standard: Geneva, Switzerland, 2007.

26. Richelli, A. EMI Susceptibility Issue in Analog Front-End for Sensor Applications. *J. Sens.* **2016**, *2016*, 1082454. [CrossRef]

27. Poulton, A.S. Effect of Conducted EMI on the DC Performance of Operational Amplifiers. *Electron. Lett.* **1994**, *30*, 282–284. [CrossRef]

28. Setti, G.; Speciale, N. Design of a Low EMI Susceptibility CMOS Transimpedance Operational Amplifier. *Microelectron. Reliab.* **1998**, *38*, 1143–1148. [CrossRef]

29. *NVNA Help.General Hardware Configuration*; Keysight Technologies: Santa Rosa, CA, USA, 2023; pp. 1–8.

Article

A Cross-Process Signal Integrity Analysis (CPSIA) Method and Design Optimization for Wafer-on-Wafer Stacked DRAM

Xiping Jiang [1,2,3], Xuerong Jia [3,4], Song Wang [3], Yixin Guo [3], Fuzhi Guo [3], Xiaodong Long [3], Li Geng [4], Jianguo Yang [2] and Ming Liu [1,2,*]

1   Institute of Microelectronics of the Chinese Academy of Sciences, Beijing 100029, China; xiping.jiang@unisemicon.com
2   School of Integrated Circuits, University of Chinese Academy of Sciences, Beijing 100029, China; yangjianguo@ime.ac.cn
3   Xi'an UniIC Semiconductors, Xi'an 710075, China; xuerong.jia@unisemicon.com (X.J.); song.wang@unisemicon.com (S.W.); yixin.guo@unisemicon.com (Y.G.); fuzhi.guo@unisemicon.com (F.G.); xiaodong.long@unisemicon.com (X.L.)
4   School of Microelectronics, Xi'an Jiaotong University, Xianning West Road 28#, Xi'an 710049, China; gengli@xjtu.edu.cn
*   Correspondence: liuming@ime.ac.cn

Abstract: A multi-layer stacked Dynamic Random Access Memory (DRAM) platform is introduced to address the memory wall issue. This platform features high-density vertical interconnects established between DRAM units for high-capacity memory and logic units for computation, utilizing Wafer-on-Wafer (WoW) hybrid bonding and mini Through-Silicon Via (TSV) technologies. This 3DIC architecture includes commercial DRAM, logic, and 3DIC manufacturing processes. Their design documents typically come from different foundries, presenting challenges for signal integrity design and analysis. This paper establishes a lumped circuit based on 3DIC physical structure and calculates all values of the lumped elements in the circuit model with the transmission line model. A Cross-Process Signal Integrity Analysis (CPSIA) method is introduced, which integrates three different manufacturing processes by modeling vertical stacking cells and connecting DRAM and logic netlists in one simulation environment. In combination with the dedicated buffer driving method, the CPSIA method is used to analyze 3DIC impacts. Simulation results show that the timing uncertainty introduced by 3DIC crosstalk ranges from 31 ps to 62 ps. This analysis result explains the stable slight variation in the maximum frequency observed in vertically stacked memory arrays from different DRAM layers in the physical testing results, demonstrating the effectiveness of this CPSIA method.

Keywords: stacked DRAM; WoW; cross-process analysis methodology; signal integrity

Citation: Jiang, X.; Jia, X.; Wang, S.; Guo, Y.; Guo, F.; Long, X.; Geng, L.; Yang, J.; Liu, M. A Cross-Process Signal Integrity Analysis (CPSIA) Method and Design Optimization for Wafer-on-Wafer Stacked DRAM. *Micromachines* **2024**, *15*, 557. https://doi.org/10.3390/mi15050557

Academic Editors: Zeheng Wang and Jing-Kai Huang

Received: 21 March 2024
Revised: 16 April 2024
Accepted: 22 April 2024
Published: 23 April 2024

1. Introduction

Today's computing systems are primarily built on the von Neumann architecture, which reflects a clear separation of processing and memory units [1]. In data processing, a significant amount of data shuttles back and forth between the processing unit and the memory unit, resulting in significant latency and energy costs [2–4], forming a critical performance bottleneck [1,4]. The cost of performing a single multiply−accumulate operation by the processing unit is much smaller compared to the cost of moving the associated data [5,6]. The incompatibility between high-density memory processes, such as Dynamic Random Access Memory (DRAM), and logic processes, along with the increasing gap between the performance of memory and processing units, collectively contribute to the memory wall [1–4,7–10]. Several near-memory architectures have been proposed to address the memory wall problem by reducing the distance between computation and memory [11–13]. In particular, in the near-memory architectures where standard high-density memory and logic process components are integrated into a single package [14–18], cross-process design and analysis methods become a popular research topic [19–21].

In near-memory architectures with high-density memory and logic process components integrated into a single package, cross-process Signal Integrity (SI) design and analysis methods depend on the specific stacking architecture, as shown in Figure 1 and Table 1.

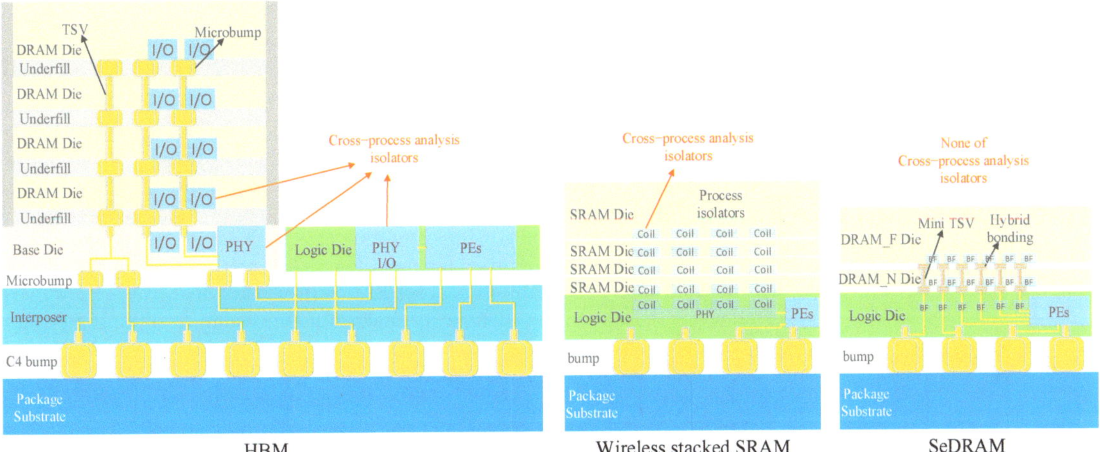

Figure 1. Three near-memory architectures with high-density memory and logic process components integrated into a single package.

Table 1. Comparison of cross-process structures.

	HBM [14,15]	Wireless Stacked SRAM [16]	SeDRAM [17,18]
Integration	2.5D	3D	3D
Memory	DRAM	High-density SRAM	DRAM
Stacked Structure	4~8 memory + 1 logic	4 memory + 1 logic	2~8 memory + 1 logic
Vertical Interconnection	TSV + microbump + interposer	None	HB + mini-TSV
	Metal	Wireless	Metal
	Package	None	WoW BEOL
Related Processes	DRAM and logic	Logic 1 and logic 2	DRAM, logic, and 3DIC
Interface between Stacked Memory and Logic	I/O	Coil on logic die	Buffer
Separation of Processes	Yes, by I/O circuit	Yes, by magnetic field	No
SI Analysis Method	Package-based validation	Virtual model	CPSIA

- High Bandwidth Memory (HBM) improves the memory access performance by a Through-Silicon Via (TSV) structured 2.5D near-memory architecture [22,23]. The DRAM dies in HBM are stacked through TSVs and microbumps, forming a DRAM stack. The DRAM stack is horizontally interconnected with logic through an interposer. All DRAM dies and logic dies are independently designed with their respective I/O circuits. These I/O circuits act as isolators for cross-process analysis, segmenting the design and analysis within the HBM package into DRAM, logic, and package-based interconnections. The SI design of constructing an HBM stack is fundamentally packaging design [20].
- Ref. [16] reports a wireless stacked Static Random Access Memory (SRAM) which utilizes semiconductor process coils to establish a vertical data path between four SRAM dies and a logic die, creating a 3D near-memory architecture. In ref. [16], there is

no metal-based signal interconnect between SRAM dies and the logic die ("Power supplies are provided via bonded wires"). The interconnection between the two different semiconductor processes is achieved through a magnetic field model within the package of this structure. As a virtual model, the magnetic field model is not constrained by any stacking manufacturing process, simplifying the SI analysis of this structure.

- Refs. [17,18] report a Stacked Embedded DRAM (SeDRAM) architecture, a noteworthy technology in the industry in recent years and the study target of this paper. SeDRAM vertically stacks DRAM dies and a logic die into a hybrid 3DIC package, resulting in the shortest physical distance for memory access at the micron level [24]. Unlike HBM's packaging integration technology, SeDRAM utilizes a Wafer-on-Wafer (WoW) Back-End-of-the-Line (BEOL) 3DIC process for manufacturing mini-TSV and Hybrid Bonding (HB) to establish high-density vertical memory access interconnects between memory and computing units, significantly enhancing memory access efficiency [25]. In this 3DIC package, a substantial number of mini-TSV and HB cells are used for interconnecting data paths. 3DIC path of SeDRAM is driven by DRAM and logic buffers, creating a cross-process signal integrity analysis environment. As a result, three different semiconductor manufacturing processes, namely, DRAM, logic, and 3DIC, are integrated into the overall design, making it challenging to distinguish boundaries of signal integrity design and analysis. Addressing the aforementioned issues, this paper proposes the Cross-Process Signal Integrity Analysis (CPSIA) method.

In Table 1, among the three near-memory architectures, HBM offers the most convenient simulation framework because the signals across stacks are isolated by I/O. However, HBM has the lowest vertical stacking density. Wireless stacked SRAM achieves an overlapping layout between the coils of the vertical channel and the memory media, achieving an area efficiency of 1162 GB/s/mm^2 [16], surpassing HBM. The SI analysis of the wireless stacked SRAM structure is conducted on a unit of stacked chips, with interconnections between stacks facilitated by virtual models. SeDRAM, leveraging WoW BEOL, greatly enhances the interconnect density across stacks. However, the I/O-less structure of SeDRAM requires a cross-process SI analysis environment that includes DRAM logic and 3DIC processes.

The SI analysis of HBM and wireless stacked SRAM among the three near-memory architectures listed in Table 1 was conducted on a unit of stacked chips, with system-level simulation implemented between the stacks. The I/O-less structure of SeDRAM requires a cross-process SI analysis environment that includes DRAM logic and 3DIC processes. The signal integrity design and analysis of the SeDRAM architecture presents a significant challenge due to its cross-process nature, encompassing the DRAM, logic, and 3DIC processes. In this hybrid architecture, the memory and computing devices are interconnected through Hybrid Bonding (HB) and mini-TSV cells, with the physical data path across different manufacturing processes in terms of libraries and design rules provided by multiple foundries. Standard Electronics Design Automation (EDA) tools do not support comprehensive SI analysis for these cross-process architectures. To establish sub-micron vertical interconnections between devices of different manufacturing processes, this cross-process vertical interconnection employs buffer drivers for the vertical interconnect units, rather than I/O circuits of HBMs [14,15] or a virtual model of wireless stacked SRAM [16]. Because of the absence of I/O circuits or a virtual model for segmenting the cross-process structure, the SI analysis of SeDRAM is geared towards buffers, essentially following the design requirements of a standard 2D chip. However, this takes place in a 3D cross-process structure. This hybrid architecture demands unique SI design approaches.

This paper addresses the cross-process design and analysis requirements for 3D vertical stacking that are compatible with three different manufacturing processes and proposes the CPSIA method for SeDRAM. This paper formulates lumped circuit models based on the 3DIC physical structure for vertical data paths, facilitating a mixed design and analysis approach that operates independently of 3DIC manufacturing processes.

Based on the lumped circuit model, a CPSIA methodology is introduced. It involves the extraction of buffer netlists based on commercial DRAM and logic foundries and the use of the combination of lumped circuits to equivalently represent the vertical stacking paths. A cross-process simulation environment is established, encompassing three commercial processes in terms of DRAM logic and 3DIC. The consistency of the comparative analysis between the simulation results and the silicon results demonstrates the effectiveness of this CPSIA method.

2. Study of the 3DIC Model

This section introduces the physical structure of the vertical stacking path used to construct the multi-layer vertical stacked DRAM platform. Following the 3DIC physical structure, a lumped circuit model is proposed, and all values of the lumped elements in the circuit model are calculated with the transmission line model. A 3DIC frequency-domain analysis is demonstrated using the circuit model.

2.1. Introduction of Study Target

As shown in Figure 2, the stack of the SeDRAM is the study target of the SI analysis methodology presented in this paper. The DRAM_Near (DRAM_N), DRAM_Far (DRAM_F), and logic components are interconnected through HB and mini-TSV technologies based on the BEOL process, with DRAM_N and DRAM_F representing the DRAM dies located near and far from the logic die, respectively. The HB cell facilitates face-to-back interconnection between DRAM_N and DRAM_F, as well as face-to-face interconnection between DRAM_N and DRAM_F. Mini-TSVs are used to establish interconnections passing through the DRAM_N substrate.

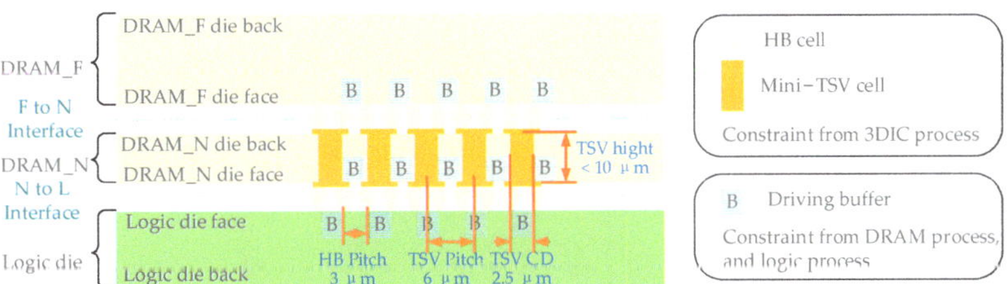

Figure 2. The stacking structure and vertical stacking cells of the SeDRAM.

The SI analysis goals for HBM and SeDRAM differ; HBM involves system-level SI analysis, while SeDRAM focuses on cross-process SI analysis. System-level SI analysis is conducted after the completion of chip design. HBM achieves DRAM design based on fixed design targets derived from the system level. It performs system-level SI analysis, including 3DIC, focusing on centralized I/O as the chip-to-chip boundary, resulting in lower analysis precision. In contrast, SeDRAM requires SI analysis for interconnections between different stacks during the design process. It involves chip design optimization based on this SI analysis, necessitating the establishment of a cross-process SI analysis environment that includes DRAM logic and 3DIC processes. The CPSIA method enables higher-precision design, simulation, and optimization during the SeDRAM stacking chip design processes, leading to improved overall system performance.

The physical vertical interconnection is driven by DRAM and logic buffers, and this vertical stacking path involves constraints from the DRAM logic and 3DIC processes separately. The 3DIC is expressed by a lumped circuit model, which helps reduce the complexity of the cross-process simulation environment.

2.2. Lumped Circuit Model of Vertical Stacking Paths

Figure 3a presents a physical model of vertical stacking paths. In this model, a 2HB+1TSV+2HB structure is employed to connect memory access signals, which is the primary focus of this work. DRAM_N is interconnected face-to-face with logic, connected by the lower Inter-Metal Dielectric 2 (IMD2) through HB cells. DRAM_F is interconnected back-to-back with DRAM_N, connected by upper IMD2 through HB cells. Mini-TSVs traverse DRAM_N to establish metal connections between the upper and lower HB layers. The circuit in DRAM_F is interconnected with the circuit in logic through the 2HB+TSV+2HB path. This vertical data path is the most complex in SeDRAM and serves as the analysis target because, in SeDRAM, the vertical data paths for the data inputs/outputs (DQs)/command and address inputs (CAs) of the two DRAM dies are individually interconnected with the logic die. For yield and impedance considerations, every two HB cells are interconnected with one mini-TSV, forming the 2HB+1TSV+2HB data path structure.

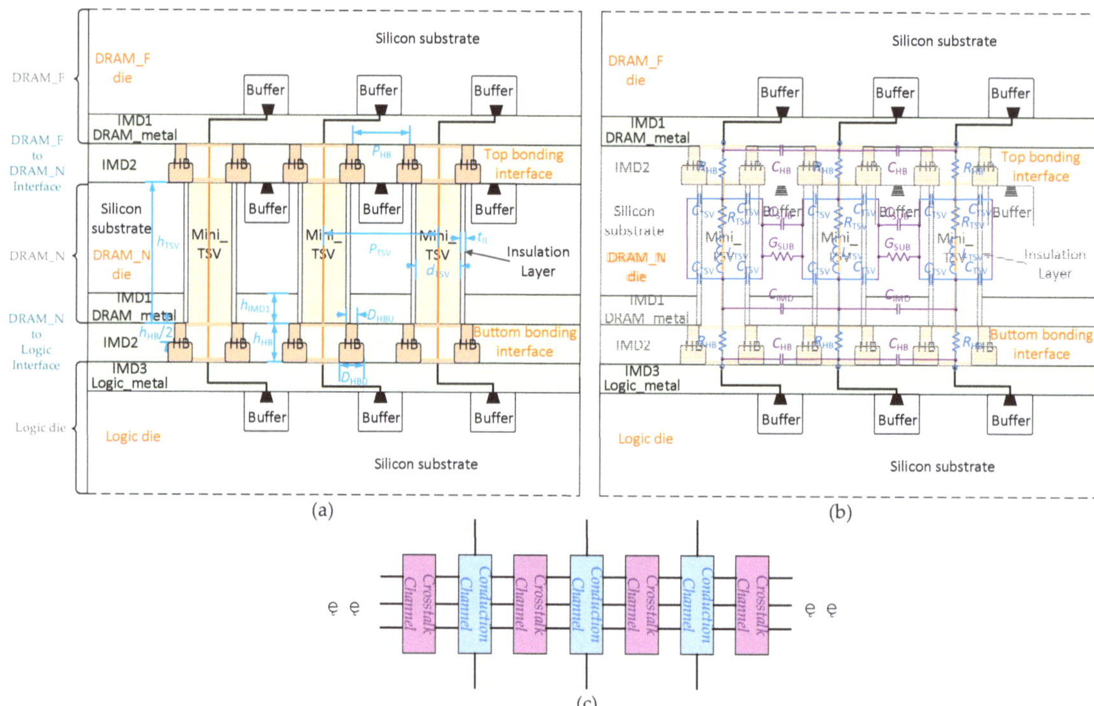

Figure 3. (**a**) Physical structure of vertical stacking paths. (**b**) Lumped circuit model of vertical stacking paths. (**c**) Simplified structures of the conduction channel and crosstalk channel.

According to the rule of thumb in transmission line theory, when the length of a transmission line is smaller than 1/20 of the target wavelength (λ), lumped elements can accurately represent the electrical behavior of the transmission line [26]. The target electromagnetic wave of this paper ranges from 1 GHz to 10 GHz, and the wavelength in silicon is from 162,000 μm to 16,200 μm. The lumped model studied in this paper consists of transmission lines in the sub-10 μm range, which is much smaller than the wavelength of the target frequency, satisfying the 1/20 λ condition. Therefore, lumped elements are used to model the vertical stacking cells.

The size of vertical stacking cells is in the sub-10 μm range, as shown in Table 2, and their electrical behaviors can be approximated using lumped elements, as illustrated in Figure 3b. Lumped elements can be categorized into two groups. The first corresponds to the lumped elements resulting from the vertical stacking cells themselves, marked in blue, while the second corresponds to the lumped elements resulting from interactions between vertical stacking cells, marked in purple. The blue lumped elements form a conduction channel, and the purple lumped elements form a crosstalk channel. The simplified structures of the conduction channel and crosstalk channel are depicted in Figure 3c.

Table 2. Physical dimensions of the model indicated in Figure 3a.

	Dimensions	Value	Unit	Description
TSV	h_{TSV}	10	μm	TSV height
	d_{TSV}	2.5	μm	TSV diameter
	P_{TSV}	6	μm	TSV pitch
HB	h_{HB}	2	μm	HB height
	d_{HBU}	0.6	μm	Up part of HB diameter
	d_{HBD}	1.5	μm	Down part of HB diameter
	P_{HB}	3	μm	HB pitch
Insulation Layer	t_{IL}	0.2	μm	Insulation layer thick
IMD1	h_{IMD1}	3	μm	IMD1 height

In Figure 3b, the blue lumped elements primarily represent the signal path between DRAM_F and logic dies, running from top to bottom, and include the following:

- R_{HB}, the equivalent resistance of the dual HB cell structure;
- R_{TSV} and L_{TSV}, the equivalent resistance and inductance of the mini-TSV cell connecting the backside and top metal layers of DRAM_N;
- C_{TSV}, the distributed capacitance formed by the outer surface of the mini-TSV copper pillar and the DRAM_N substrate, enclosed by the insulation layer (SiO2) surrounding the TSV.

The purple lumped elements in Figure 3b arise from the medium between vertical stacking cells and manifest in two types of structures. The IMD2 is formed by the BEOL process to create HB cells, exhibiting good insulating properties but possessing a significant relative dielectric constant (see Table 3). Due to the thinning of the DRAM_N substrate, the DRAM_N substrate in the structure of Figure 3a consists solely of the p-type substrate, which has both non-ideal conductivity and a significant relative dielectric constant (see Table 3). Distributed parameters exist in the IMD2 and DRAM_N substrate media, serving as coupling channels for crosstalk between adjacent vertical stacking paths, and their equivalent lumped elements are as follows:

- C_{HB}, the distributed capacitance formed by the adjacent dual HB cell structures through the IMD2 medium;
- C_{IMD}, the distributed capacitance formed by the adjacent mini-TSV cells through the Inter-Metal Dielectric 1 (IMD1) medium (the metal layer of DRAM_N);
- C_{Sub} and G_{Sub}, the equivalent capacitance and conductance formed by the adjacent mini-TSV cells through the medium of the DRAM_N substrate.

Table 3. Material parameters of the model indicated in Figure 3a.

	Parameters	Value	Unit	Description
TSV	ρ_{TSV}	1.68×10^{-8}	Ω·m	Resistivity of TSV (Cu)
	μ_{r_TSV}	1	\	Relative permeability of TSV
HB	ρ_{HB}	1.68×10^{-8}	Ω·m	Resistivity of HB (Cu)
Si Sub	σ_{Sub}	10	S/m	Conductivity of Si Sub
	ε_{r_Sub}	11.9	\	Relative permittivity of Si Sub
IMD1/IMD2	ε_{r_IMD}	4.1	\	Relative permittivity of IMD
Insulation Layer	ε_{r_IL}	4.1	\	Relative permittivity of insulator
Free Space	ε_0	8.854×10^{-12}	F/m	Permittivity of free space
	μ_0	1.257×10^{-6}	H/m	Permeability of free space

The lumped elements in the lumped circuit model can be determined using the formulas of the transmission line model, including coaxial line, two-wire line, and planar line models [27]. The calculation of these lumped elements is dependent on physical dimensions and material parameters, which are detailed in Tables 2 and 3, respectively.

R_{TSV} is calculated using the cylindrical resistor formula:

$$R_{\text{TSV}} = \rho_{\text{TSV}} \times \frac{h_{\text{TSV}}}{\pi \times (d_{\text{TSV}}/2)^2} \ . \tag{1}$$

L_{TSV} is determined through the coaxial cable model formula:

$$L_{\text{TSV}} = \frac{\mu_0 \times \mu_{\text{r\_TSV}}}{2\pi} \times \ln\left(\frac{P_{\text{TSV}}}{(d_{\text{TSV}}/2)}\right) \times h_{\text{TSV}} \ . \tag{2}$$

R_{HB} represents the parallel resistance of two HB cells, where the resistance of each HB cell is computed in two parts based on the HB structure and using the cylindrical resistor formula:

$$R_{\text{HB}} = \frac{1}{2}(R_{\text{HBU}} + R_{\text{HBD}}) = \frac{1}{2}\left(\rho_{\text{HB}} \times \frac{h_{\text{HB}}/2}{\pi \times (d_{\text{HBU}}/2)^2} + \rho_{\text{HB}} \times \frac{h_{\text{HB}}/2}{\pi \times (d_{\text{HBD}}/2)^2}\right). \tag{3}$$

C_{TSV} represents the distributed capacitance of the mini-TSV in the insulation layer, and it is calculated using the coaxial line capacitance formula:

$$C_{\text{TSV}} = \frac{1}{4}\left(\frac{2\pi \times \varepsilon_0 \times \varepsilon_{\text{r\_IL}}}{\ln\left(\frac{(d_{\text{TSV}}/2+t_{\text{IL}})}{(d_{\text{TSV}}/2)}\right)} \times (h_{\text{TSV}} - h_{\text{IMD1}})\right) \ . \tag{4}$$

C_{HB} denotes the distributed capacitance of the dual copper pillar structure in IMD2 and is calculated in two parts using the two-wire line capacitance formula:

$$C_{\text{HB}} = C_{\text{HBU}} + C_{\text{HBD}} = \frac{\pi \times \varepsilon_0 \times \varepsilon_{\text{r\_IMD}}}{\cosh^{-1}\left(\frac{P_{\text{HB}}}{d_{\text{HBU}}}\right)} \times \frac{h_{\text{HB}}}{2} + \frac{\pi \times \varepsilon_0 \times \varepsilon_{\text{r\_IMD}}}{\cosh^{-1}\left(\frac{P_{\text{HB}}}{d_{\text{HBD}}}\right)} \times \frac{h_{\text{HB}}}{2} \ . \tag{5}$$

C_{IMD} represents the distributed capacitance of adjacent mini-TSVs in IMD1, calculated using the two-wire line capacitance formula:

$$C_{\text{IMD}} = \frac{\pi \times \varepsilon_0 \times \varepsilon_{\text{r\_IMD}}}{\cosh^{-1}\left(\frac{P_{\text{TSV}}}{d_{\text{TSV}}}\right)} \times h_{\text{IMD1}}. \tag{6}$$

G_{Sub} is the distributed conductance of adjacent mini-TSVs in the DRAM_N substrate, determined by the two-wire line capacitance formula:

$$G_{\text{Sub}} = \frac{\pi \times \sigma_{\text{Sub}}}{\cosh^{-1}\left(\frac{P_{\text{TSV}}}{d_{\text{TSV}}}\right)} \times (h_{\text{TSV}} - h_{\text{IMD1}}) \ . \tag{7}$$

C_{Sub} represents the equivalent capacitance of adjacent mini-TSVs in the DRAM_N substrate and is calculated using the two-wire line capacitance formula:

$$C_{\text{Sub}} = \frac{\pi \times \varepsilon_0 \times \varepsilon_{\text{r\_Sub}}}{\cosh^{-1}\left(\frac{P_{\text{TSV}}}{d_{\text{TSV}}}\right)} \times (h_{\text{TSV}} - h_{\text{IMD1}}) \ . \tag{8}$$

The values of all lumped elements in Figure 3b are determined using the method described above, as summarized in Table 4. Among these elements, R_{TSV}, L_{TSV}, R_{HB}, and C_{TSV} impact conduction along the signal path, whereas C_{HB}, C_{IMD}, G_{Sub}, and C_{Sub} constitute coupling channels for crosstalk between adjacent vertical stacking paths.

Table 4. Lumped element values of the model indicated in Figure 3b.

Function	Medium	Parameters	Value	Unit	Description
Conduction	TSV	R_{TSV}	34.22	mΩ	Equivalent resistance of mini-TSV
		L_{TSV}	3.138	pH	Equivalent inductance of the dual HB structure
	HB	R_{HB}	34.46	mΩ	Equivalent resistance of the dual HB structure
	Insulation Layer	C_{TSV}	2.689	fF	Distributed capacitance between the mini-TSV copper pillar and the DRAM_N substrate
Crosstalk	IMD2	C_{HB}	0.136	fF	Distributed capacitance between adjacent dual HBs
	IMD1	C_{IMD}	0.225	fF	Distributed capacitance between adjacent mini-TSVs
	Si Sub	G_{Sub}	0.144	mS	Equivalent conductance of SI Sub
		C_{Sub}	1.522	fF	Equivalent capacitance of SI Sub

The lumped elements associated with conduction functionality primarily consist of, approximately, a 100 mΩ resistor and a 10 fF capacitor, affecting the data channels, which are comparable to the distributed parameters of metal layers within 2D chips. The lumped elements related to crosstalk functionality primarily consist of, approximately, a 0.1 mS conductance and a 1.5 fF capacitance, establishing crosstalk between data channels, which is comparable to the distributed parameters of metal layers within 2D chips. The scale of distributed numerical values introduced by 3DIC is not fundamentally different from 2D chip designs. Therefore, SeDRAM employs buffers to drive the vertical stacking paths and operates at the DRAM core frequency. The driving units and speeds of 3DIC are the essential differences between SeDRAM and HBM. It is necessary to conduct cross-process signal integrity analysis in combination with SeDRAM's dedicated driving methods.

2.3. Frequency-Domain Analysis

The framework for frequency-domain analysis introduces seven channels of lumped circuits for vertical stacking paths, aiming to analyze the channel characteristics of vertical stacking paths, including 3DIC crosstalk, as illustrated in Figure 4. Channel 3 in the middle is considered the victim, while the three outer pairs of channels act as the aggressors. Fourteen terminations are distributed on both sides of the seven channels for S-parameter analysis. Each channel incorporates components such as R_{TSV}, L_{TSV}, R_{HB}, and C_{TSV} for conduction, as well as C_{HB}, C_{IMD}, G_{Sub}, and C_{Sub} for crosstalk. In this setup, channel 3 near the center of the framework is chosen as the subject of analysis.

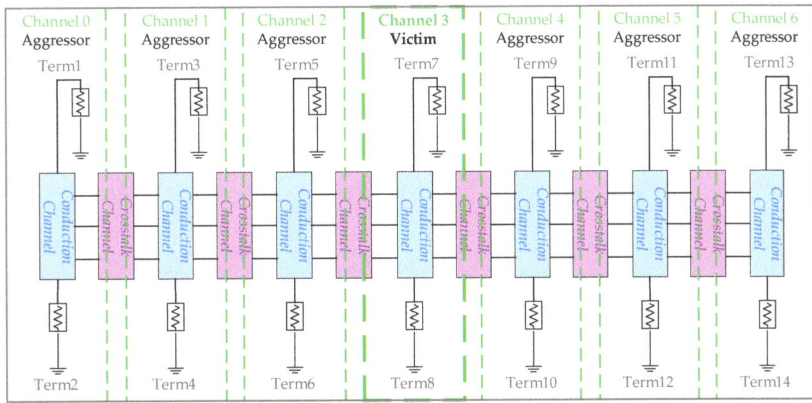

Figure 4. Frequency-domain analysis frame.

Figure 5 illustrates the frequency-domain analysis for channel 3. In Figure 5a, the insertion loss on channel 3 is displayed in terms of the frequency response ratio between termination 7 to 8 and termination 8 to 7. Insertion loss represents the proportion of signal loss from the input to the output caused by the 3DIC path, with values closer to zero indicating better performance. The insertion loss on channel 3 is greater than -0.2 dB below 10 GHz, indicating that insertion loss is not the primary factor affecting 3DIC signal integrity. The return loss on channel 3 is depicted in Figure 5b, including the frequency response ratio between termination 7 to 7 and termination 8 to 8. Return loss characterizes the proportion of the input signal reflected back to the input terminal through the 3DIC path compared to the input signal. Typically, a value lower than -30 dB does not significantly affect channel performance. In SeDRAM, vertical stacking paths operate at the DRAM core frequency, which is lower than 1 GHz, resulting in a return loss below -34 dB, which has no significant impact on signal integrity. However, above 2.5 GHz, the return loss exceeds -25 dB, becoming a major challenge for signal integrity.

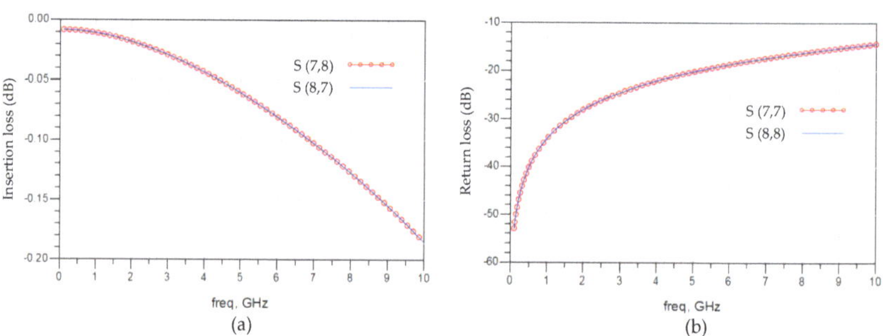

Figure 5. Frequency-domain analysis. (**a**) Insertion loss of channel 3. (**b**) Return loss of channel 3.

Figure 6a,b demonstrate the impacts of near-end crosstalk and far-end crosstalk on channel 3. Taking Figure 6a as an example, six frequency response ratios are overlaid, representing the near-end crosstalk impacts of the six aggressors from terminations 1, 3, 5, 9, 11, and 13. Terminations 5 and 9 are closest to termination 7 and have the greatest crosstalk impacts on termination 7, with an impact of -75 dB at 1 GHz. The farther the near-end terminations are from relative termination 7, the less their impact on termination 7. The patterns and numerical values of far-end crosstalk on channel 3 in Figure 6b are similar to near-end crosstalk. Therefore, crosstalk is not a significant challenge to signal integrity.

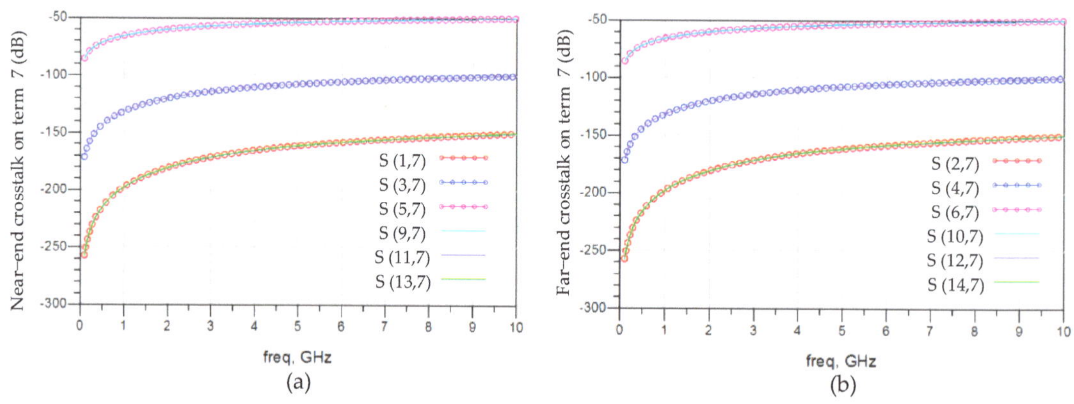

Figure 6. Frequency-domain analysis for the impacts of crosstalk on channel 3. (**a**) Impacts of near-end crosstalk on termination 7. (**b**) Impacts of far-end crosstalk on termination 7.

Figure 7 demonstrates the crosstalk on channel 3 with the impact of 3DIC lumped element variation, considering the statistical variations of the vertical stacking path. Under the 3DIC slow condition, all lumped element values of the vertical stacking path are increased by 40%, corresponding to the lowest slew rate of digital signals. Conversely, the 3DIC fast condition involves reducing all lumped element values by 40%, reflecting the highest slew rate. Among the near-end crosstalk response and the far-end crosstalk response, the 3DIC fast condition introduces the least crosstalk. Both the 3DIC fast and slow conditions introduce crosstalk response deviations of less than 6% on the basis of -65 dB at 1 GHz.

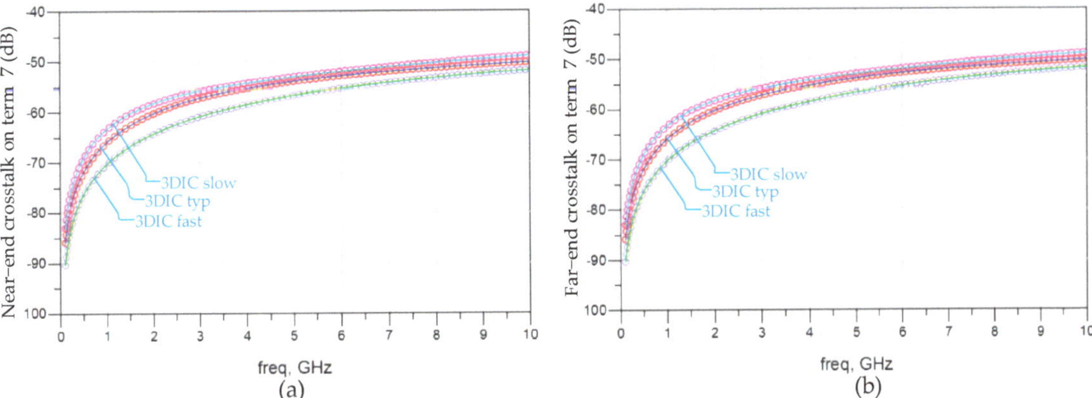

Figure 7. Frequency-domain analysis for the crosstalk on channel 3 with the impact of 3DIC lumped element variation. (**a**) Near-end crosstalk on termination 7. (**b**) Far-end crosstalk on termination 7.

This section, in conjunction with the stacking structure and the vertical stacking cell features of the vertical stacked DRAM platform, highlights the distinct nature of WoW SI analysis: cross-process and the absence of process segmentation by I/O circuit. The lumped circuit based on the 2HB+1TSV+2HB structure is introduced to establish a modeling methodology for the vertical stacked DRAM platform. All values of the lumped elements in the circuit model are calculated with reference to the transmission line model. Frequency-domain analysis of vertical stacking paths based on lumped circuits is presented, highlighting that the impact of 3DIC channels increases with the frequency, and the influence of 3DIC channels below 1 GHz meets the design requirements of SeDRAM.

3. Cross-Process Analysis

To address the buffer driving (I/O less) method, a cross-process timing-domain analysis method is established, where the 3DIC is represented in the form of a lumped circuit; the DRAM and logic buffers are represented in netlist form, and they are integrated into one simulation environment. Employing this method, an impact analysis introduced by 3DIC crosstalk is demonstrated, coupled with memory access behavior across the 3DIC.

3.1. CPSIA Method

In the SeDRAM architecture, the vertical stacking path has a high density, and I/O-driven chip-to-chip interconnect technologies are neither necessary nor feasible. Instead, the vertical stacking path is directly driven by a buffer cell within the DRAM and logic chip. The area overhead of the driving circuit is minimal, aligning well with the high-density interconnect characteristics of the vertical stacking path. Unlike the channel analysis method with 50 ohm terminations in Section 2.3, this sub-section combines SeDRAM's dedicated driving method for cross-process signal integrity analysis.

As shown in Figure 8, a CPSIA framework consists of three parts: two kinds of netlists of 25 nm DRAM and 28 nm logic driving buffer based on commercial foundries and the lumped circuit model of the vertical stacking path. The combination of these three simulation elements forms an integrated simulation environment, which includes three processes. This CPSIA environment establishes a 3D SI analysis method equivalent to standard 2D chip design, meeting the requirements of the I/O-less driving structure of SeDRAM and providing greater accuracy than the I/O-based SI analysis. The netlists of DRAM and logic driving buffers include the Transceivers/Receivers (TX/RX) of DRAM_N, DRAM_F, and logic, along with impedances of ZimD0, ZimD1, ..., and ZimD6. These impedances represent the inner connecting metal layers between the TX/RX ports and the 3DIC logic interface. The netlists of the DRAM and logic driving buffers include analog behavior described in the DRAM and logic process libraries for signal analysis. The circuit model of the vertical stacking path consists of lumped elements corresponding to mini-TSV and HB cells, enabling cross-process simulations without the need for 3DIC process libraries.

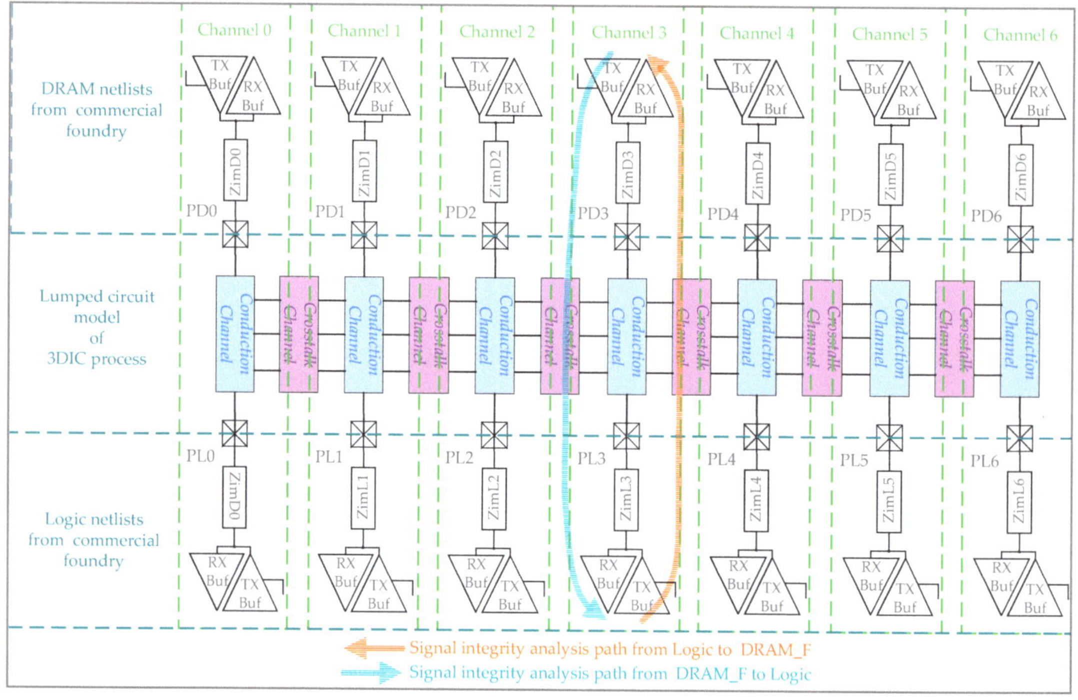

Figure 8. CPSIA frame.

In SeDRAM, all memory access data signals of DRAM_F and DRAM_N are independently designed, with DRAM_F having a longer vertical stacking path, which is the focus of this analysis. In this CPSIA frame, there are two SI analysis paths, as shown in Figure 8. This approach closely approximates the actual circuit's driver and load responses, avoiding rough evaluation with a 50 ohm driver impedance.

Pseudo-Random Binary Sequence (PRBS) excitation is applied to seven channels. Due to the synchronous design in the DRAM circuit, eye diagrams of seven channels with the same direction are overlaid, resulting in two sets of eye diagrams. The first set is obtained by overlaying eye diagrams collected with seven logic buffers as the TXs on PD0−PD6, and the second set is obtained by overlaying eye diagrams collected with seven DRAM_F buffers as the TXs on PL0−PL6. The two sets of eye diagrams are shown in Figure 9, corresponding to the two signal integrity paths in Figure 9. The two eye diagrams indicating writing and reading data paths of memory access have high quality and are easily recoverable by

RXs, resulting in minimal impacts on DRAM timing. Since the speed of the vertical data paths is below 1 Gbps, the impact of 3DIC is within the tolerance of the vertical stacked DRAM platform.

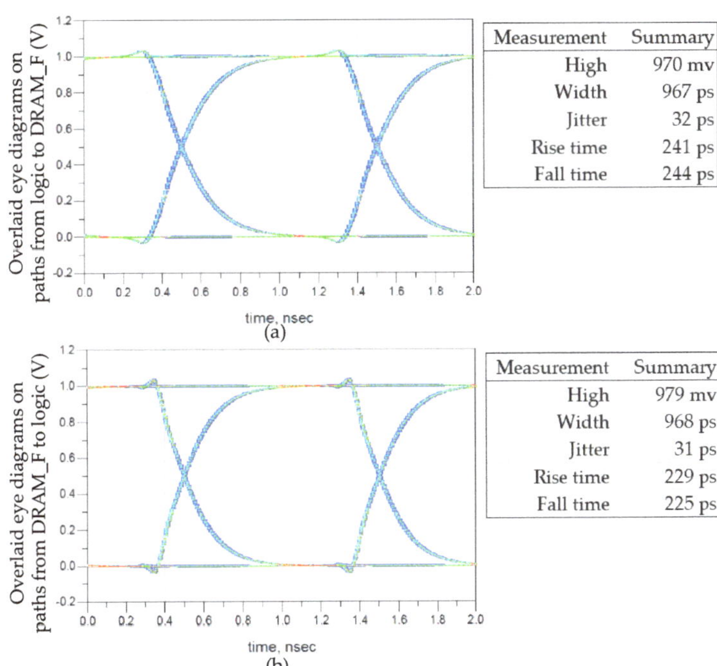

Figure 9. Overlaid eye diagrams at 1 Gbps for the signal integrity analysis. (**a**) Overlaid eye diagrams from logic to DRAM_F. (**b**) Overlaid eye diagrams from DRAM_F to logic.

3.2. Impact Analysis Introduced by 3DIC

This sub-section analyzes the impact of 3DIC on memory access. Figure 10 illustrates the Logic Memory Access (LMA) path for logic reading and writing data from and to DRAM_N and DRAM_F. The red, green, and blue lines represent CAs, data writing paths (from the DRAM perspective), and data reading paths of DRAM_F. The yellow, orange, and cyan lines represent CAs, data writing paths (from the DRAM perspective), and data reading paths of DRAM_N. CAs include the command address and tCK, which are unidirectional signals from logic to DRAM, where the address goes through a decoder. Data writing and reading paths connect the DRAM array and interface through the DRAM internal data path. The design and layout of DRAM_F and DRAM_N are identical; in fact, there is only one type of DRAM used in manufacturing the 3DIC wafer, without distinction between stack layers. The design differences in the 3DIC layers enable DRAM_F to extend the data path to the logic interface through the 3DIC structure. The only distinction in the LMA path between DRAM_F and DRAM_N lies in the 3DIC path. The LMA paths of DRAM_N pass through HBs, while the LMA paths of DRAM_F go through a longer 3DIC connection (2HB+1TSV+2HB structure). The degradation of SI in LMA due to 3DIC is evident in terms of signal jitter and signal delay introduced by 3DIC.

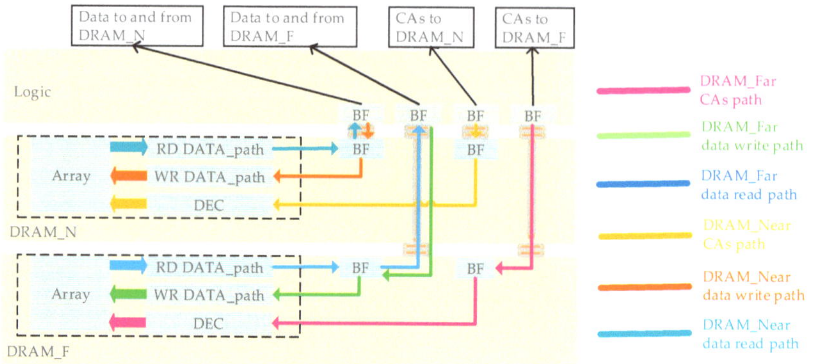

Figure 10. Logic memory access paths.

Figure 11 expands on the focus of jitter in the overlaid eye diagrams from logic to DRAM_F. A jitter of 32 ps is observed in the channel response, including a background noise of 1 ps, leading to uncertainty in the sampling time on the DRAM and a reduction in the timing margin for sampling frequency. To isolate the impact of factors other than crosstalk on signal jitter, only the excitation of channel 3 is retained, while the remaining logic TXs are set to zero. A jitter of 1.0 ps is discovered, which is not caused by the flipping of adjacent 3DIC channels. The jitter on the channel response originates from the 3DIC crosstalk channels and is determined by the random encoding of aggressor channels.

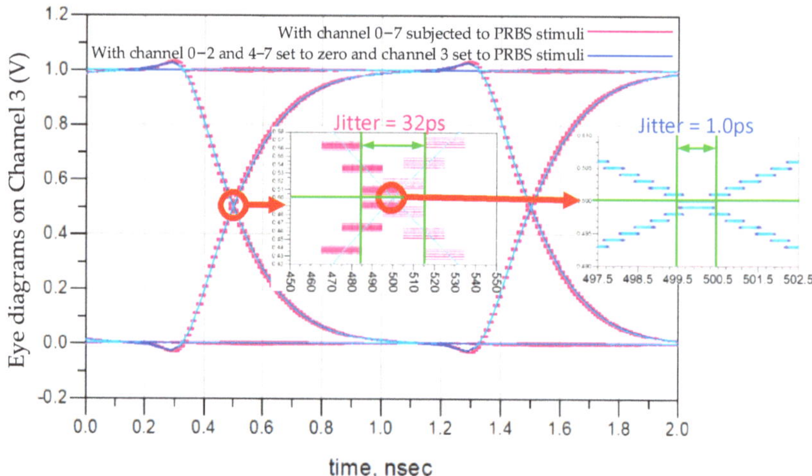

Figure 11. Jitter analysis of data path.

Figure 12 presents the jitter analysis, considering the impact of driver Process Voltage and Temperature (PVT) deviations as well as 3DIC variation (see Section 2.2). The Driver FF/TT/SS conditions utilize netlists extracted from both the DRAM and logic, featuring the fastest, typical, and slowest combinations of the P-Channel Metal Oxide Semiconductor (PMOS) and the N-Channel Metal Oxide Semiconductor (NMOS). The combination of driver FF and 3DIC fast corresponds to the fastest 3DIC channel, while the combination of driver SS and 3DIC slow corresponds to the slowest 3DIC channel. Figure 12 illustrates the deviations in signal setup time under various conditions. Notably, the absolute values of time-domain jitter remain the same across the three conditions, exhibiting a phase deviation of 4 ps.

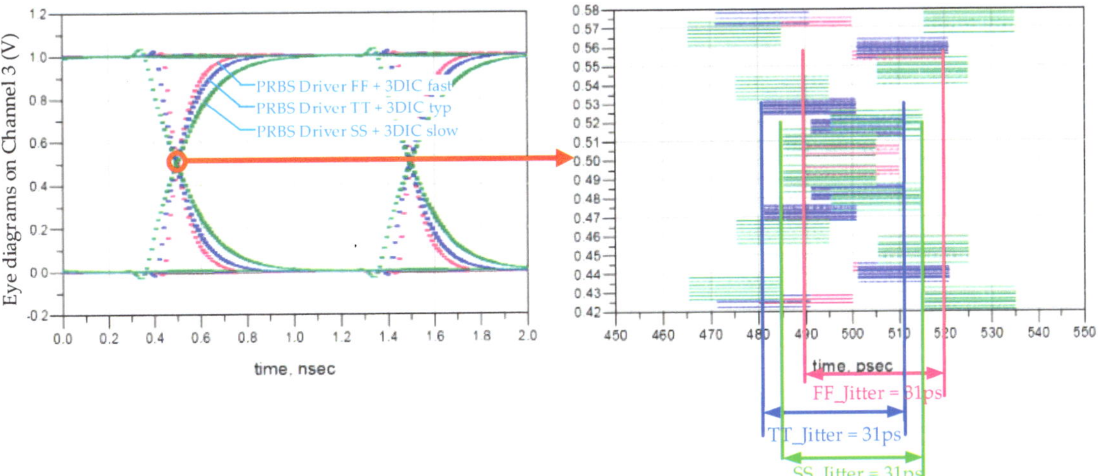

Figure 12. Jitter analysis on data paths with the impact of driver PVT deviation and 3DIC variation.

The jitter introduced by 3DIC reduces the timing margin of the sampling circuits, becoming a leading cause of the reduced maximum frequency for DRAM_F. Specifically, when writing data from logic to DRAM_F, the jitter on the DRAM data writing path (the green path in Figure 10) decreases the timing margin of the first-level data sampling in the DRAM. When reading data from DRAM_F to logic, the jitter on the DRAM data reading path (the blue path in Figure 10) reduces the timing margin of the first-level data sampling in the logic. The CAs (including clock signals) are driven from logic into DRAM_F through the red path in Figure 10; thus, the jitter on CAs reduces the timing margin of all data sampling in DRAM_F. The 31 ps jitter demonstrated in Figure 11 is a random jitter introduced by crosstalk, which is present not only on the data path but also on the clock net, resulting in a maximum timing uncertainty ranging from 31 ps to 62 ps.

The 3DIC also introduces signal transmission delay. Figure 13 includes the 3DIC driving sources of the buffers, the response of PRBS excitation in the 3DIC channel, and a response without a 3DIC channel between logic and DRAM buffers. There is a 0.7 ns delay between the two responses from the source. In particular, when zooming in on the figure, there is a 9 ps difference between the two buffer responses, using an 80% VDD threshold as the transition from low to high.

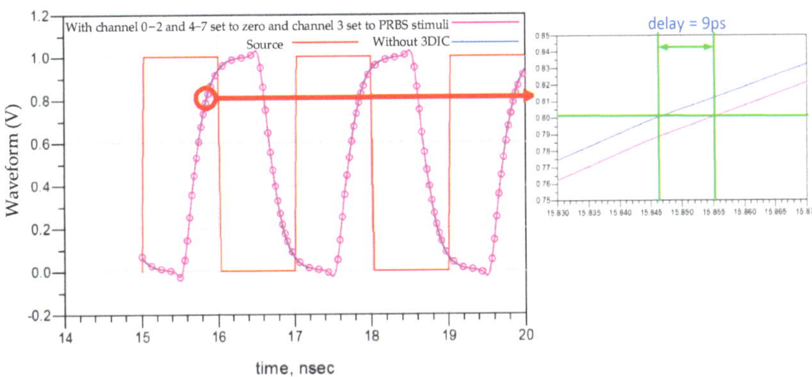

Figure 13. The impact of transmission delay on DRAM_N.

The 3DIC load introduces an additional 9 ps delay in the driving response in DRAM_F. The 3DIC delay does not impact writing data to the memory but affects reading data from memory. The data reading circuit belongs to the internal tCK clock domain of DRAM_F, while the sampling data circuit on the logic DIE belongs to the logic tCK clock domain. Only the former undergoes transmission delay introduced by the 3DIC, reducing the timing margin of the first-level sampling in logic.

The jitters introduced by the 3DIC are associated with the behavior of aggressor channels; they are random during the memory access process. Random jitters contribute to the timing uncertainty of DRAM_F memory access, ranging from 31 ps to 62 ps. Furthermore, the delay in the 3DIC path affects the timing of the first-level sampling of DRAM in logic.

3.3. Design Optimization of Vertical Stacking

The diverse combinations of mini-TSV and HB form the vertical stacking path between DRAM_F and logic, as shown in Figure 14. The eye diagrams of the 1HB+1TSV+1HB, 2HB+1TSV+2HB, and 4HB+1TSV+4HB structures are shown separately in Figure 15a, Figure 9a, and Figure 15b. The proportions of HB and TSV cells have a minimal impact on the channel. The performance of these three structures in signal transmission is similar, but they differ in terms of design resource utilization. Mini-TSV cells connect the internal metal layers of the DRAM_N to the HB layer on the backside of the DRAM_N silicon substrate, resulting in an active layer footprint on DRAM_N. Unlike HB cells that do not impact the active layer layout, the number of mini-TSV cells is constrained in DRAM design. The failure ratio of the signal HB is less than 0.1 ppm. In the case of the maximum 64Gb DRAM, there exists a requirement for 300k HB connections carrying critical signals. The employment of the 2HB+1TSV+2HB structure for critical signal connections results in a 3% yield improvement in the 64Gb near-memory product. To prevent the entire DRAM failing due to the bonding failure of a single HB cell, using the 2HB+1TSV+2HB structure to establish the vertical stacking data path represents an excellent tradeoff. Yield is a crucial focus in the large-scale production of SeDRAM. In the collaborative design, a diverse combination of mini-TSV and HB is utilized to create vertical signal and power interconnects.

- A 1HB+1TSV+1HB structure is employed for testing signal interconnects.
- A 2HB+1TSV+2HB structure is employed for interconnecting memory access data signals, such as DQs and CAs. Its advantages include reducing the contact resistance of HB cells in the data path and enhancing the product yield targets.
- A 4HB+1TSV+4HB structure is utilized for the power network. Four sets of HBs in parallel are used to address the high contact resistance issue in HBs, reducing voltage drop and current density in HB cells.

Along with the dedicated buffer driving method, a CPSIA approach is proposed and utilized to analyze the 3DIC jitters, integrating DRAM logic and 3DIC designs in a simulation environment. This approach quantifies the timing uncertainty introduced by 3DIC crosstalk, ranging from 31 ps to 62 ps.

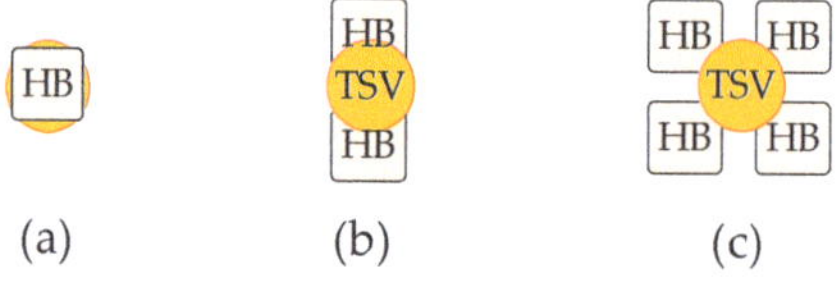

Figure 14. The physical structures of vertical stacking paths: (**a**) the 1HB+1TSV+1HB structure; (**b**) the 2HB+1TSV+2HB structure; (**c**) the 4HB+1TSV+4HB structure.

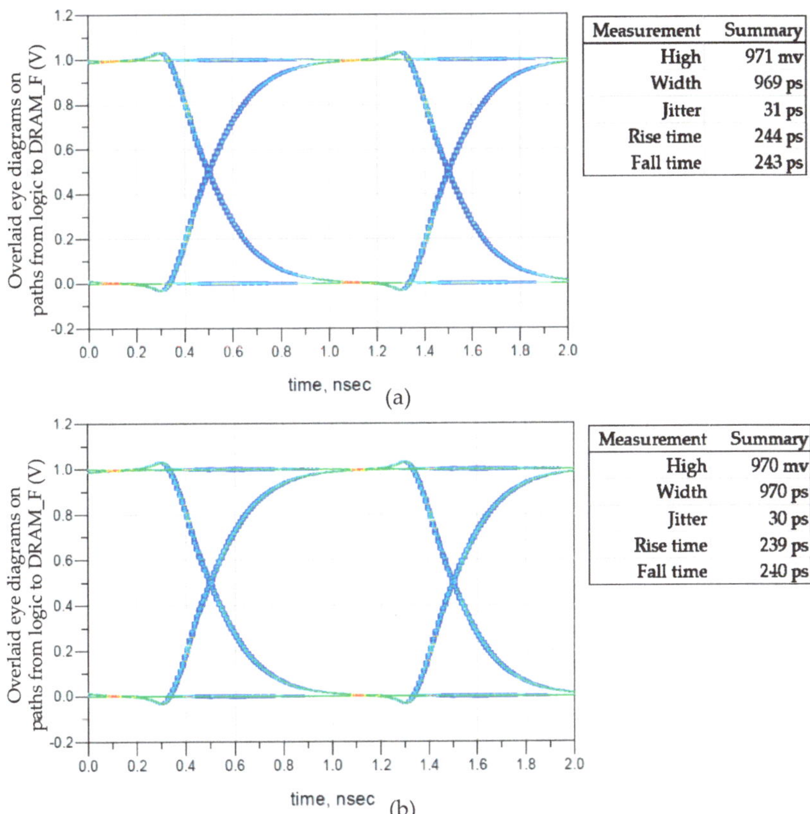

Figure 15. Overlaid eye diagrams from logic to DRAM_F: (**a**) the 1HB+1TSV+1HB structure; (**b**) the 4HB+1TSV+4HB structure.

4. Physical Testing and Result Analyses

The timing uncertainty, ranging from 31 ps to 62 ps, introduced by the random behavior of aggressor channels coupled through 3DIC crosstalk, was determined. The 3DIC path represents the only difference between DRAM_F and DRAM_N, considering their identical design and layout. Therefore, the quantified 3DIC impact should manifest in the physical testing of SeDRAM. This section provides the physical testing results of the tCK shmoo in a cross-process test structure with commercial DRAM logic and 3DIC manufacturing processes. DRAM_F and DRAM_N from the same 3DIC wafer exhibit an unsymmetric distribution in maximum frequency. Subsequently, a study was conducted to explore the relationship between this phenomenon and the analysis presented in Section 3.

4.1. The Test Chip

A physical testing wafer is established with a DRAM_N, DRAM_F, and logic stacking structure, as shown in Figure 16. The logic includes DRAM test circuits and test pads used for interconnection with the test tooling. DRAM_N is vertically interconnected with the logic through HB cells; DRAM_F is vertically interconnected with the logic through HB and mini-TSV cells. The vertically interconnected units corresponding to functionally identical signals for DRAM_N and DRAM_F are physically arranged adjacently to reduce channel differences in signals with the same function across the two DRAM stacks.

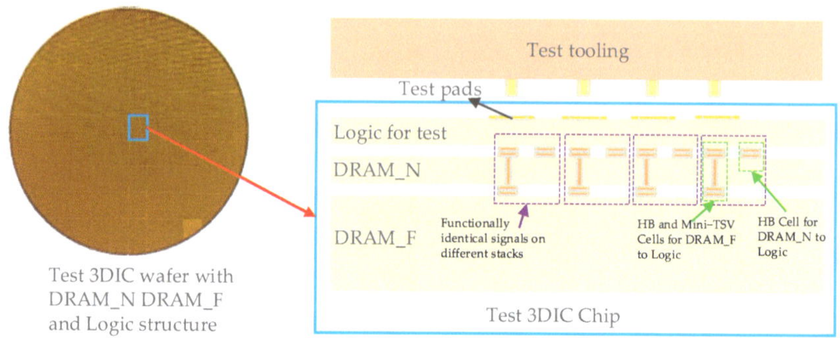

Figure 16. Testing chip environment.

The Logic, DRAM_N, and DRAM_F dies are organized into the testing chip structure through the 3DIC process, as shown in Figure 17. The Logic die includes a Design-for-Test (DFT) circuit used to test the DRAM arrays on DRAM_N and DRAM_F through their LMA interfaces. The memory access path in the test structure is consistent with the memory−compute integration application, including the 3DIC data paths from logic buffers to DRAM_F and DRAM_N buffers. This alignment is also consistent with the cross-process structure shown in Figure 8. The only distinction between the LMA interfaces of DRAM_F and DRAM_N is that the DRAM_F includes a more complex 3DIC path with an HB mini-TSV and HB structure. Under the same 2D chip design, the additional impact of 3DIC on DRAM_F exists in LMA interfaces. This leads to deviations in test results in terms of DRAM_F and DRAM_N.

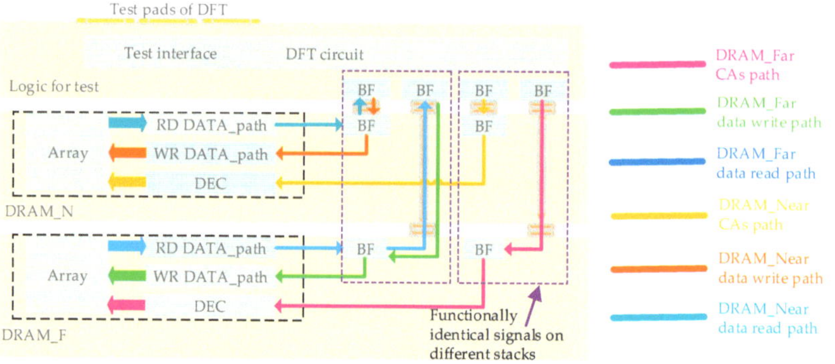

Figure 17. Testing chip structure.

tCK is the synchronous core clock of the DRAM. The tCK shmoo test follows the standard DRAM testing procedure: a fixed frequency (tCK) is set to perform read and write operations on SeDRAM. Diverse data are written into the SeDRAM, including two DRAM arrays on both DRAM_F and DRAM_N. After reading the operations, the data bus is checked at each Access Time (tAC) step. After scanning through multiple patterns, if all the DRAM arrays pass the write and read loops, the tCK shmoo is marked as a pass (in green); otherwise, it is marked as a fail (in red). The tCK shmoo is a result of extensive scanning of the DRAM arrays with multiple patterns.

Figure 18 shows the tCK shmoo test results of the double-layered DRAM test chip. The shortest tCKs for DRAM_N and DRAM_F are 1.64 ns and 1.68 ns, respectively, with DRAM_F having a slightly lower maximum frequency than DRAM_N. DRAM_F and DRAM_N are two stacked DRAM arrays on the same 3DIC wafer under the same temperature. This tCK Shmoo comparison shows that the minimum tCK (maximum frequency) of DRAM_N is better than that of DRAM_F by 40 ps. The 3DIC is the only distinction between the two DUTs of DRAM_F and DRAM_N. The impact of 3DIC is speculated to be the primary factor influencing this phenomenon.

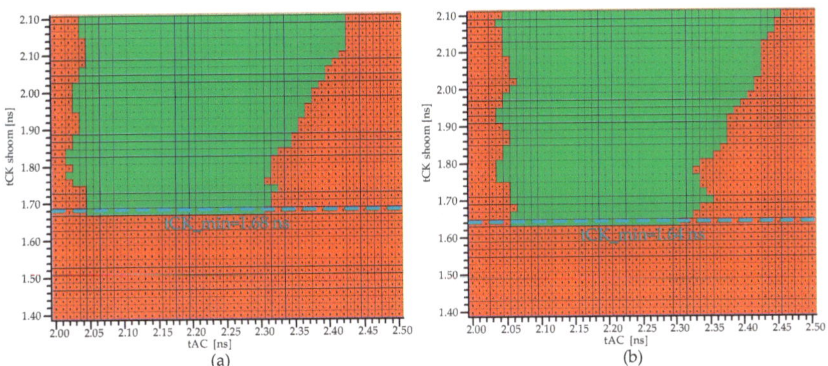

Figure 18. DRAM tCK test shmoo. (**a**) DRAM_F_N. (**b**) DRAM.

To illustrate the performance gap between DRAM_N and DRAM_F, 12 sets of samples are depicted in Figure 19. The histogram displays the distribution of the minimum tCK differences between DRAM_F and DRAM_N for the 12 sets of samples. The test results include manufacturing deviations and testing errors, indicating that DRAM_N has a speed advantage over DRAM_F. This is reflected in two aspects: a predominance of positive values over negative values in the distribution of the difference between the tCK_min of DRAM_F and the tCK_min of DRAM_N and the average tCK_min differences, which indicate that the average tCK_min of DRAM_F is smaller than that of the tCK_min of DRAM_N by 26.67 ps. The impact of 3DIC is speculated to be the primary factor influencing this phenomenon.

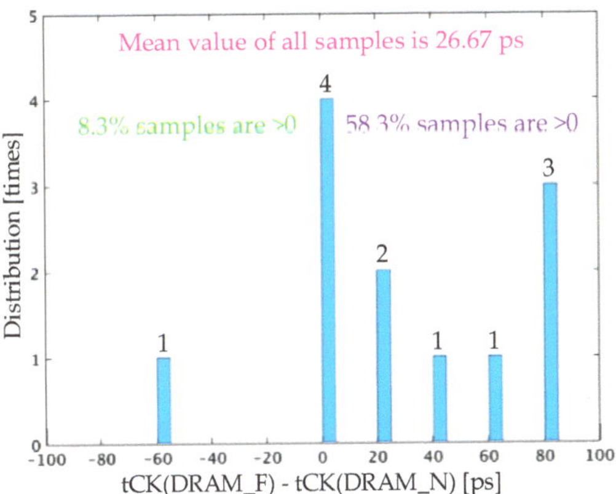

Figure 19. The distribution of tCK_min differences between DRAM_F and DRAM_N.

4.2. Analysis of Test Results

The cross-process jitter analysis in Figure 11 shows that 3DIC crosstalk contributes to 31 ps of jitter on the LMA path, with the timing uncertainty of 31 ps introduced by the random behavior of aggressor channels coupled through 3DIC crosstalk, which is the uncertainty in both data and tCK. In the testing environment, the memory access path in DRAM_F, which includes the vertical stacking path formed by HB mini-TSV and HB cells, is distinct from DRAM_N. The tCK path uncertainty of 31 ps reduces the timing margin for all DFF/latch samplings within the DRAM, while the data path uncertainty of 31 ps reduces the timing margin for the first-level sampling in the writing path and the last-level sampling in the reading path. Therefore, this study attributes the 31–62 ps tCK period deviation observed in the tCK shmoo test results to the impact of 3DIC crosstalk.

4.3. Model Extension

While the impact of 3DIC on signal integrity meets the requirement of the analysis target below 1 Gbps, this method plays a significant role in determining the evolutionary path of this vertical stacked DRAM platform, reflected in the expansion of stacking structures and the enhancement of LMA speed.

The combination of HB and mini-TSV enables us to design higher-stacked DRAM platforms, thereby increasing DRAM density. Lumped circuit models of an eight DRAM and one logic stacking structure and of a four DRAM and one logic stacking structure were established for comparison with the two DRAM and one logic stacking structure analyzed in this paper. Following the frequency analysis method of Figure 4, it was assumed that channel 3 was the victim and channel 2 was the aggressor. The near-end and far-end crosstalk responses of term 5 and term 6 of channel 2 to term 7 of channel 3 are shown in Figure 20. As the stacking structure becomes more complex, the crosstalk introduced by 3DIC gradually increases.

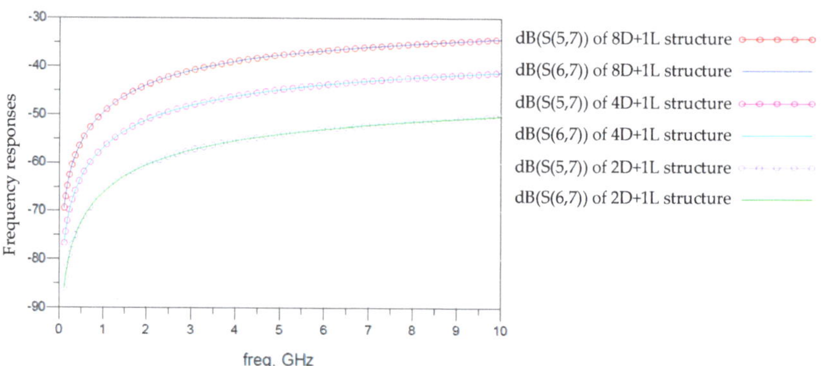

Figure 20. Near-end and far-end crosstalk on termination 8 in the 2D+1L 4D+1L and 8D+1L structures.

According to the objective of this study, the DRAM to logic interface frequency matches the internal clock of the DRAM array. On the roadmap of SeDRAM, we will employ prefetching techniques to fetch data at 8 or 16 times the DRAM array frequency to the DRAM to logic interface, thus enabling data to flow through the DRAM to logic interface at 8 or 16 times the DRAM array frequency. Figure 21 illustrates the eye diagrams for a frequency increase to 2 Gbps and 4 Gbps in stacking structures of 4D+1L (four DRAM layers and one logic layer) and 8D+1L. Above 2 Gbps, the eye diagrams gradually degrade. Combining frequency-domain analysis, the primary cause is return loss. In this next-level structure, the impact of 3DIC cannot be ignored, and quantitative analysis and optimization using this method are necessary.

Based on the relationship between the lumped circuit model and the corresponding physical structure, it is easy to identify three quantitative optimization methods for 3DIC SI, aiming to meet the advancement of SeDRAM, in terms of the expansion of stacking structures and the enhancement of LMA speed. One approach is to increase the pitch of the vertical stacking paths. Another is to introduce direct current channels in the signal vertical stacking path array. The third method focuses on optimizing the 3DIC process through the analysis of key factors leading to 3DIC responses, including structural and material enhancements, such as C_{TSV} sensitive to the insulation layer thickness.

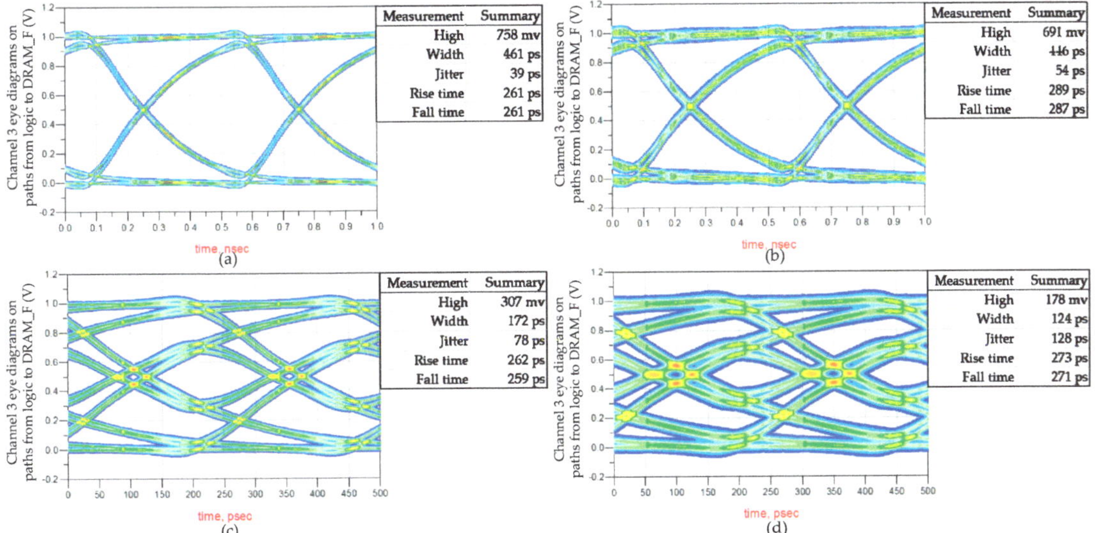

Figure 21. Overlaid eye diagrams from logic to DRAM_F: (**a**) the 2 Gbps on 4D+1L structure; (**b**) the 2 Gbps 8D+1L structure; (**c**) the 4 Gbps on 4D+1L structure; (**d**) the 4 Gbps 8D+1L structure.

The jitter analysis result provides an explanation for this interesting physical testing phenomenon, demonstrating the effectiveness of this CPSIA method. The model extension analysis for higher speeds and increased stacking structures illustrates that this method will play a crucial role in SeDRAM's technological advancements as channel degradation progresses.

5. Conclusions

This paper highlights the distinct nature of WoW 3D multi-layer vertical stacked DRAM Platform SI analysis in terms of cross-process and the absence of process segmentation by I/O circuit. A lumped circuit based on the 3DIC physical structure is introduced to establish a modeling methodology for the vertical stacked DRAM platform. All values of the lumped elements in the circuit model are calculated with the transmission line model. In combination with the dedicated buffer driving method, the CPSIA method is proposed and used for the analysis of 3DIC jitters, integrating DRAM logic and 3DIC designs in a simulation environment, determining the timing uncertainty introduced by 3DIC crosstalk ranging from 31 ps to 62 ps. The silicon results show that the distribution of DRAM_N's maximum frequency is better than that of DRAM_F, with the average of the tCK_min differences being 26.67 ps, demonstrating the effectiveness of this CPSIA method.

Author Contributions: Conceptualization, M.L. and X.J.; investigation, X.J. (Xiping Jiang), X.J. (Xuerong Jia), S.W. and L.G.; methodology, X.J. (Xuerong Jia), S.W., L.G. and Y.G.; validation, F.G. and X.L.; resources, F.G. and X.L.; writing—original draft preparation, X.J. (Xiping Jiang) and Y.G.; writing—review and editing, J.Y., X.J. (Xiping Jiang), X.J. (Xuerong Jia), Y.G. and S.W.; supervision, M.L., J.Y., X.J. (Xiping Jiang) and L.G. All authors have read and agreed to the published version of the manuscript.

Funding: This research was funded by the National Key R&D Program of China under grant no. 2019YFB2204800.

Data Availability Statement: Simulation and measurement data are available on request from the corresponding author.

Conflicts of Interest: Authors Xiping Jiang, Xuerong Jia, Song Wang, Yixin Guo, Fuzhi Guo, and Xiaodong Long were employed by the company Xi'an UniIC Semiconductors. The remaining authors declare that the research was conducted in the absence of any commercial or financial relationships that could be construed as a potential conflict of interest.

References

1. Sebastian, S.; Gallo, M.L.; Khaddam-Aljameh, R.; Eleftheriou, E. Memory devices and applications for in-memory computing. *Nat. Nanotechnol.* **2020**, *15*, 529–544. [CrossRef] [PubMed]
2. Khan, K.; Pasricha, S.; Kim, R.G. A Survey of Resource Management for Processing-in-Memory and Near-Memory Processing Architectures. *J. Low Power Electron. Appl.* **2020**, *10*, 30. [CrossRef]
3. Santoro, G.; Turvani, G.; Graziano, H. New Logic-in-Memory Paradigms: An Architectural and Technological Perspective. *Micromachines* **2019**, *10*, 368. [CrossRef] [PubMed]
4. Tian, W.; Li, B.; Li, Z.; Cui, H.; Shi, J.; Wang, Y.; Zhao, J. Using Chiplet Encapsulation Technology to Achieve Processing-in-Memory Functions. *Micromachines* **2022**, *13*, 1790. [CrossRef] [PubMed]
5. Horowitz, M. Computing's Energy Problem (and what we can do about it). In Proceedings of the IEEE International Solid-State Circuits Conference (ISSCC), San Francisco, CA, USA, 9–13 February 2014.
6. Ke, L.; Zhang, X.; So, J.; Lee, J.G.; Kang, S.H.; Lee, S.; Han, S.; Cho, Y.G.; Kim, J.H.; Kwon, Y.; et al. Near-Memory Processing in Action: Accelerating Personalized Recommendation With AxDIMM. *IEEE Micro* **2022**, *12*, 116–127. [CrossRef]
7. Mutlu, O.; Ghose, S.; Gómez-Luna, J.; Ausavarungnirun, R. Processing data where it makes sense: Enabling in-memory computation. *Microprocess. Microsyst.* **2019**, *67*, 28–41. [CrossRef]
8. Spessot, A.; Oh, H. 1T-1C Dynamic Random Access Memory Status, Challenges, and Prospects. *IEEE Trans. Electron. Devices* **2020**, *67*, 1382–1393. [CrossRef]
9. Lee, S.; Cho, H.; Son, Y.H.; Ro, Y.; Kim, N.S.; Ahn, J.U. Leveraging Power-Performance Relationship of Energy-Efficient Modern DRAM Devices. *IEEE Access* **2018**, *6*, 31387–31398. [CrossRef]
10. Park, N.; Ryu, S.; Kung, J.; Kim, J.-J. High-throughput Near-Memory Processing on CNNs with 3D HBM-like Memory. *ACM Trans. Des. Autom. Electron. Syst.* **2021**, *26*, 1–20. [CrossRef]
11. Bernhardt, A.; Koch, A.; Petrov, I. pimDB: From Main-Memory DBMS to Processing-In-Memory DBMS-Engines on Intelligent Memories. In Proceedings of the 19th International Workshop on Data Management on New Hardware, DaMoN '23, Seattle, WA, USA, 18–23 June 2023; pp. 44–52.
12. Qureshi, Y.M.; Simon, W.A.; Zapater, M.; Olcoz, K.; Atienza, D. Gem5-X: A Many-core Heterogeneous Simulation Platform for Architectural Exploration and Optimization. *ACM Trans. Archit. Code Optim.* **2021**, *18*, 1–27. [CrossRef]
13. Zhou, M.; Xu, W.; Kang, J.; Rosing, T. TransPIM: A Memory-based Acceleration via Software-Hardware Co-Design for Transformer. In Proceedings of the IEEE International Symposium on High-Performance Computer Architecture (HPCA), Seoul, Republic of Korea, 2–6 April 2022; pp. 1071–1085.
14. Lee, D.U.; Cho, H.S.; Kim, J.; Ku, Y.J.; Oh, S.; Kim, C.D.; Kim, H.W.; Lee, W.Y.; Kim, T.K.; Yun, T.S.; et al. A 128 Gb 8-High 512 GB/s HBM2E DRAM with a Pseudo Quarter Bank Structure, Power Dispersion and an Instruction-Based At-Speed PMBIST. In Proceedings of the IEEE International Solid-State Circuits Conference (ISSCC), San Francisco, CA, USA, 16–20 February 2020.
15. Park, M.J.; Cho, H.S.; Yun, T.S.; Byeon, S.; Koo, Y.J.; Yoon, S.; Lee, D.U.; Choi, S.; Park, J.; Lee, J.; et al. A 192-Gb 12-High 896-GB/s HBM3 DRAM with a TSV Auto-Calibration Scheme and Machine-Learning-Based Layout Optimization. In Proceedings of the IEEE International Solid-State Circuits Conference (ISSCC), San Francisco, CA, USA, 20–26 February 2022.
16. Shiba, K.; Okada, M. A 7-nm FinFET 1.2-TB/s/mm^2 3D-Stacked SRAM Module with 0.7-pJ/b Inductive Coupling Interface Using Over-SRAM Coil and Manchester-Encoded Synchronous Transceiver. *IEEE J. Solid-State Circuits* **2023**, *58*, 2075–2086. [CrossRef]
17. Wang, S.; Jiang, X.; Bai, F.; Xiao, W.; Long, X.; Ren, Q.; Kang, Y. A True Process-Heterogeneous Stacked Embedded DRAM Structure Based on Wafer-Level Hybrid Bonding. *Electronics* **2023**, *12*, 1077. [CrossRef]
18. Wang, S.; Yu, B.; Xiao, W.; Bai, F.; Long, X.; Bai, L.; Jia, X.; Zuo, F.; Tan, J.; Guo, Y.; et al. An 85 GBps/Gbit 0.66 pJ/bit Stacked DRAM with Multilayer Arrays by Fine Pitch Hybrid Bonding and Mini-TSV. In Proceedings of the IEEE Symposium on VLSI Technology and Circuits (VLSI Technology and Circuits), Kyoto, Japan, 11–16 June 2023.
19. Beyene, W.; Juneja, N.; Hahm, Y.-C.; Kollipara, R.; Kim, J. Signal and Power Integrity Analysis of High-Speed Links with Silicon Interposer. In Proceedings of the IEEE 67th Electronic Components and Technology Conference (ECTC), Orlando, FL, USA, 30 May–2 June 2017; pp. 2377–5726.
20. Kim, J.; Pak, J.P.; Cho, J.; Song, E.; Cho, J.; Kim, H.; Song, T.; Lee, J.; Lee, H.; Park, K.; et al. High-Frequency Scalable Electrical Model and Analysis of a Through Silicon Via (TSV). *IEEE Trans. Compon. Packag. Manuf. Technol.* **2011**, *1*, 181–195.
21. Tsai, Y.C.; Lee, C.H.; Chang, H.C.; Liu, J.H.; Hu, H.W.; Ito, H.; Kim, Y.S.; Ohba, T.; Chen, K.-N. Electrical Characteristics and Reliability of Wafer-on-Wafer (WOW) Bumpless Through-Silicon Via. *IEEE Trans. Electron. Devices* **2021**, *68*, 3520–3525. [CrossRef]
22. Amin Farmahini-Farahani, A.; Gurumurthi, S.; Loh, G.; Ignatowski, M. Challenges of High-Capacity DRAM Stacks and Potential Directions. In Proceedings of the MCHPC'18: Proceedings of the Workshop on Memory Centric High Performance Computing, Dallas, TX, USA, 4–13 November 2018.
23. Sakui, K.; Ohba, T. High Bandwidth Memory (HBM) and High Bandwidth NAND (HBN) with the Bumpless TSV Technology. In Proceedings of the International 3D Systems Integration Conference (3DIC), Sendai, Japan, 8–10 October 2019.

24. Bai, F.; Jiang, X.; Wang, S.; Yu, B.; Tan, J.; Zuo, F.; Wang, C.; Wang, F.; Long, X.; Yu, G.; et al. A Stacked Embedded DRAM Array for LPDDR4/4X using Hybrid Bonding 3D Integration with 34 GB/s/1 Gb 0.88 pJ/b Logic-to-Memory Interface. In Proceedings of the IEEE International Electron Devices Meeting (IEDM), San Francisco, CA, USA, 12–18 December 2020.
25. Niu, D.; Li, S.; Wang, Y.; Han, W.; Zhang, Z.; Guan, Y.; Guan, T.; Sun, F.; Xue, F.; Duan, L.; et al. 184QPSW 64 Mb/mm^2 3D Logic-to-DRAM Hybrid Bonding with Process-Near-Memory Engine for Recommendation System. In Proceedings of the IEEE International Solid-State Circuits Conference (ISSCC), San Francisco, CA, USA, 20–26 February 2022.
26. Cho, J.; Song, E.; Yoon, K.; Pak, J.S.; Kim, J.; Lee, W.; Song, T.; Kim, K.; Lee, J.; Lee, H.; et al. Modeling and Analysis of Through-Silicon Via (TSV) Noise Coupling and Suppression Using a Guard Ring. *IEEE Trans. Compon. Packag. Manuf. Technol.* **2011**, *1*, 220–233. [CrossRef]
27. Sadiku, M.N.O. Transmission Lines. In *Book Elements of Electromagnetics*, 7th ed.; Sedra, A.S., Ed.; Oxford University Press: New York, NY, USA, 2018; Volume II, pp. 554–555.

micromachines

Article

Compact Modeling of Advanced Gate-All-Around Nanosheet FETs Using Artificial Neural Network

Yage Zhao [1], Zhongshan Xu [1], Huawei Tang [1], Yusi Zhao [1], Peishun Tang [1], Rongzheng Ding [1], Xiaona Zhu [1], David Wei Zhang [1,2] and Shaofeng Yu [1,2,*]

1   School of Microelectronics, Fudan University, Shanghai 200433, China; 20112020044@fudan.edu.cn (Y.Z.); 21112020147@m.fudan.edu.cn (Z.X.); 23112020147@m.fudan.edu.cn (H.T.); 22212020147@m.fudan.edu.cn (P.T.); xnzhu@fudan.edu.cn (X.Z.); dwzhang@fudan.edu.cn (D.W.Z.)
2   National Integrated Circuit Innovation Center, Shanghai 201203, China
*   Correspondence: shaofeng_yu@fudan.edu.cn

Abstract: As the architecture of logic devices is evolving towards gate-all-around (GAA) structure, research efforts on advanced transistors are increasingly desired. In order to rapidly perform accurate compact modeling for these ultra-scaled transistors with the capability to cover dimensional variations, neural networks are considered. In this paper, a compact model generation methodology based on artificial neural network (ANN) is developed for GAA nanosheet FETs (NSFETs) at advanced technology nodes. The DC and AC characteristics of GAA NSFETs with various physical gate lengths (L_g), nanosheet widths (W_{sh}) and thicknesses (T_{sh}), as well as different gate voltages (V_{gs}) and drain voltages (V_{ds}) are obtained through TCAD simulations. Subsequently, a high-precision ANN model architecture is evaluated. A systematical study on the impacts of ANN size, activation function, learning rate, and epoch (the times of complete pass through the entire training dataset) on the accuracy of ANN models is conducted, and a shallow neural network configuration for generating optimal ANN models is proposed. The results clearly show that the optimized ANN model can reproduce the DC and AC characteristics of NSFETs very accurately with a fitting error (MSE) of 0.01.

Keywords: gate-all-around (GAA) Nanosheet FETs (NSFETs); compact model; artificial neural network (ANN); TCAD simulation

Citation: Zhao, Y.; Xu, Z.; Tang, H.; Zhao, Y.; Tang, P.; Ding, R.; Zhu, X.; Zhang, D.W.; Yu, S. Compact Modeling of Advanced Gate-All-Around Nanosheet FETs Using Artificial Neural Network. *Micromachines* **2024**, *15*, 218. https://doi.org/10.3390/mi15020218

Academic Editors: Zeheng Wang and Jing-Kai Huang

Received: 1 January 2024
Revised: 28 January 2024
Accepted: 29 January 2024
Published: 31 January 2024

1. Introduction

In response to market demands, the transistor dimensions have been scaled down proportionally according to Moore's Law. As an alternative to planar metal-oxide-semiconductor field-effect transistors (MOSFETs), fin field-effect transistors (FinFETs), which utilize a three-dimensional architecture with the gate wrapping around vertical fins on top and sides, have been developed and commercialized in 22 nm CMOS technology [1–3]. During the past decades, FinFET technology has been successfully applied to 5 nm and even 3 nm technology nodes through higher aspect ratio and layout optimization [4–8]. However, the scaling of Fin-FETs has also encountered fabrication- and performance-related obstacles due to fundamental physical limitations and difficulties in developing the required process. The performance improvement is constrained by the severe short channel effects (SCEs), while it is difficult to populate multiple fins in a limited space as CGP is further reduced. As the most feasible solution to extend Moore's Law and Dennard's Law, gate-all-around (GAA) nanosheet FETs (NSFETs) are poised to become a mainstream device architecture for 2 nm node and beyond [9–12]. Compared to traditional FinFETs or planar MOSFETs, GAA NSFETs offer superior electrostatic control, higher driving capability, lower leakage current, and more effective footprint [10]. This is because they not only have the gates surrounding the channel, but also have wider effective widths in the same footprint. Nevertheless, this advancement imposes a challenge to semiconductor device models.

Semiconductor device models are regarded as a bridge between foundry, EDA vendor, and design house, as well as a key-enabler for accurate integrated circuit (IC) simulations. The conventional semiconductor device models include macro models [13,14], compact models [15–18], and look-up table (LUT) models [19,20]. In particular, compact models are the mainstream ones and are composed of physics-based equations, which have been developed for decades. The first industry standard compact model is BSIM (Berkeley short-channel insulated-gate field-effect transistor model), whose genesis can be traced to the 1980s [21], and several versions have been developed and remain in use today [16,22,23]. Generally, analytical equations are used to describe device I–V and C–V characteristics in the subthreshold, linear, and saturation regions in a unified way. The accuracy of the compact models is crucial for efficient analysis and design of ICs. However, for advanced transistors, the underlying physics becomes much more complicated, making the models more difficult to fit. In addition, the actual electrical properties of miniaturized transistors are case sensitive due to dimension variations. Since developing suitable analytical compact models is complex and often takes several years, it requires novel modeling methodology to circumvent the high costs of time and labor.

The need for a new technique brings the artificial neural network (ANN) method to the attention of researchers, which has been attempted for planar MOSFETs modeling since the early 1990s and showed good precision [24]. ANNs represent a class of machine learning models inspired by the neuromorphic architecture, and use a set of multilayered perceptrons/neurons, also known as feed-forward neural networks, consisting of an input layer, multiple hidden layers, and an output layer [25,26]. Because of the robust learning capability, they have once been a powerful tool used in the computer science to deal with machine learning issues. The primary objective of ANNs is to learn complex mappings between inputs and outputs by adjusting the weights and biases of interconnected neurons, in a nutshell, is to achieve a good means of solving data fitting problems. This learning process involves the application of mathematical principles, particularly the chain rule in calculus, to update the network parameters and minimize the error between predicted and actual outcomes. In other words, with a reasonable network configuration, ANNs can fit arbitrary nonlinear functions and hence can also be developed as black-box models to address nonlinear systems or more sophisticated internal expressions, such as the compact modeling of semiconductor devices in advanced nodes mentioned earlier. Although the ANN models seemed to be a simple black box, there are many parameters within the neural network that have an impact on the accuracy of models, which will further affect the subsequent circuit simulations. Thereby, an in-depth study of ANN-based compact modeling methodology is necessary for the development and application of GAA devices and even complementary FET (CFET) devices, which are more sophisticated architectures with n-FET folded onto p-FET, in advanced technologies [27,28]. Actually, there are some interesting and meaningful studies on ANN-based device modeling that have been published in recent years [29–32]. However, most of the literature in this field have only superficially studied ANN modeling, focusing instead on its implementation in subsequent circuits or on the unique electrical properties under investigation, and lacking an in-depth understanding and full exploration of the ANNs used for modeling.

In this work, we conduct a comprehensive evaluation of the compact modeling of advanced GAA NSFETs based on ANN, with the datasets from finely calibrated TCAD simulations. Referring to [10] and IRDS 2022 [33], an N-channel GAA NSFET was built as the nominal transistor for the modeling study. The applied voltages on terminals and 3-D nanosheet dimensions were set as input parameters and varied to obtain datasets, some of which were used for training data feeding into the ANN and the others were used for testing data for the final test. Appropriate data preprocessing and neural network configurations, as well as $L2$ regularization were adopted to improve model accuracy. Without considering the physical characteristics of real transistors, high fitting accuracy can be achieved by using transistor data for model training. The DC and AC characteristics are well mapped with the five input variants, including applied voltages and geometrical dimensions.

2. Device Structure, TCAD Simulation Calibration, and Dataset Generation

2.1. Device Structure

The Sentaurus Technology Computer-Aided Design (TCAD) [34] tool is exploited to construct the GAA NSFET devices and generate physical electric characteristic data for subsequent studies. Figure 1a–c shows the 3-D schematic of nominal GAA NSFET structure and 2-D cross-sectional along and across the channel views, respectively. Detailed parameters of a nominal highly scaled device at 2 nm technology node are specifically listed in Table 1 following IRDS 2022 [33], where the physical gate length (L_g) of 14 nm, nanosheet width (W_{sh}) of 15 nm, nanosheet thickness (T_{sh}) of 6 nm, the spacer length (L_{sp}) of 6 nm, and the sheet-to-sheet spacing (T_{sp}) of 10 nm are adopted. For n-type MOS, the in-situ uniform doping profiles for channels and source/drain regions were performed with 1×10^{10} cm^{-3} of boron doping concentration and 5×10^{20} cm^{-3} of arsenic doping concentration, respectively. As for the high-k/metal gate (HKMG) stack, the equivalent oxide thickness (EOT) is 1.35 nm, which consists of HfO$_2$ of 2 nm and interfacial oxide SiO$_2$ of 1 nm. The work-function metal used in the gate stack is TiN and the effective work-function (WF) is set to 4.4 eV. Note that for high-performance devices, the geometric parameters are the same as above except for the nanosheet width being wider.

Table 1. Detailed parameters of nominal device at 2 nm technology node [33].

Parameters	Value
Physical gate length (L_g)	14 nm
Source/drain length (L_{sd})	12 nm
Spacer length (L_{sp})	6 nm
Nanosheet width (W_{sh})	15 nm
Nanosheet thickness (T_{sh})	6 nm
Sheet-to-sheet spacing (T_{sp})	10 nm
Equivalent oxide thickness (EOT)	1.35 nm
Source/drain doping concentration (N_{sd})	5×10^{20} cm^{-3}
Channel doping concentration (N_{ch})	1×10^{10} cm^{-3}
Metal gate work-function (WF)	4.4 eV

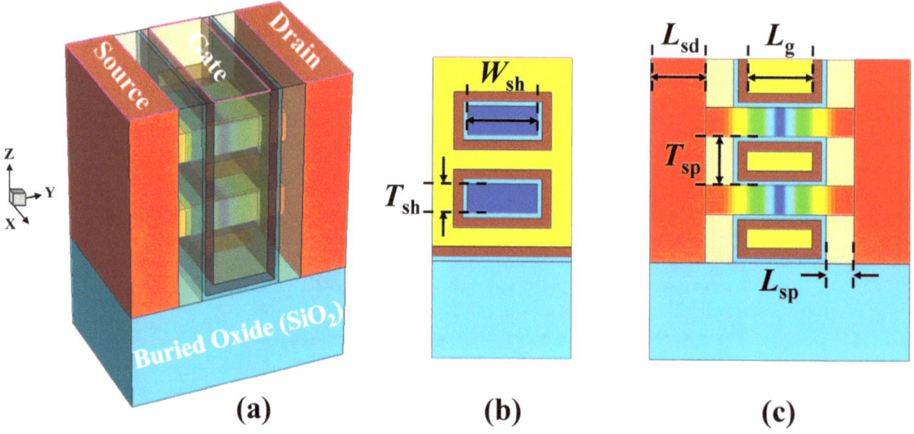

Figure 1. An illustration of nominal gate-all-around nanosheet FET (GAA NSFET) and details of device structure : (**a**) Entire 3-D schematic; (**b**) X–Z cut plane and (**c**) Y–Z cut plane of the nominal device structure.

2.2. TCAD Simulation Calibration

Since nanoscale devices typically exhibit size-dependent behavior, the corresponding physical model parameters built-in in the TCAD simulator may not be accurate enough with the scale shrinking, which affects the validity of the device characteristics resulted from TCAD simulations. Therefore, in order to ensure the accuracy of the subsequent simulations to generate more physically accurate datasets for the subsequent ANN model, it is essential to calibrate the simulator against experimental data to lay a solid ground for the ANN modeling work. In this calibration work, both DC and AC characteristics were covered, comprehensively demonstrating the exactitude of the simulation platform.

Besides the nominal GAA NSFET structure illustrated in the previous section for DC calibration, an n-type MOS capacitor was generated according to the device description for AC calibration [35]. TCAD calibrations against experimental data of Refs. [10,35] andwere performed in the framework of drift-diffusion (DD) transport model with quantum correction in electrostatics. The results are shown in Figure 2a,b, where the calibrated simulator closely matches the experimental I_{ds}–V_{gs} and C–V characteristics after adjustments of the relevant model parameters. The physical models used include the Philip unified mobility, thin-layer mobility and high-field saturation models, as well as Shockley–Read–Hall (SRH) recombination, Auger recombination and band-to-band tunneling models in the drift-diffusion (DD) framework. Physically more correct, Fermi-Dirac statistics are used for high doping concentrations. Furthermore, the density-gradient and kinetic velocity models are considered to account for quantum confinement and ballistic effects.

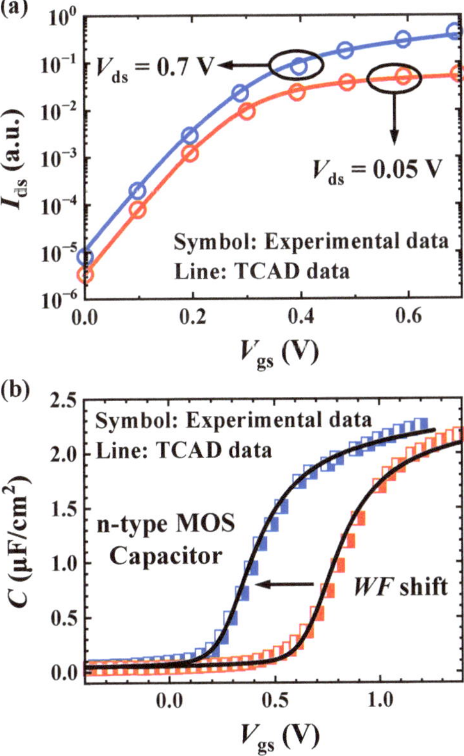

Figure 2. TCAD simulation calibrations against the experimental data under the same simulation environment. (**a**) Calibrated I_{ds}–V_{gs} characteristics of n-type NSFET versus experiment data from Ref. [10], and (**b**) C–V characteristics of n-type MOS capacitor versus experimental data from Ref. [35].

2.3. Dataset Generation

Based on the previously calibrated simulation environment, a number of GAA NSFETs were designed by altering the nanosheet dimensions (L_g, W_{sh}, and T_{sh}) of the nominal device. And the ranges of dimensional variants were designed to cover the specifications of IRDS roadmap organized for 3 nm to 1 nm nodes [33]. Here, L_g ranges from 10 to 20 nm, W_{sh} ranges from 15 to 30 nm, while T_{sh} has a smaller movable range between 4–7 nm. Then, the DC and AC characteristics were extracted to create dataset when V_{ds} and V_{gs} were set at 0–0.7 V. For the C-V model, the AC characteristics were obtained with a frequency of 10^6 Hz. In the practical simulation experiments, we can flexibly control the number of points taken in the electric characteristic curves. Considering that too much data generated by TCAD are very likely to cause overfitting and waste of computing resource, we finally randomly selected 4000 sets of data to form the dataset used for the subsequent study.

3. Development and Optimization of ANN Model

3.1. Development of ANN Model

Figure 3 shows the proposed schematic diagram of developing a regression ANN model, which is executed in the following steps: (1) accepting the input data, (2) fine-tuning the input and output parameters while training the model, (3) testing and (4) evaluating the trained model using the testing data. A complete five parameters are used as input variants, including gate-to-source voltage (V_{gs}), drain-to-source voltage (V_{ds}), physical gate length (L_g), nanosheet width (W_{sh}), and nanosheet thickness (T_{sh}). The training/testing data, which comprising DC and AC characteristics for various input parameters, is obtained from physical TCAD simulations. The hidden layers consist of two layers, with k ($k = 10$) and s ($s = 5$) neurons respectively. The number of neurons in the output layer is p ($p = 4$), one is used for the I–V model, and the other three are used for the C–V model. Besides, we define the conversion function for mapping the output values of the ANN model to the real current I_{ds} and the capacitance $C_{g,g}$, $C_{g,d}$ and $C_{g,s}$. The training of the ANN model is realized using python with the assistance of the PyTorch package.

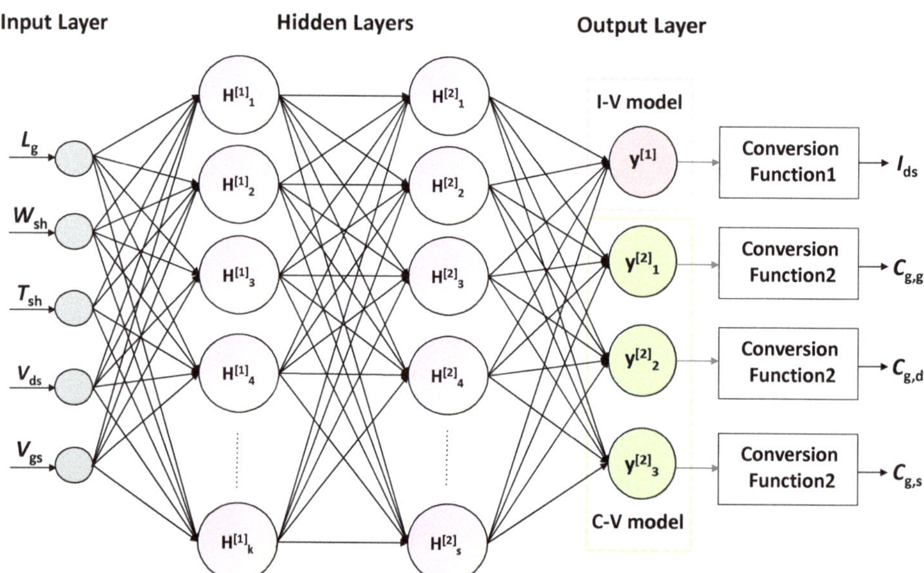

Figure 3. The regression neural network topology framework.

In our ANN model, each hidden layer consists of multiple neurons, and the connections between neurons are characterized by weights w and biases b. The mathematical foundation of ANNs relies on the activation function, often denoted as f, which introduces non-linearity into the model. The training of the network includes two steps: the forward process and the backward process. The forward process of the network involves calculating the weighted sum of inputs, applying the activation function, and passing the result to the next layer. This process is repeated layer by layer until the final output is obtained. The mathematical representation of the forward process can be expressed as follows:

$$net_j^{(k)} = \sum_{i=1}^{n^{(k-1)}} w_{j,i}^{(k)} * y_i^{(k-1)} + b_j^{(k)} \tag{1}$$

$$y_j^{(k)} = f(net_j^{(k)}) \tag{2}$$

Here, $net_j^{(k)}$ represents the weighted sum of inputs for neuron j in layer k, $w_{j,i}^{(k)}$ denotes the weight connecting neuron i in layer $k - 1$ to neuron j in layer k, $y_i^{(k-1)}$ is the output of neuron i in layer $k - 1$, $b_j^{(k)}$ is the bias for neuron j in layer k, and $f(net_j^{(k)})$ is the activation function. Especially, the hyperbolic tangent function $\tanh(x)$ was used as the activation function [36,37]. The output of $\tanh(x)$ lies within the range of $[-1, 1]$, which, compared to the $[0, 1]$ range of the sigmoid function, makes $\tanh(x)$ advantageous in zero-centering. This helps mitigate the exploding gradient problem during gradient descent.

The training process involves minimizing a predefined loss function, typically the mean squared error (MSE) , which measures the discrepancy between the predicted and actual outputs. MSE is calculated by taking the average of the squared differences between predicted and actual values, which is a simple and easily differentiable form. This simplicity facilitates the updating of weights in optimization algorithms like gradient descent. In addition, as MSE involves squaring the errors, it is less sensitive to outliers (samples with significantly different actual values). This means that individual outliers do not have a disproportionately large impact on the overall loss function, enhancing the robustness of the model.

The backward process, also known as backpropagation, is a crucial step in updating the network parameters. The gradients are propagated backward through the network, and the weights and biases are adjusted using optimization algorithms such as stochastic gradient descent (SGD) [38]. The chain rule is applied iteratively to compute the gradients of the loss (L) with respect to the network parameters:

$$\frac{\partial L}{\partial w_{j,i}^{(k)}} = \frac{\partial L}{\partial net_j^{(k)}} * \frac{\partial net_j^{(k)}}{\partial w_{j,i}^{(k)}} \tag{3}$$

$$\frac{\partial L}{\partial b_j^{(k)}} = \frac{\partial L}{\partial net_j^{(k)}} * \frac{\partial net_j^{(k)}}{\partial b_j^{(k)}} \tag{4}$$

These gradients guide the parameters update during the training process, gradually optimizing the network to improve its predictive capabilities. The iterative nature of backpropagation allows the network to learn complex patterns and relationships within the data.

Thus, a four-layered regression ANN involves intricate mathematical formulations, including the forward pass equations for computing neuron activations and the backward pass equations for updating weights and biases during training. The application of the chain rule in calculus is fundamental to these computations, enabling the network to learn and adapt to complex patterns in the data.

3.2. Optimization of ANN Model

Before the training process, we noticed that the orders of magnitude of the outputs are too small, $10^{-13} \sim 10^{-3}$ for the I–V model and $10^{-18} \sim 10^{-17}$ for the C–V model, which are not favorable for data fitting. So, we preprocessed the outputs (I_{ds}, $C_{g,g}$, $C_{g,d}$, and $C_{g,s}$) in order to achieve the accurate fitting through a linear preprocessing method. Here, we multiplied the output currents and capacitances by factors of 1×10^6 and 1×10^{18}, respectively, thereby converting the units from A and F to μA and aF.

Since then, the processed dataset was utilized for training, but another problem was identified, namely, the ANN model had overfitting, which means it performs well during training but fails to generalize effectively to the test samples. In other words, our model has a significant gap between the model's performance during training and its performance when making predictions on new data. The network excels in fitting the training data but struggles to make accurate predictions on unseen examples.

Overfitting often leads to excessively complex neural network models. These models tend to capture noise and outliers in the training data, making them less suitable for generalization. Moreover, the loss function used during training may not accurately reflect the network's performance on new data. The model might minimize the training loss, giving a false sense of success, while failing to minimize the loss on validation or test data. Let L_{train} denotes the training loss, L_{val} the validation loss, and L_{test} the test loss. Overfitting occurs when L_{train} is significantly smaller than both L_{val} and L_{test}.

$$L_{train} \ll L_{val}, L_{test} \tag{5}$$

To address this issue, we adopt $L2$ regularization (also known as weight decay) [39], which is a widely adopted technique to address overfitting by adding a penalty term to the loss function. The regularized loss function is given by:

$$L = \frac{1}{2}\|Xw - y\|^2 + \lambda\|w\|^2 \tag{6}$$

Here, X is the input matrix, w is the weight vector, y is the target vector, and λ is the regularization parameter that controls the strength of the regularization. The first term $\frac{1}{2}\|Xw - y\|^2$ represents the MSE (described in Section 3.1), aiming to minimize the difference between the predicted and actual values. The second term $\lambda\|w\|^2$ is the $L2$ regularization term. It penalizes large weights by adding the squared magnitude of the weight vector. The regularization parameter λ controls the trade-off between fitting the training data and preventing overfitting.

From the viewpoint of convex optimization, the introduction of the $L2$ regularization term transforms the optimization problem into a constrained optimization problem. The regularization term induces a constraint on the magnitude of the weight vector, effectively defining a hypersphere in the weight space. This transformation has a smoothing effect on the optimization landscape, making it more convex. The regularization term adds a regularization force that discourages the weights from reaching extreme values, leading to a more stable and generalizable model.

In summary, $L2$ regularization mitigates overfitting by penalizing large weights in a linear regression model. The mathematical formulation introduces a balance between fitting the training data and controlling the complexity of the model. From a convex optimization perspective, the regularization term induces a constraint that shapes a more well-behaved optimization landscape.

4. Results and Discussion

A total of 4000 samples for ANN training and testing are obtained from I_{ds}–V_{gs} and C–V data generated by previous TCAD simulations. We randomly split these samples into a training set (a total of 3200 samples) and a testing set (a total of 800 samples) in a 4:1 ratio. Theoretically, as the number of hidden layers and neurons increases, the ANN model becomes more capable of extracting the non-linear mapping relationship between input and output. However, in practice, too many hidden layers or number of neurons can also bring about overfitting problems. And in most cases, the fitting accuracy is determined synergistically by both the number of hidden layers and neurons. For most fitting cases with limited input and output variants, a shallow neural network is sufficient, which is easier to be trained and converges to the optimal solution faster , with a more favorable computational and memory footprint. Thus, to obtain an optimal network, we studied the impact of network sizes on the errors (MSE) for shallow neural networks with two hidden layers. As shown in Figure 4, we find that the MSE of the testing set tends to decrease and then increase as the number of neurons increases. The minimum MSE is 0.01 with ten neurons in the first hidden layer and five neurons in the second hidden layer.

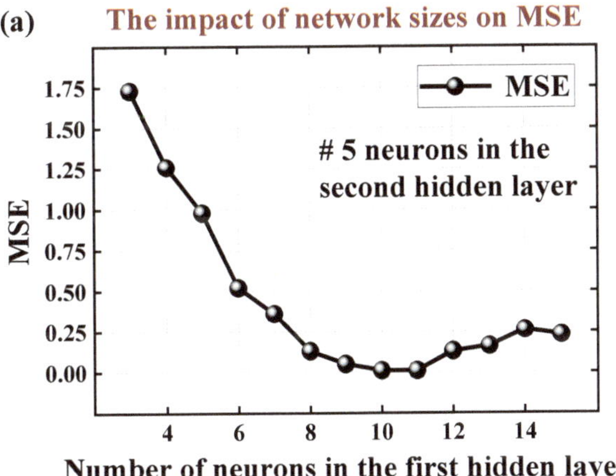

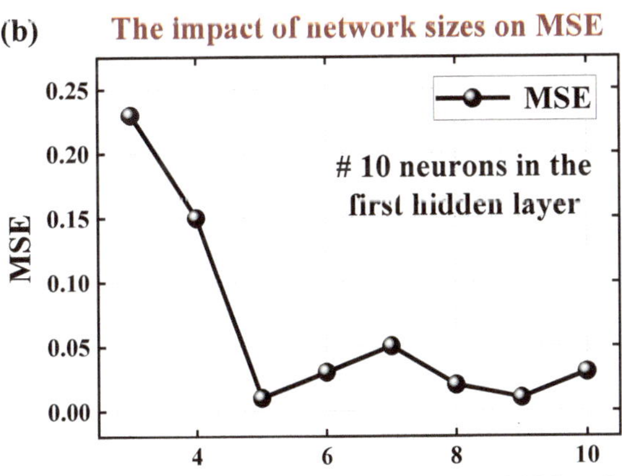

Figure 4. The MSE for all the test samples with different numbers of neurons in the first hidden layer (a) and second hidden layer (b).

Besides, we investigated the impact of different types of activation functions of the neurons on MSE for the test dataset, as depicted in Figure 5. Using the hyperbolic tangent function $\tanh(x)$ has the lowest MSE. Through our analysis, the $\tanh(x)$ function is zero-centered, meaning its mean is zero. This is beneficial for optimization algorithms such as gradient descent, as it helps prevent the gradient updates from consistently favoring a particular direction, thus improving the convergence speed of the model. The derivative of the $\tanh(x)$ function is non-zero in most regions, aiding in the propagation of gradients during backpropagation. Unlike the sigmoid function, the gradient of $\tanh(x)$ does not approach zero in regions of large or small inputs, reducing the risk of the vanishing gradient problem.

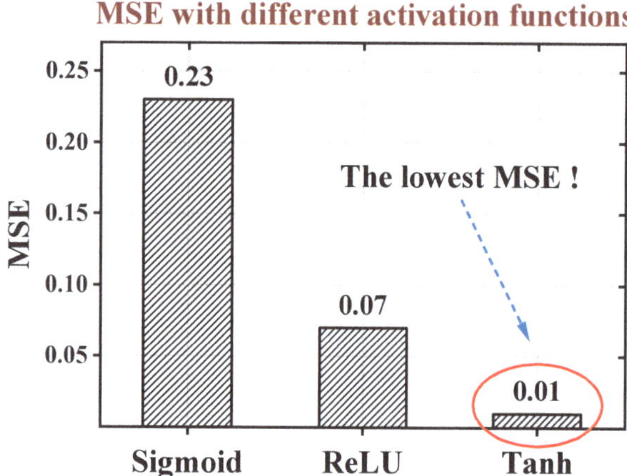

Figure 5. The MSE with different activation functions of the neurons. Popular activation functions: sigmoid, relu, and tanh.

Moreover, we have studied the impact of the learning rate on MSE. In our ANN model, the learning rate is a crucial hyper parameter in training neural networks. It controls the magnitude of updates applied to the weights during the training process. A higher learning rate means larger updates, leading to faster convergence but with the risk of overshooting the optimal weights. Conversely, a lower learning rate allows for smaller weight updates, potentially resulting in slower convergence but increased precision in finding the global minimum. Through our experiments, we think learning rate of 0.02 is the best solution for the ANN model as shown in Figure 6.

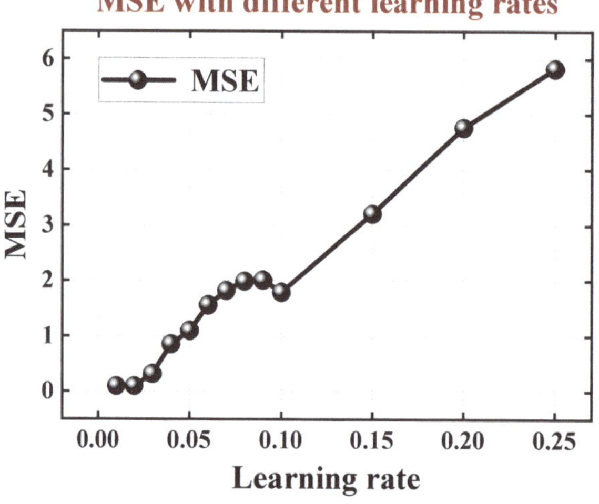

Figure 6. The MSE with different learning rates.

Epochs represent the number of times the entire dataset is fed forward and backward through the neural network during the training process. The choice of the number of epochs plays a pivotal role in determining how well the model generalizes to unseen data. Too few epochs may result in under fitting, where the model fails to capture the underlying patterns in the data. On the other hand, an excessive number of epochs may lead to overfitting, causing the model to memorize the training data but perform poorly on new, unseen data. Moreover, the relationship between epoch and learning rate is interdependent. A higher learning rate may require fewer epochs to converge, as each iteration leads to more substantial weight updates. Conversely, a lower learning rate might necessitate a higher number of epochs to allow the model to converge gradually. Finally, as shown in Figure 7, we choose the lowest MSE (0.01) scheme with the epoch of 5000 and the learning rate of 0.02.

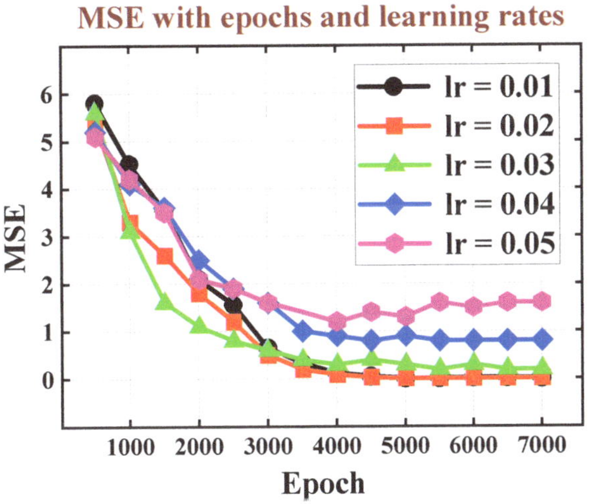

Figure 7. The MSE decline process with different learning rates as epoch increases.

Here, we summarize the primary parameters of the proposed ANN model, as shown in Table 2. Based on the experiments and analysis, we can utilize the ANN model to predict the current and capacitance output based on input data (L_g, W_{sh}, T_{sh}, V_{gs} and V_{ds}) with MSE of 0.01. As shown in Figure 8, the example results of both the DC and AC characteristics of ANN model are fitted against the TCAD data of high-density and high-performance GAA NSFETs at 2 nm technology node. It can be seen that the output $I–V$ and $C–V$ performances generated by ANN fit well with the TCAD results. In addition, we predicted the $I–V$ performance of the nominal NSFET device at high gate bias ($V_{gs} = 0.7–0.8$ V) to examine the model scalability. The extrapolation behavior also fits well to simulation results. The optimistic results reveal that the proposed network is capable of handling the electrical characterization of advanced GAA NSFETs with great accuracy.

Table 2. The primary parameters of the ANN model.

Parameters	Features
Network size	5-10-5-4
Activation function	Hyperbolic tangent function
Learning rate	0.02
Epoch	5000
#Training samples	3200
#Test samples	800
Task	Regression
MSE	0.01
Regularization	$L2$ Regularization

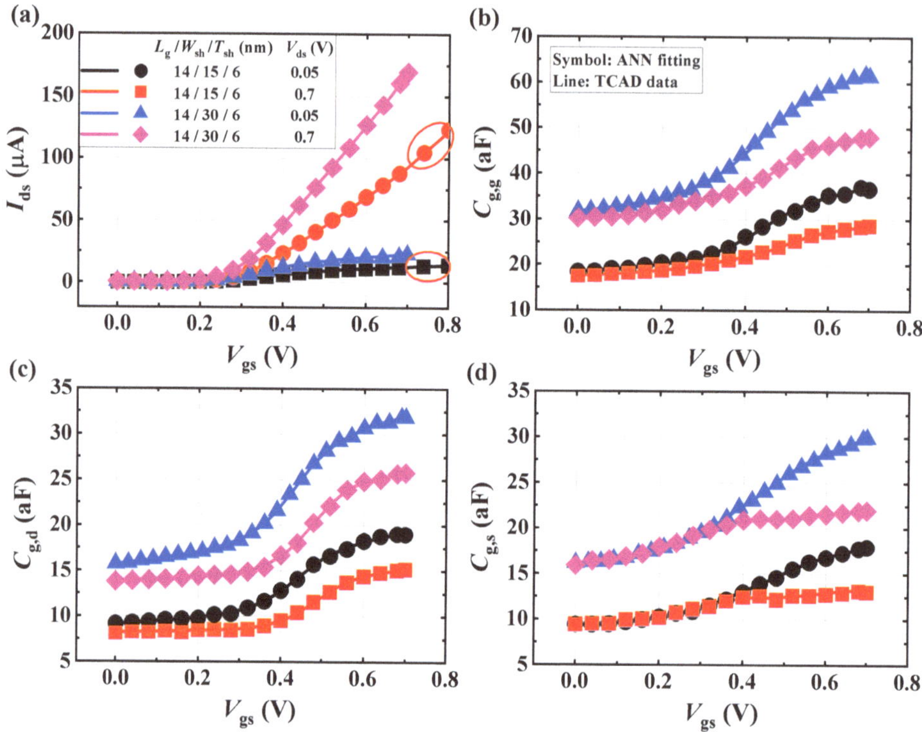

Figure 8. Example ANN model fitting results of the simulated DC and AC characteristics for high-density and high-performance GAA NSFETs at 2 nm technology node. (**a**) I_{ds}–V_{gs}. (**b**) $C_{g,g}$–V_{gs}. (**c**) $C_{g,d}$–V_{gs}. (**d**) $C_{g,s}$–V_{gs}.

5. Conclusions

In summary, the ANN-based compact modeling methodology has been thoroughly investigated for advanced GAA NSFETs. Here, the impacts of ANN size, activation function, learning rate, and epoch on the accuracy of ANN models were systematically evaluated. Based on the precisely calibrated simulation environment, various GAA NSFET devices were constructed by varying the nanosheet dimensions, and their DC as well as AC characteristics were extracted. The generated dataset contains five input variants (V_{gs}, V_{ds}, L_g, W_{sh}, and T_{sh}) and four output quantities (I_{ds}, $C_{g,g}$, $C_{g,d}$, and $C_{g,s}$). Before the training process, the output data were preprocessed to circumvent unnecessary fitting mistakes using a linear preprocessing method. By adopting the $L2$ regularization, the overfitting issue was perfectly resolved with the addition of a penalty term to the loss function. The optimized ANN model fully demonstrates its superior fitting properties under various conditions with a low fitting MSE error of 0.01. Furthermore, the scalability was also validated. This work contributes to the development of ANN-based compact models, holding great promise for adoption in advanced fast turn-around design and technology co-optimization (DTCO) as well as large-scale product-design-oriented circuit simulations.

Author Contributions: Conceptualization, Y.Z. (Yage Zhao); methodology, Y.Z. (Yage Zhao) and S.Y.; investigation, Y.Z. (Yage Zhao), S.Y. and Z.X.; writing–original draft preparation, Y.Z. (Yage Zhao); writing—review and editing, Y.Z. (Yage Zhao), S.Y., Z.X., H.T., Y.Z. (Yusi Zhao), P.T., R.D., X.Z. and D.W.Z.; supervision, S.Y. All authors have read and agreed to the published version of the manuscript.

Funding: This research was funded by a platform for the development of next generation integrated circuit technology and Shanghai Sailing Program, 20YF1401700.

Data Availability Statement: The data and code are available from the corresponding authors upon reasonable request.

Conflicts of Interest: The authors declare no conflict of interest.

References

1. Hisamoto, D.; Lee, W.C.; Kedzierski, J.; Takeuchi, H.; Asano, K.; Kuo, C.; Anderson, E.; King, T.J.; Bokor, J.; Hu, C. FinFET-a self-aligned double-gate MOSFET scalable to 20 nm. *IEEE Trans. Electron Devices* **2000**, *47*, 2320–2325. [CrossRef]
2. Jan, C.H.; Bhattacharya, U.; Brain, R.; Choi, S.J.; Curello, G.; Gupta, G.; Hafez, W.; Jang, M.; Kang, M.; Komeyli, K.; et al. A 22 nm SoC platform technology featuring 3-D tri-gate and high-k/metal gate, optimized for ultra low power, high performance and high density SoC applications. In Proceedings of the 2012 International Electron Devices Meeting (IEDM), San Francisco, CA, USA, 10–13 December 2012; pp. 3.1.1–3.1.4.
3. Sell, B.; Bigwood, B.; Cha, S.; Chen, Z.; Dhage, P.; Fan, P.; Giraud-Carrier, M.; Kar, A.; Karl, E.; Ku, C.J.; et al. 22FFL: A high performance and ultra low power FinFET technology for mobile and RF applications. In Proceedings of the 2017 IEEE International Electron Devices Meeting (IEDM), San Francisco, CA, USA, 2–6 December 2017; pp. 29.24.21–29.24.24.
4. Yeap, G.; Lin, S.S.; Chen, Y.M.; Shang, H.L.; Wang, P.W.; Lin, H.C.; Peng, Y.C.; Sheu, J.Y.; Wang, M.; Chen, X.; et al. 5 nm CMOS Production Technology Platform featuring full-fledged EUV, and High Mobility Channel FinFETs with densest 0.021 μm^2 SRAM cells for Mobile SoC and High Performance Computing Applications. In Proceedings of the 2019 IEEE International Electron Devices Meeting (IEDM), San Francisco, CA, USA, 7–11 December 2019; pp. 36.37.31–36.37.34.
5. Liu, J.C.; Mukhopadhyay, S.; Kundu, A.; Chen, S.H.; Wang, H.C.; Huang, D.S.; Lee, J.H.; Wang, M.I.; Lu, R.; Lin, S.S.; et al. A Reliability Enhanced 5nm CMOS Technology Featuring 5th Generation FinFET with Fully-Developed EUV and High Mobility Channel for Mobile SoC and High Performance Computing Application. In Proceedings of the 2020 IEEE International Electron Devices Meeting (IEDM), San Francisco, CA, USA, 12–18 December 2020; pp. 9.2.1–9.2.4.
6. Ding, Y.; Luo, X.; Shang, E.; Hu, S.; Chen, S.; Zhao, Y. A Device Design for 5 nm Logic FinFET Technology. In Proceedings of the 2020 China Semiconductor Technology International Conference (CSTIC), Shanghai, China, 26 June–17 July 2020; pp. 1–5.
7. Chang, C.H.; Chang, V.S.; Pan, K.H.; Lai, K.T.; Lu, J.H.; Ng, J.A.; Chen, C.Y.; Wu, B.F.; Lin, C.J.; Liang, C.S.; et al. Critical Process Features Enabling Aggressive Contacted Gate Pitch Scaling for 3 nm CMOS Technology and Beyond. In Proceedings of the 2022 International Electron Devices Meeting (IEDM), San Francisco, CA, USA, 3–7 December 2022; pp. 27.21.21–27.21.24.
8. Chung, S.S.; Chiang, C.K.; Pai, H.; Hsieh, E.R.; Guo, J.C. The Extension of the FinFET Generation Towards Sub-3 nm: The Strategy and Guidelines. In Proceedings of the 2022 6th IEEE Electron Devices Technology & Manufacturing Conference (EDTM), Oita, Japan, 6–9 March 2022; pp. 15–17.
9. Feng, P.; Song, S.; Nallapati, G.; Zhu, J.; Bao, J.; Moroz, V.; Choi, M.; Lin, X.; Lu, Q.; Colombeau, B.; et al. Comparative Analysis of Semiconductor Device Architectures for 5-nm Node and beyond. *IEEE Electron Device Lett.* **2017**, *38*, 1657–1660. [CrossRef]
10. Loubet, N.; Hook, T.; Montanini, P.; Yeung, C.W.; Kanakasabapathy, S.; Guillom, M.; Yamashita, T.; Zhang, J.; Miao, X.; Wang, J.; et al. Stacked nanosheet gate-all-around transistor to enable scaling beyond FinFET. In Proceedings of the 2017 Symposium on VLSI Technology, Kyoto, Japan, 5–8 June 2017; pp. T230–T231.
11. Das, U.K.; Bhattacharyya, T.K. Opportunities in Device Scaling for 3-nm Node and Beyond: FinFET versus GAA-FET versus UFET. *IEEE Trans. Electron Devices* **2020**, *67*, 2633–2638. [CrossRef]
12. Ritzenthaler, R.; Mertens, H.; Eneman, G.; Simoen, E.; Bury, E.; Eyben, P.; Bufler, F.M.; Oniki, Y.; Briggs, B.; Chan, B.T.; et al. Comparison of Electrical Performance of Co-Integrated Forksheets and Nanosheets Transistors for the 2nm Technological Node and Beyond. In Proceedings of the 2021 IEEE International Electron Devices Meeting (IEDM), San Francisco, CA, USA, 11–16 December 2021; pp. 26.22.21–26.22.24.
13. Strohbehn, K.; Martin, M.N. SPICE macro models for annular MOSFETs. In Proceedings of the 2004 IEEE Aerospace Conference Proceedings (IEEE Cat. No.04TH8720), Big Sky, MT, USA, 6–13 March 2004; pp. 2370–2377.
14. Oh, J.H.; Yu, Y.S. Macro-Modeling for N-Type Feedback Field-Effect Transistor for Circuit Simulation. *Micromachines* **2021**, *12*, 1174. [CrossRef] [PubMed]
15. Song, J.; Yuan, Y.; Yu, B.; Xiong, W.; Taur, Y. Compact Modeling of Experimental n- and p-Channel FinFETs. *IEEE Trans. Electron Devices* **2010**, *57*, 1369–1374. [CrossRef]
16. Ding, J.; Asenov, A. Reliability-Aware Statistical BSIM Compact Model Parameter Generation Methodology. *IEEE Trans. Electron Devices* **2020**, *67*, 4777–4783. [CrossRef]
17. Wu, T.; Luo, H.; Wang, X.; Asenov, A.; Miao, X. A Predictive 3-D Source/Drain Resistance Compact Model and the Impact on 7 nm and Scaled FinFETs. *IEEE Trans. Electron Devices* **2020**, *67*, 2255–2262. [CrossRef]
18. Jung, S.G.; Kim, J.K.; Yu, H.Y. Analytical Model of Contact Resistance in Vertically Stacked Nanosheet FETs for Sub-3-nm Technology Node. *IEEE Trans. Electron Devices* **2022**, *69*, 930–935. [CrossRef]
19. Wang, J.; Xu, N.; Woosung, C.; Keun-Ho, L.; Youngkwan, P. A generic approach for capturing process variations in lookup-table-based FET models. In Proceedings of the 2015 International Conference on Simulation of Semiconductor Processes and Devices (SISPAD), Washington, DC, USA, 9–11 September 2015; pp. 309–312.
20. Thakker, R.A.; Sathe, C.; Sachid, A.B.; Baghini, M.S.; Rao, V.R.; Patil, M.B. A Novel Table-Based Approach for Design of FinFET Circuits. *IEEE Trans. Comput.-Aided Des. Integr. Circuits Syst.* **2009**, *28*, 1061–1070. [CrossRef]

21. Sheu, B.; Scharfetter, D.; Hu, C.; Pederson, D. A compact IGFET charge model. *IEEE Trans. Circuits Syst.* **1984**, *31*, 745–748. [CrossRef]

22. Duarte, J.P.; Khandelwal, S.; Medury, A.; Hu, C.; Kushwaha, P.; Agarwal, H.; Dasgupta, A.; Chauhan, Y.S. BSIM-CMG: Standard FinFET compact model for advanced circuit design. In Proceedings of the ESSCIRC Conference 2015-41st European Solid-State Circuits Conference (ESSCIRC), Graz, Austria, 14–18 September 2015; pp. 196–201.

23. Singh, S.K.; Gupta, S.; Vega, R.A.; Dixit, A. Accurate Modeling of Cryogenic Temperature Effects in 10-nm Bulk CMOS FinFETs Using the BSIM-CMG Model. *IEEE Electron Device Lett.* **2022**, *43*, 689–692. [CrossRef]

24. Litovski, V.B.; Radjenović, J.I.; Mrčarica, Ž.M.; Milenković, S.L. MOS transistor modelling using neural network. *Electron. Lett.* **1992**, *28*, 1766–1768. [CrossRef]

25. Jain, A.K.; Jianchang, M.; Mohiuddin, K.M. Artificial neural networks: A tutorial. *Computer* **1996**, *29*, 31–44. [CrossRef]

26. Saha, N.; Swetapadma, A.; Mondal, M. A Brief Review on Artificial Neural Network: Network Structures and Applications. In Proceedings of the 2023 9th International Conference on Advanced Computing and Communication Systems (ICACCS), Coimbatore, India, 17–18 March 2023; pp. 1974–1979.

27. Ryckaert, J.; Schuddinck, P.; Weckx, P.; Bouche, G.; Vincent, B.; Smith, J.; Sherazi, Y.; Mallik, A.; Mertens, H.; Demuynck, S.; et al. The Complementary FET (CFET) for CMOS scaling beyond N3. In Proceedings of the 2018 IEEE Symposium on VLSI Technology, Honolulu, HI, USA, 18–22 June 2018; pp. 141–142.

28. Subramanian, S.; Hosseini, M.; Chiarella, T.; Sarkar, S.; Schuddinck, P.; Chan, B.T.; Radisic, D.; Mannaert, G.; Hikavyy, A.; Rosseel, E.; et al. First Monolithic Integration of 3D Complementary FET (CFET) on 300 mm Wafers. In Proceedings of the 2020 IEEE Symposium on VLSI Technology, Honolulu, HI, USA, 16–19 June 2020; pp. 1–2.

29. Ko, K.; Lee, J.K.; Kang, M.; Jeon, J.; Shin, H. Prediction of Process Variation Effect for Ultrascaled GAA Vertical FET Devices Using a Machine Learning Approach. *IEEE Trans. Electron Devices* **2019**, *66*, 4474–4477. [CrossRef]

30. Butola, R.; Li, Y.; Kola, S.R.; Chen, C.Y.; Chuang, M.H. Artificial Neural Network-Based Modeling for Estimating the Effects of Various Random Fluctuations on DC/Analog/RF Characteristics of GAA Si Nanosheet FETs. *IEEE Trans. Microw. Theory Tech.* **2022**, *70*, 4835–4848. [CrossRef]

31. Qi, G.; Chen, X.; Hu, G.; Zhou, P.; Bao, W.; Lu, Y. Knowledge-based neural network SPICE modeling for MOSFETs and its application on 2D material field-effect transistors. *Sci. China Inf. Sci.* **2023**, *66*, 122405. [CrossRef]

32. Wei, J.; Wang, H.; Zhao, T.; Jiang, Y.L.; Wan, J. A New Compact MOSFET Model Based on Artificial Neural Network with Unique Data Preprocessing and Sampling Techniques. *IEEE Trans. Comput.-Aided Des. Integr. Circuits Syst.* **2023**, *42*, 1250–1254. [CrossRef]

33. IRDS. International Roadmap for Devices and Systems 2022 (IRDS 2022). Available online: https://irds.ieee.org/ (accessed on 31 July 2022).

34. Synopsys, Inc. *Sentaurus TCAD User's Manual*; Synopsys Inc.: Mountain View, CA, USA, 2019.

35. Arimura, H.; Ragnarsson, L.Å.; Oniki, Y.; Franco, J.; Vandooren, A.; Brus, S.; Leonhardt, A.; Sippola, P.; Ivanova, T.; Verni, G.A.; et al. Dipole-First Gate Stack as a Scalable and Thermal Budget Flexible Multi-Vt Solution for Nanosheet/CFET Devices. In Proceedings of the 2021 IEEE International Electron Devices Meeting (IEDM), San Francisco, CA, USA, 11–16 December 2021; pp. 13.15.11–13.15.14.

36. Zamanlooy, B.; Mirhassani, M. Efficient VLSI Implementation of Neural Networks With Hyperbolic Tangent Activation Function. *IEEE Trans. Very Large Scale Integr. (VLSI) Syst.* **2014**, *22*, 39–48. [CrossRef]

37. Lau, M.M.; Lim, K.H. Investigation of activation functions in deep belief network. In Proceedings of the 2017 2nd International Conference on Control and Robotics Engineering (ICCRE), Bangkok, Thailand, 1–3 April 2017; pp. 201–206.

38. Ruder, S. An overview of gradient descent optimization algorithms. *arXiv* **2016**, arXiv:1609.04747.

39. Gupta, S.; Gupta, R.; Ojha, M.; Singh, K.P. A Comparative Analysis of Various Regularization Techniques to Solve Overfitting Problem in Artificial Neural Network. In Proceedings of the Data Science and Analytics, Singapore, 8 March 2018; pp. 363–371.

 *micromachines*

Article

Comprehensive Comparison of MOCVD- and LPCVD-SiN$_x$ Surface Passivation for AlGaN/GaN HEMTs for 5G RF Applications

Longge Deng [1,†], Likun Zhou [2,†], Hao Lu [1,*], Ling Yang [1,*], Qian Yu [1], Meng Zhang [1], Mei Wu [1], Bin Hou [1], Xiaohua Ma [1] and Yue Hao [1]

1   State Key Discipline Laboratory of Wide Bandgap Semiconductor Technology, School of Microelectronics, Xidian University, Xi'an 710071, China
2   Advanced Materials and Nanotechnology, Xidian University, Xi'an 710071, China
*   Correspondence: luhao@xidian.edu.cn (H.L.); yangling@xidian.edu.cn (L.Y.)
†   These authors contributed equally to this work.

Abstract: Passivation is commonly used to suppress current collapse in AlGaN/GaN HEMTs. However, the conventional PECV-fabricated SiN$_x$ passivation layer is incompatible with the latest process, like the "passivation-prior-to-ohmic" method. Research attention has therefore turned to high-temperature passivation schemes. In this paper, we systematically investigated the differences between the SiN$_x$/GaN interface of two high-temperature passivation schemes, MOCVD-SiN$_x$ and LPCVD-SiN$_x$, and investigated their effects on the ohmic contact mechanism. By characterizing the device interface using TEM, we reveal that during the process of MOCVD-SiN$_x$, etching damage and Si diffuses into the semiconductor to form a leakage path and reduce the breakdown voltage of the AlGaN/GaN HEMTs. Moreover, N enrichment at the edge of the ohmic region of the LPCVD-SiN$_x$ device indicates that the device is more favorable for TiN formation, thus reducing the ohmic contact resistance, which is beneficial to improving the PAE of the device. Through the CW load-pull test with drain voltage V_{DS} = 20V, LPCVD-SiN$_x$ devices obtain a high PAE of 66.35%, which is about 6% higher than MOCVD-SiN$_x$ devices. This excellent result indicates that the prospect of LPCVD-SiN$_x$ passivation devices used in 5G small terminals will be attractive.

Keywords: AlGaN/GaN; high electron mobility transistors (HEMTs); SiN$_x$ passivation; low pressure chemical vapor deposition (LPCVD); ohmic contact; SiN$_x$/GaN interface

 check for updates

Citation: Deng, L.; Zhou, L.; Lu, H.; Yang, L.; Yu, Q.; Zhang, M.; Wu, M.; Hou, B.; Ma, X.; Hao, Y. Comprehensive Comparison of MOCVD- and LPCVD-SiN$_x$ Surface Passivation for AlGaN/GaN HEMTs for 5G RF Applications. *Micromachines* **2023**, *14*, 2104. https://doi.org/10.3390/mi14112104

Academic Editor: Ha Duong Ngo

Received: 15 September 2023
Revised: 22 October 2023
Accepted: 6 November 2023
Published: 16 November 2023

1. Introduction

With the rapid growth of mobile data volume, traditional 4G networks are no longer able to meet the demands for higher speeds, lower latency, and more reliable connections. As a result, the development of 5G as the next generation of mobile communication technology aims to provide more efficient, flexible, and innovative communication solutions to support various novel applications and the expanding needs of the digital society. In 5G applications, GaN semiconductors offer unique advantages including wide bandgap and high operation frequency compared to Si and group III–V material [1]. Devices based on GaN materials exhibit higher output power density and energy conversion efficiency, enabling system miniaturization and lightweight design. Today, GaN devices have been widely used in radio frequency (RF) applications [2–4] and power electronics [5–7]. However, during the process of GaN HEMT epitaxial growth, surface states like dangling bonds or defects will exist on the interrupted surfaces [8,9]. These surface states will lead to some adverse phenomena such as current collapse, thereby significantly impacting the RF output performance of GaN HEMT devices [10,11]. There are currently various methods available for suppressing surface states, such as growth-controlling conditions [12,13] or surface passivation [14]. Surface passivation has become an effective and common method due to its simplicity. It has a relatively simple process and can reduce the possibility of chemical contamination or mechanical damage to the surface during subsequent packaging processes. The passivation material mostly used on GaN HEMTs is SiN_x [15] and the most commonly used deposition method for it is plasma-enhanced chemical vapor deposition (PECVD) with a growth temperature below 350 °C [16]. However, due to its tendency to crack under high-temperature conditions [17], it is not possible to carry out the "passivation-prior-to-ohmic" process [18]. In addition, SiN_x growth using PECVD has lower film compactness compared to the SiN_x deposited by metal organic chemical vapor deposition (MOCVD) and low-pressure chemical vapor deposition (LPCVD) [18,19]. The active plasma sources in PECVD can also potentially cause damage to the surface of AlGaN or GaN and further degrade the quality of the deposited SiN_x itself [20]. Therefore, improved high-temperature solutions, including MOCVD (over 900 °C) and LPCVD (over 750 °C), are used for depositing the passivation layer [21–23]. MOCVD, an in-situ SiNx passivation technique, is employed in which an epitaxial layer is deposited using MOCVD, followed by the deposition of an additional SiNx passivation layer within the same chamber. This in-situ growth approach helps to prevent some negative impacts during the process chamber transfer, such as oxidation reactions. During the LPCVD process for SiNx deposition, the pressure is typically maintained at around 200 mTorr. At a specific temperature, the lower operating pressure results in an increased mean free path of gas molecules within the chamber, leading to a significant reduction in diffusion rate. Consequently, the reaction time is prolonged, thereby achieving the objective of enhancing thin film quality and density. The research above has revealed that SiN_x passivation layers deposited at high temperatures exhibit higher thermal stability and better growth quality, which can significantly suppress surface state density and enhance the output characteristics of devices.

However, it is noted that a high-temperature deposited SiN_x layer is often used in power electronic devices to form MIS gate structures [23,24], while there are limited evaluations in the field of RF applications [25] and there has been little research on the interface under SiN_x layers and the mechanism of ohmic contact, which are important concerns for enhancing the RF performance of AlGaN/GaN HEMTs. The impact of the passivation layer on ohmic alloying is not yet clear, and there is no relevant research on the interface quality of the passivation layer and its interdiffusion with the barrier layer. Therefore, this work systematically compares MOCVD-SiN_x and LPCVD-SiN_x, which are known for their high-temperature processes to evaluate interface quality and the impact on ohmic contact. Finally, this study will assess the performance between the two passivation layer approaches in RF applications.

2. Device Structure and Fabrication

The structures of sample A and sample B used in this work are shown in Figure 1a,b. The AlGaN/GaN heterostructure of both samples was grown on a three-inch SiC substrate using MOCVD. The epilayer, from the bottom to the top, included a GaN buffer (a 400 nm unintentionally doped GaN channel layer). On top of the GaN channel layer, there was a 1 nm AlN insertion layer, facilitating the growth of a 20 nm AlGaN layer with 25% Al composition. A 2 nm GaN cap layer was deposited on the barrier layer. Considering the growth quality of the passivation layer and the stress management, the difference between sample A and sample B is that there was a 20 nm in situ SiN_x passivation layer utilized using MOCVD and a 40 nm ex situ SiN_x passivation layer utilized by LPCVD. The device fabrication process followed the "passivation-prior-to-ohmic" strategy. The fabrication process flow is shown in Figure 1c. The SiN_x passivation layer above the ohmic region was first etched away by F-based etching. The F-based etching gas mixture consists of CF4/O2 = 25/5 sccm, and the etching pressure is set to 5 mTorr. Then, another ohmic photolithography process was performed followed by ohmic contact formation. The ohmic contact metal stack was Ti/Al/Ni/Au = 20/160/45/55 nm, with annealing performed at 860 °C/60 s under ambient N_2 conditions. The electrical isolation of the devices was achieved using nitrogen ion implantation. To form the T-gate structure, on sample A, the 20 nm SiN_x passivation layer was supplemented with 100 nm SiN_x grown by PECVD, and the 40 nm SiN_x layer was supplemented with an additional 80 nm SiN_x grown by PECVD on sample B. The T-gate electrode was formed by Ni/Au = 45/400 nm stacks with a 0.5 μm foot length and a 1 μm cap length, while the gate width was 100 μm. Finally, interconnection metal fabrication was completed by evaporating Ti/Au = 20/400 nm.

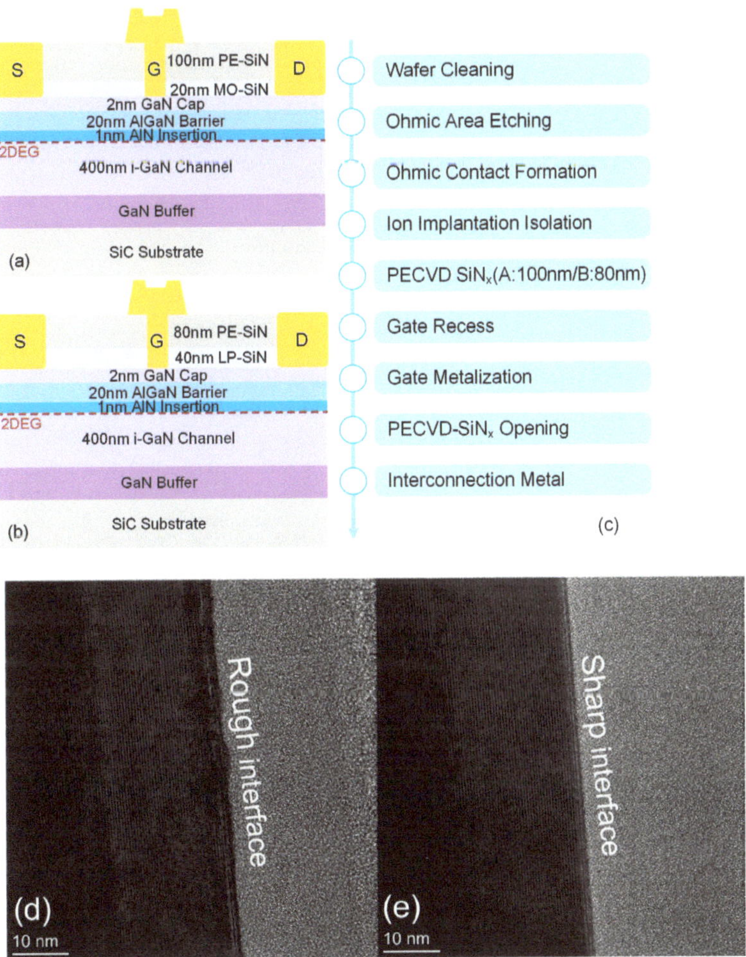

Figure 1. The device structure diagram of (**a**) sample A and (**b**) sample B. (**c**) The fabrication process flow of the samples. The TEM image of the interface quality of SiN_x/GaN in (**d**) sample A and (**e**) sample B.

3. Results and Discussion

After the fabrication, the samples were tested using a Keysight B1500A semiconductor parameter analyzer to obtain the result of the transmission line model (TLM) and the DC characteristics. The pulse I-V characteristics were tested using a Keithley 4200A-SCS parameter analyzer.

A. The Epitaxial Growth of Device

Figure 1d,e compare the epitaxial growth quality of sample A and sample B after ohmic annealing. It can be observed from the transmission electron microscopy (TEM) that Si was significantly diffused at the interface of sample A, resulting in a rough SiN_x/GaN boundary. In contrast, in sample B, there was a clear boundary between the LPCVD-SiN_x and the semiconductor interface, indicating the good suppression of Si diffusion. This is because, during the process of depositing the SiN_x film using MOCVD, SiH_4 had an etching effect on the semiconductor layer, allowing Si to diffuse into the semiconductor as an impurity. According to the results, we suppose that LPCVD can effectively prevent the diffusion of Si. Zhu et al. have demonstrated that the surface diffusion of Si is insufficient to compensate for the electron concentration and enhance performance [26]. On the contrary, etching damage accompanied by the diffusion of Si into the semiconductor will generate a leakage current path through the gate, affecting the reverse characteristics of the device [27].

B. Formation Mechanism of Ohmic Contact

Figures 2 and 3 compare the morphology of ohmic regions in samples A and B. The formation of TiN in high-temperature annealing can promote ohmic metal formation by ensuring direct contact with 2DEG or by forming N vacancies reported by many reports [28–31]. There was no apparent low-work function TiN alloy incorporation in the edge of the ohmic region of sample A, while there was large TiN alloy incorporation in sample B. Since there was sidewall contact between the ohmic metal stack and the passivation layer in the "passivation-prior-to-ohmic" scheme, some different alloy reactions occurred under the same annealing conditions. Based on an analysis of the distribution of Ti and N elements at the edge of the ohmic region for sample A (shown in the red box in Figure 2g,j) and sample B (shown in the red box in Figure 3g,j), we infer that this is because the deposition temperature of MOCVD-SiN_x was above 900 °C while the annealing temperature of sample A was lower than 900 °C, resulting in N being difficult to participate in the alloy reaction in order to form TiN from the side walls during the annealing alloying process. In contrast, the deposition temperature of LPCVD-SiN_x was 780 °C, which is less than the annealing temperature of 860 °C. This may make it easier for N to combine with Ti to form TiN in sample B. This enrichment of TiN at the edge of the ohmic region, which is also compared in Figures 2k and 3k, effectively reduced the ohmic contact resistance.

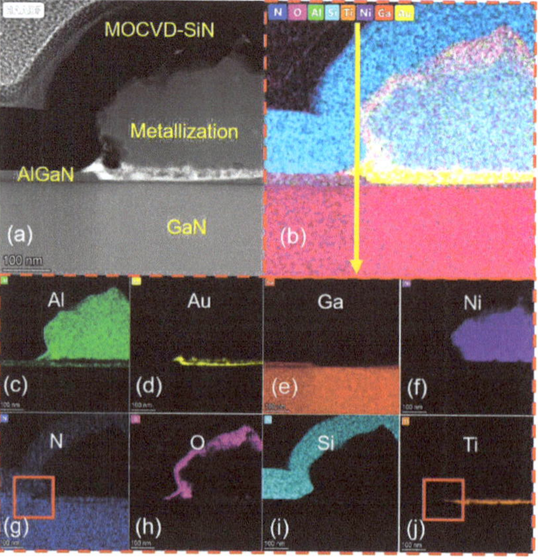

Figure 2. *Cont.*

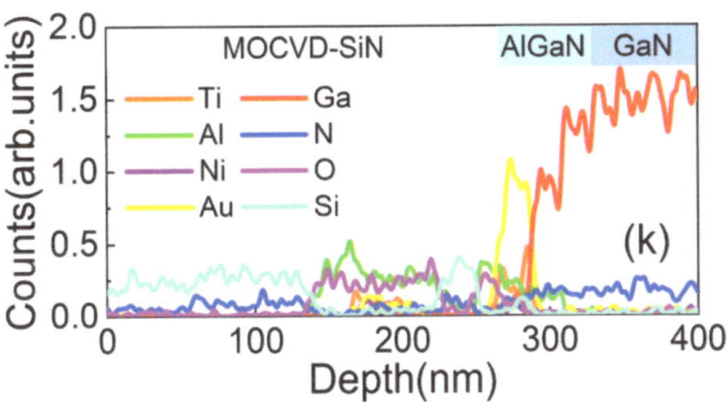

Figure 2. (**a**) The HAADF-STEM micrograph of the sidewall ohmic region for sample A. (**b**) EDX mapping of all elements. (**c–j**) EDX mapping of Al, Au, Ga, Ni, N, O, Si, and Ti. (**k**) EDS line scan profile of yellow arrow presented in (**b**).

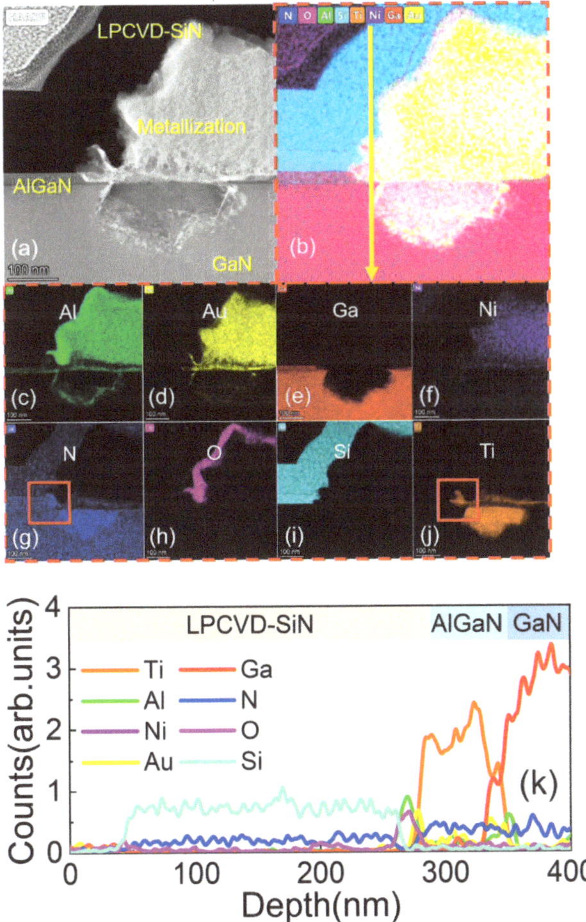

Figure 3. (**a**) The HAADF-STEM micrograph of the sidewall ohmic region for sample B. (**b**) EDX mapping of all elements. (**c–j**) EDX mapping of Al, Au, Ga, Ni, N, O, Si, and Ti. (**k**) EDS line scan profile of yellow arrow presented in (**b**).

C. Ohmic Contact Resistance and DC I-V Characteristics

The ohmic contact results obtained by the transmission line model (TLM) test are shown in Figure 4a. The ohmic contact resistance (R_c) for sample B with apparent side wall effect was 0.32 ± 0.05 $\Omega\cdot$mm and the sheet resistance (R_{sh}) was 291 Ω/sq, whereas the R_c for sample A was 0.39 ± 0.05 $\Omega\cdot$mm, with an R_{sh} of 302 Ω/sq. Figure 4b displays the DC characteristic curves of samples A and B. From the transfer I–V curves, it can be observed that the threshold voltage for sample A was -3.7 V with a maximum transconductance ($g_{m,max}$) of 294 mS/mm, while sample B had a threshold voltage of -3.8 V and $g_{m,max}$ of 346 mS/mm, which was 50 mS/mm higher than sample A.

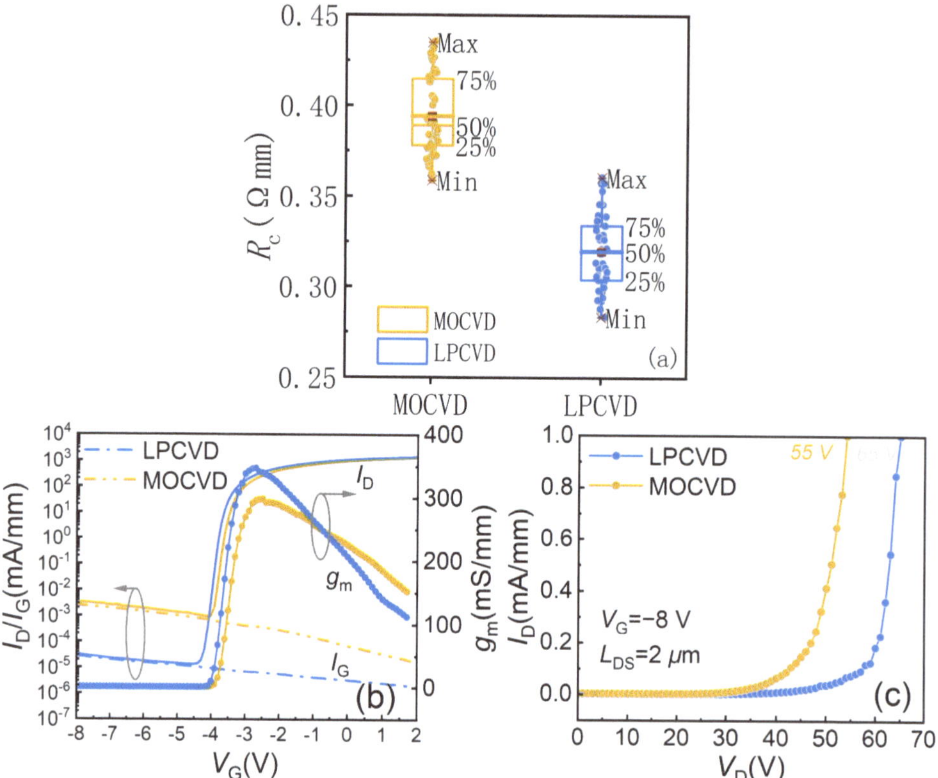

Figure 4. (**a**) TLM result of the MOCVD-SiN$_x$ and LPCVD-SiN$_x$ device. (**b**) Transfer I–V characteristics in semi-log scale of the MOCVD-SiN$_x$ and LPCVD-SiN$_x$ device. (**c**) Breakdown characteristics of the MOCVD-SiN$_x$ and LPCVD-SiN$_x$ devices.

Under the conditions of gate voltage (V_{gs}) = 2 V and drain voltage (V_{ds}) = 10 V, the maximum output currents were 1.28 A/mm and 1.34 A/mm for samples A and B, respectively. To minimize the influence of thermal resistance on on-resistance (R_{on}), the pulse test at a quiescent voltage of (0 V, 0 V) was performed with a 2 μm drain source spacing (L_{DS}) device. The measured R_{on} value for sample A was 1.75 $\Omega\cdot$mm and that for sample B was 1.54 $\Omega\cdot$mm. It can be seen that sample B had a lower R_{on}, which is consistent with the superior TLM results on sample B. Moreover, sample B, which lacks the diffusion of Si and etching damage, did not form an additional leakage path in passivation layers, resulting in relatively lower gate leakage current by 1–2 orders of magnitude below 10^{-5} mA/mm compared to sample A.

Figure 4c depicts the breakdown characteristics for the samples. At $V_{gs} = -8$ V, the breakdown voltage of sample A was measured to be 55 V for a 2 μm L_{DS} device, while the breakdown voltage of sample B was 65 V. This indicates that the excellent growth interface of LPCVD-SiN$_x$ can generate a smoother electric field, thereby enhancing the breakdown characteristics of the device.

D.   Pulsed I-V Characteristics

In RF applications, especially for an RF power output, the temperature of the working environment usually increases due to device heat dissipation. Therefore, to characterize the stability of the passivation layer, pulse I-V measurements were performed at room temperature of 25 °C and high temperature of 85 °C with a pulse width and a duty cycle of 500 ns and 0.05%, respectively. The voltage range of gate voltage (V_{GS}) was −4 to 2 V with a step of 3 V. The gate and drain quiescent bias (V_{GQ} and V_{DQ}) used for the pulse test is shown in Figure 5, ranging from (V_{GQ}, V_{DQ}) = (0 V, 0 V)~(−8 V, 20 V). Under room temperature conditions, the current collapse shown in Figure 5a of sample A was 7.8% for the high reverse gate voltage and high drain voltage condition of (−8 V, 20 V), while that of sample B shown in Figure 5b was 5.9%. At 85 °C, the current collapse of sample A was 7.4% (Figure 5c), while that of sample B was 6% (Figure 5d). Comparing the passivation effects between high and low temperatures, the variations observed were relatively insignificant and we thought that the discrepancy was a consequence of different surface trapping states at the interface to the SiN layer. The different growth methods and recipes for the passivation layer had a significant impact on the introduction of traps [32,33], and, in the future, we will focus on more accurate trap testing, like deep-level transient spectroscopy (DLTS), to demonstrate the differences in traps related to the two passivation layers. Nevertheless, both samples show excellent thermal stability and also show that sample B has a better passivation effect at either room temperature or at high temperatures. These results prove that the high-temperature process of SiN$_x$ deposition is beneficial to improving the thermal stability of the passivation.

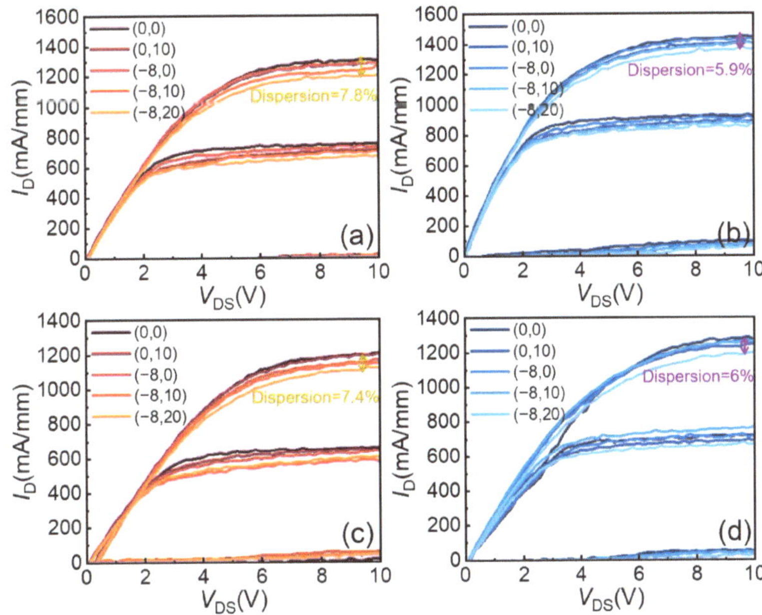

Figure 5. Pulsed output characteristic of (**a**) sample A and (**b**) sample B at 25 °C. Pulsed output characteristics of (**c**) sample A and (**d**) sample B at 85 °C.

E.   Small- and Large Signal Power Characteristic

Small-signal measurements of the two samples are shown in Figure 6a,b. S-parameters were measured from 0.1 GHz to 40 GHz using a Keysight network analyzer. The current gain cutoff frequency f_T for sample A at $V_{ds} = 20$ V was 22 GHz, while that of sample B was 24 GHz. The maximum oscillation frequency f_{max} was 64 GHz in sample A and 72 GHz in sample B. The improvement of f_{max} may be ascribed to the better transconductance and reduced parasitic resistance deriving from low R_c in sample B [34].

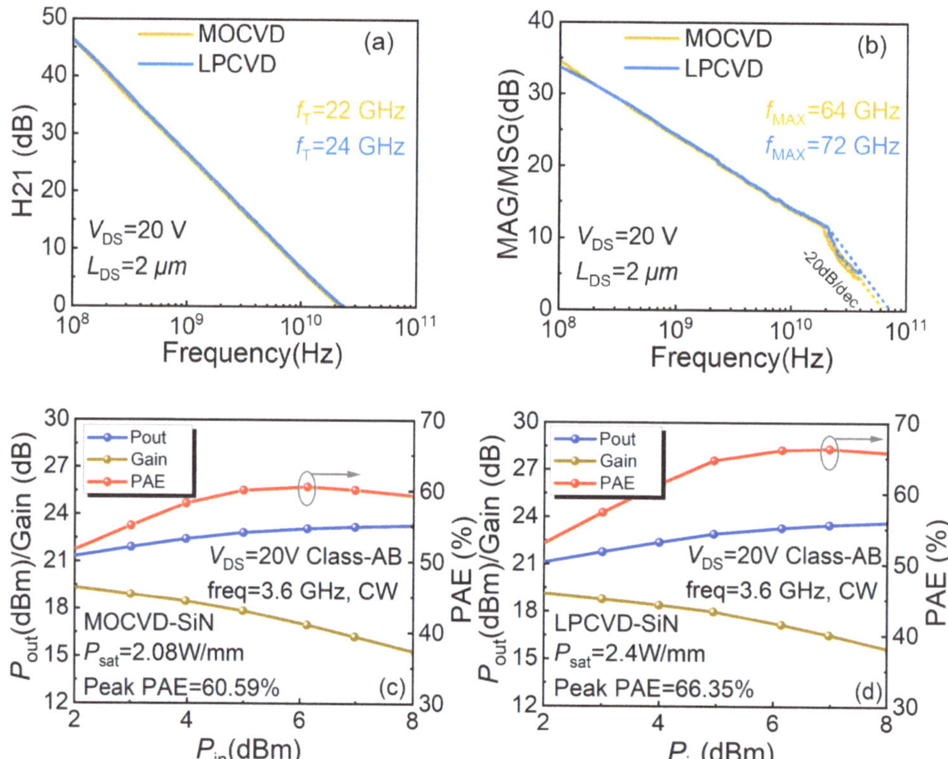

Figure 6. Small signal comparison between MOCVD- and LPCVD-SiN$_x$ of (**a**) f_T and (**b**) f_{max}. Large-signal results of (**c**) the MOCVD-SiN$_x$ device and (**d**) the LPCVD-SiN$_x$ device.

CW load–pull measurements were measured at 3.6 GHz using a Focus load–pull system with class-AB bias. The impedance matching in the load-pull test is PAE optimum tuning. These results are shown in Figure 6c,d. It can be seen that sample B has better output power density and power-added efficiency (PAE). This is attributed to lower leakage current [35] and lower parasitic resistance [34] in sample B. According to the large-signal test, sample B exhibited a PAE of 66.35% under a drain voltage bias of 20 V, which is 6% higher than 60.59% of sample A. The sample B fabricated in this study exhibited significant efficiency advantages in low-voltage power output compared to other state-of-the-art LPCVD-SiN$_x$ AlGaN/GaN HEMTs measured at a sub-6G condition, as shown in Figure 7 [18,25,36–38].

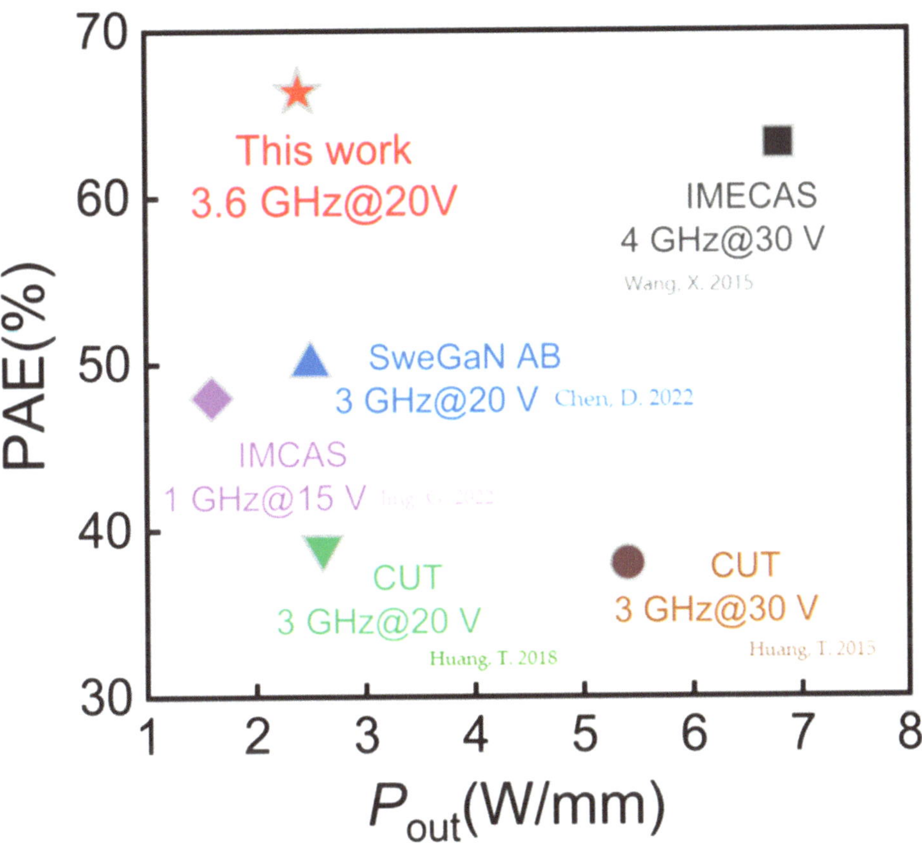

Figure 7. Benchmark of sub-6G high-signal power characteristics for state-of-the-art AlGaN/GaN HEMTs [18,25,36–38].

4. Conclusions

Due to the low-temperature characteristics of PECVD-SiNx, it cannot meet the requirements of an advanced "passivation-prior-to-ohmic" process. This study investigates SiNx passivation fabricated using two high-temperature deposition methods. In this paper, we compared the differences between MOCVD-SiN$_x$ and LPCVD-SiN$_x$ in terms of ohmic contact and related interface. We discovered that the growth interface of LPCVD-SiN$_x$ was smoother than that of MOCVD-SiN$_x$, resulting in better leakage suppression. Additionally, LPCVD-SiN$_x$ devices achieved improved RF output performance by forming lower Ohmic contact resistance. The maximum current of the LPCVD-SiN$_x$ device exceeded 1.3 A/mm at V_{gs} = 2 V, with gate leakage below 10^{-5} mA/mm. The f_{max} of the 2 μm L_{DS} device reached 72 GHz. Under a drain voltage of 20 V, the output power exceeded 2.4 W/mm with PAE greater than 66.35%. The results presented in this study demonstrates significant efficiency advantages in low-voltage power output compared to other state-of-the-art LPCVD-SiNx AlGaN/GaN HEMT operated at 5G frequency spectrum. These results demonstrate the excellent performance of LPCVD-SiN$_x$ devices in small-sized modules working in low-voltage applications and highlight its prospects in 5G small terminals.

Author Contributions: L.D.: investigation (equal); methodology (equal); writing—original draft (equal); and formal analysis (equal). L.Z.: investigation (equal); data curation; visualization (equal); and writing—review and editing (equal). H.L.: writing—review and editing (supporting). L.Y.: methodology (equal). Q.Y.: investigation (equal). M.Z.: methodology (equal). M.W.: validation (supporting). B.H.: methodology (equal). X.M.: funding acquisition (lead); project administration (lead); and supervision (lead). Y.H.: funding acquisition (lead) and supervision (lead). All authors have read and agreed to the published version of the manuscript.

Funding: This work was supported by the National Natural Science Foundation of China under grant nos. 62234009 and 62090014; in part by the China Postdoctoral Science Foundation under grant 2023M732730; and in part by the Fundamental Research Funds for the Central Universities of China under grant XJSJ23056.

Conflicts of Interest: The authors declare no conflict of interest.

References

1. Hao, Y.; Ma, X.; Mi, M.; Yang, L. Research on GaN-Based RF Devices: High-Frequency Gate Structure Design, Submicrometer-Length Gate Fabrication, Suppressed SCE, Low Parasitic Resistance, Minimized Current Collapse, and Lower Gate Leakage. *IEEE Microw. Mag.* **2021**, *22*, 34. [CrossRef]
2. Mishra, U.K.; Parikh, P.; Wu, Y.F. AlGaN/GaN HEMTs—An Overview of Device Operation and Applications. *Proc. IEEE* **2002**, *90*, 1022. [CrossRef]
3. Lu, H.; Hou, B.; Yang, L.; Zhang, M.; Deng, L.; Wu, M.; Si, Z.; Hunag, S.; Ma, X.; Hao, Y. High RF performance GaN-on-Si HEMTs with passivation implanted termination. *IEEE Electron Device Lett.* **2022**, *43*, 188. [CrossRef]
4. Wu, M.; Zhang, M.; Yang, L.; Hou, B.; Yu, Q.; Li, S.; Shi, C.; Zhao, W.; Lu, H.; Chen, W.; et al. First Demonstration of State-of-the-art GaN HEMTs for Power and RF Applications on A Unified Platform with Free-standing GaN Substrate and Fe/C Co-doped Buffer. In Proceedings of the 2022 International Electron Devices Meeting (IEDM), San Francisco, CA, USA, 3–7 December 2022; pp. 11.3.1–11.3.4.
5. Wang, Z.; Zhang, Z.; Wang, S.; Chen, C.; Wang, Z.; Yao, Y. Design and optimization on a novel high-performance ultra-thin barrier AlGaN/GaN power HEMT with local charge compensation trench. *Appl. Sci.* **2019**, *9*, 3054. [CrossRef]
6. Wang, Z.; Yang, D.; Cao, J.; Wang, F.; Yao, Y. A novel technology for turn-on voltage reduction of high-performance lateral heterojunction diode with source-gate shorted anode. *Superlattices Microstruct.* **2019**, *125*, 144. [CrossRef]
7. Chen, K.J.; Häberlen, O.; Lidow, A.; lin Tsai, C.; Ueda, T.; Uemoto, Y.; Wu, Y. GaN-on-Si power technology: Devices and applications. *IEEE Trans. Electron Devices* **2017**, *64*, 779–795. [CrossRef]
8. Mitrofanov, O.; Manfra, M. Mechanisms of gate lag in GaN/AlGaN/GaN high electron mobility transistors. *Superlattices Microstruct.* **2003**, *34*, 33. [CrossRef]
9. Makiyama, K.; Ohki, T.; Okamoto, N.; Kanamura, M.; Masuda, S.; Nakasha, Y.; Joshin, K.; Imanishi, K.; Hara, N.; Ozaki, S.; et al. High-power GaN-HEMT with low current collapse for millimeter-wave amplifier. *Phys. Status Solidi C* **2011**, *8*, 2442. [CrossRef]
10. Huang, S.; Jiang, Q.; Yang, S.; Zhou, C.; Chen, K.J. Effective passivation of AlGaN/GaN HEMTs by ALD-grown AlN thin film. *IEEE Electron Device Lett.* **2012**, *33*, 516. [CrossRef]
11. Hashizume, T.; Hasegawa, H. Effects of nitrogen deficiency on electronic properties of AlGaN surfaces subjected to thermal and plasma processes. *Appl. Surf. Sci.* **2004**, *234*, 387. [CrossRef]
12. Chen, J.; Qin, J.; Xiao, W.; Wang, H. Effective Suppression of Current Collapse in AlGaN/GaN HEMT With N_2O Plasma Treatment Followed by High Temperature Annealing in N 2 Ambience. *IEEE J. Electron Devices Soc.* **2022**, *10*, 356. [CrossRef]
13. Weimann, N.G.; Manfra, M.J.; Wachtler, T. Unpassivated AlGaN-GaN HEMTs with minimal RF dispersion grown by plasma-assisted MBE on semi-insulating 6H-SiC substrates. *IEEE Electron Device Lett.* **2003**, *24*, 57. [CrossRef]
14. Ohno, Y.; Nakao, T.; Kishimoto, S.; Maezawa, K.; Mizutani, T. Effects of surface passivation on breakdown of AlGaN/GaN high-electron-mobility transistors. *Appl. Phys. Lett.* **2004**, *84*, 2184. [CrossRef]
15. Zhang, Y.; Huang, S.; Wei, K.; Zhang, S.; Wang, X.; Zheng, Y.; Liu, G.; Chen, X.; Li, Y.; Liu, X. Millimeter-wave AlGaN/GaN HEMTs with 43.6% power-added-efficiency at 40 GHz fabricated by atomic layer etching gate recess. *IEEE Electron Device Lett.* **2020**, *41*, 701. [CrossRef]
16. Pei, Y.; Rajan, S.; Higashiwaki, M.; Chen, Z.; DenBaars, S.P.; Mishra, U.K. Effect of dielectric thickness on power performance of AlGaN/GaN HEMTs. *IEEE Electron Device Lett.* **2009**, *30*, 313. [CrossRef]
17. Moon, S.W.; Lee, J.; Seo, D.; Jung, S.; Choi, H.G.; Shim, H.; Yim, J.S.; Twynam, J.; Roh, S.D. High-voltage GaN-on-Si hetero-junction FETs with reduced leakage and current collapse effects using SiNx surface passivation layer deposited by low pressure CVD. *Jpn. J. Appl. Phys.* **2014**, *53*, 08NH02. [CrossRef]
18. Wang, X.; Huang, S.; Zheng, Y.; Wei, K.; Chen, X.; Liu, G.; Yuan, T.; Luo, W.; Pang, L.; Jiang, H. Robust SiN x/AlGaN interface in GaN HEMTs passivated by thick LPCVD-grown SiN x layer. *IEEE Electron Device Lett.* **2015**, *36*, 666. [CrossRef]
19. Siddique, A.; Ahmed, R.; Anderson, J.; Nazari, M.; Yates, L.; Graham, S.; Holtz, M.; Piner, E.L. Structure and interface analysis of diamond on an AlGaN/GaN HEMT utilizing an in situ SiN x interlayer grown by MOCVD. *ACS Appl. Electron. Mater.* **2019**, *1*, 1387. [CrossRef]

20. Zhang, S.; Wei, K.; Ma, X.H.; Hou, B.; Liu, G.G.; Zhang, Y.C.; Wang, X.-H.; Zheng, Y.-K.; Huang, S.; Li, Y.-K.; et al. Reduced reverse gate leakage current for GaN HEMTs with 3 nm Al/40 nm SiN passivation layer. *Appl. Phys. Lett.* **2019**, *114*, 013503. [CrossRef]

21. Lu, X.; Ma, J.; Jiang, H.; Liu, C.; Lau, K.M. Low trap states in in situ SiNx/AlN/GaN metal-insulator-semiconductor structures grown by metal-organic chemical vapor deposition. *Appl. Phys. Lett.* **2014**, *105*, 102911. [CrossRef]

22. Lu, H.; Hou, B.; Yang, L.; Niu, X.; Si, Z.; Zhang, M.; Wu, M.; Mi, M.; Zhu, Q.; Cheng, K.; et al. Aln/GaN/InGaN coupling-channel HEMTs for improved g m and gain linearity. *IEEE Trans. Electron Devices* **2021**, *68*, 3308. [CrossRef]

23. Hua, M.; Liu, C.; Yang, S.; Liu, S.; Fu, K.; Dong, Z.; Cai, Y.; Zhang, B.; Chen, K.J. GaN-based metal-insulator-semiconductor high-electron-mobility transistors using low-pressure chemical vapor deposition SiN x as gate dielectric. *IEEE Electron Device Lett.* **2015**, *36*, 448. [CrossRef]

24. Jiang, H.; Liu, C.; Chen, Y.; Lu, X.; Tang, C.W.; Lau, K.M. Investigation of in situ SiN as gate dielectric and surface passivation for GaN MISHEMTs. *IEEE Trans. Electron Devices* **2017**, *64*, 832. [CrossRef]

25. Huang, T.; Malmros, A.; Bergsten, J.; Gustafsson, S.; Axelsson, O.; Thorsell, M.; Rorsman, N. Suppression of dispersive effects in AlGaN/GaN high-electron-mobility transistors using bilayer SiN x grown by low pressure chemical vapor deposition. *IEEE Electron Device Lett.* **2015**, *36*, 537. [CrossRef]

26. Zhu, L.; Zhou, Q.; Chen, K.; Gao, W.; Cai, Y.; Cheng, K.; Li, Z.; Zhang, B. The Modulation Effect of LPCVD-Si x N y Stoichiometry on 2-DEG Characteristic of UTB AlGaN/GaN Heterostructure. *IEEE Trans. Electron Devices* **2022**, *69*, 4828. [CrossRef]

27. Song, L.; Fu, K.; Zhang, Z.; Sun, S.; Li, W.; Yu, G.; Hao, R.; Fan, Y.; Shi, W.; Cai, Y.; et al. Interface Si donor control to improve dynamic performance of AlGaN/GaN MIS-HEMTs. *AIP Adv.* **2017**, *7*, 125023. [CrossRef]

28. Douglas, E.A.; Reza, S.; Sanchez, C.; Koleske, D.; Allerman, A.; Klein, B.; Armstrong, A.M.; Kaplar, R.J.; Baca, A.G. Ohmic contacts to Al-rich AlGaN heterostructures. *Phys. Status Solidi (A)* **2017**, *214*, 1600842. [CrossRef]

29. Wang, L.; Kim, D.H.; Adesida, I. Direct contact mechanism of Ohmic metallization to AlGaN/GaN heterostructures via Ohmic area recess etching. *Appl. Phys. Lett.* **2009**, *95*, 172107. [CrossRef]

30. Lu, H.; Hou, B.; Yang, L.; Song, F.; Zhang, M.; Wu, M.; Ma, X.; Hao, Y. Low-resistance Ta/Al/Ni/Au ohmic contact and formation mechanism on AlN/GaN HEMT. *IEEE Trans. Electron Devices* **2022**, *69*, 6023. [CrossRef]

31. Yang, L.; Lu, H.; Niu, X.; Zhang, M.; Shi, C.; Deng, L.; Hou, B.; Mi, M.; Wu, M.; Cheng, K.; et al. Investigation of contact mechanism and gate electrostatic control in multi-channel AlGaN/GaN high electron mobility transistors with deep recessed ohmic contact. *J. Appl. Phys.* **2022**, *132*, 165703. [CrossRef]

32. Guo, H.; Shao, P.; Zeng, C.; Bai, H.; Wang, R.; Pan, D.; Chen, P.; Chen, D.; Lu, H.; Zhang, R.; et al. Improved LPCVD-SiNx/AlGaN/GaN MIS-HEMTs by using in-situ MOCVD-SiNx as an interface sacrificial layer. *Appl. Surf. Sci.* **2022**, *590*, 153086. [CrossRef]

33. Liu, X.; Wang, X.; Zhang, Y.; Wei, K.; Zheng, Y.; Kang, X.; Jiang, H.; Li, J.; Wang, W.; Wu, X.; et al. Insight into the near-conduction band states at the crystallized interface between GaN and SiN x grown by low-pressure chemical vapor deposition. *ACS Appl. Mater. Interfaces* **2018**, *10*, 21721. [CrossRef] [PubMed]

34. Lu, H.; Ma, X.; Hou, B.; Yang, L.; Zhang, M.; Wu, M.; Si, Z.; Zhang, X.; Niu, X.; Hao, Y. Improved RF power performance of AlGaN/GaN HEMT using by Ti/Au/Al/Ni/Au shallow trench etching ohmic contact. *IEEE Trans. Electron Devices* **2021**, *68*, 4842. [CrossRef]

35. Jimenez, J.L.; Chowdhury, U. X-Band GaN FET reliability. In Proceedings of the 2008 IEEE International Reliability Physics Symposium, 2008, Phoenix, AZ, USA, 27 April–1 May 2008, (unpublished).

36. Chen, D.Y.; Persson, A.R.; Wen, K.H.; Sommer, D.; Grünenpütt, J.; Blanck, H.; Thorsell, M.; Kordina, O.; Darakchieva, V.; Rorsman, N. Impact of in situ NH3 pre-treatment of LPCVD SiN passivation on GaN HEMT performance. *Semicond. Sci. Technol.* **2022**, *37*, 035011. [CrossRef]

37. Huang, T.; Bergsten, J.; Thorsell, M.; Rorsman, N. Small-and large-signal analyses of different low-pressure-chemical-vapor-deposition SiN x passivations for microwave GaN HEMTs. *IEEE Trans. Electron Devices* **2018**, *65*, 908. [CrossRef]

38. Jing, G.; Wang, X.; Huang, S.; Jiang, Q.; Deng, K.; Wang, Y.; Li, Y.; Fan, J.; Wei, K.; Liu, X. Mechanism of Linearity Improvement in GaN HEMTs by Low Pressure Chemical Vapor Deposition-SiN x Passivation. *IEEE Trans. Electron Devices* **2022**, *69*, 6610. [CrossRef]

Review

Power Electronics Revolutionized: A Comprehensive Analysis of Emerging Wide and Ultrawide Bandgap Devices

S M Sajjad Hossain Rafin [1], Roni Ahmed [2], Md. Asadul Haque [3], Md. Kamal Hossain [3], Md. Asikul Haque [3] and Osama A. Mohammed [1,*]

[1] Energy Systems Research Laboratory, Department of ECE, Florida International University, Miami, FL 33174, USA; srafi010@fiu.edu or rafin@ieee.org
[2] Department of ECE, Presidency University, Dhaka 1212, Bangladesh; roniahmed355@pu.edu.bd
[3] Department of EEE, Northern University Bangladesh, Dhaka 1230, Bangladesh; mdasadeee91@gmail.com (M.A.H.); kamalbijoy25@gmail.com (M.K.H.); asikul.haque.bd@gmail.com (M.A.H.)
* Correspondence: mohammed@fiu.edu

Abstract: This article provides a comprehensive review of wide and ultrawide bandgap power electronic semiconductor devices, comparing silicon (Si), silicon carbide (SiC), gallium nitride (GaN), and the emerging device diamond technology. Key parameters examined include bandgap, critical electric field, electron mobility, voltage/current ratings, switching frequency, and device packaging. The historical evolution of each material is traced from early research devices to current commercial offerings. Significant focus is given to SiC and GaN as they are now actively competing with Si devices in the market, enabled by their higher bandgaps. The paper details advancements in material growth, device architectures, reliability, and manufacturing that have allowed SiC and GaN adoption in electric vehicles, renewable energy, aerospace, and other applications requiring high power density, efficiency, and frequency operation. Performance enhancements over Si are quantified. However, the challenges associated with the advancements of these devices are also elaborately described: material availability, thermal management, gate drive design, electrical insulation, and electromagnetic interference. Alongside the cost reduction through improved manufacturing, material availability, thermal management, gate drive design, electrical insulation, and electromagnetic interference are critical hurdles of this technology. The review analyzes these issues and emerging solutions using advanced packaging, circuit integration, novel cooling techniques, and modeling. Overall, the manuscript provides a timely, rigorous examination of the state of the art in wide bandgap power semiconductors. It balances theoretical potential and practical limitations while assessing commercial readiness and mapping trajectories for further innovation. This article will benefit researchers and professionals advancing power electronic systems.

Keywords: wide bandgap devices; ultrawide bandgap devices; silicon; silicon carbide; GaN; diamond; power semiconductor devices

check for updates

Citation: Rafin, S.M.S.H.; Ahmed, R.; Haque, M.A.; Hossain, M.K.; Haque, M.A.; Mohammed, O.A. Power Electronics Revolutionized: A Comprehensive Analysis of Emerging Wide and Ultrawide Bandgap Devices. *Micromachines* **2023**, *14*, 2045. https://doi.org/10.3390/mi14112045

Academic Editors: Zeheng Wang and Jing-Kai Huang

Received: 6 October 2023
Revised: 25 October 2023
Accepted: 27 October 2023
Published: 31 October 2023

1. Introduction

In the era of power electronics, wide and ultrawide bandgap power electronic semiconductors have become a game-changing innovation. These cutting-edge materials, such as silicon carbide (SiC), gallium nitride (GaN), and diamond, perform better than conventional Si-based products. In recent years, significant improvements have been made in wide bandgap power electronic semiconductors regarding the materials' caliber, device design, and production techniques. The creation of superior SiC and GaN substrates, advancements in crystal growth methods, and improved device production procedures have all been created by academics and business stakeholders. Wide bandgap devices are becoming more commercially viable due to these developments' increased material performance, greater device yields, and lower production costs. Electronic switching devices are essentially used in power electronic converters to control electrical energy efficiently. Higher efficiency, greater power densities, and more integrated systems have always been the direction of power electronics technology development. Like many technologies, power semiconductor technology has been growing towards this constant progress. The development of diverse Si power devices over the past 50 years has been the main driver of the advancements [1].

Si is currently, by far, the most established semiconductor material used in power devices. However, due to its limitations, engineers and academics have made significant efforts to identify alternatives to Si-based power devices for greater performance. These devices are getting close to their material limits [2]. The introduction of power devices based on wide bandgap (WBG) materials such as SiC and GaN has been a revolutionary advancement. Utilizing these new wide bandgaps (WBGs) power semiconductor devices increases the efficiency of electric energy transformations, allowing for a more logical use of electric energy and a significant reduction in power converter size and robustness. SiC and GaN are excellent trade-offs between theoretical and practical properties among the possible semiconductor materials candidates. Moreover, these materials' key advantages over Si include good performance across a wide temperature range, high dielectric strength, and high saturation drift velocity [3].

SiC is one of the most widely studied and commercially available wide bandgap materials. It possesses a bandgap energy of approximately 3.3 electron volts (eV), significantly higher than Si's 1.1 eV. SiC-based power devices offer numerous advantages, including reduced conduction and switching losses, higher temperature tolerance, and increased efficiency. These properties make SiC devices well-suited for electric vehicles, renewable energy systems, industrial motor drives, and aerospace applications [4,5]. GaN is another prominent wide bandgap material recently gaining significant attention. GaN exhibits a bandgap energy of around 3.4 eV, similar to SiC. GaN-based power devices provide exceptional performance characteristics, including high breakdown voltages, fast switching speeds, and low on-resistance. These attributes make GaN devices ideal for applications requiring high-frequency operation, such as wireless power transfer, data centers, and radar systems.

It is widely known that, when running under reverse bias in the natural environment, Si power metal oxide field effect transistors (MOSFETs) are highly capable of single-event burnout (SEB) [6,7]. The parasitic bipolar transistor built into the design may turn on due to the transitory current created by intense heavy ion penetration through the device. Voltage ranges can vary up to 600 V regarding Si MOSFETs [8]. With the support of more than 600 V voltage applications, insulated gate bipolar transistors (IGBTs) have developed necessary applications. However, due to their maximum switching losses, IGBT devices have become low-efficient at high frequencies [9].

Over the past 20 years, a great deal of research has been carried out on SiC power devices, and many are currently in the market. In particular, SiC is utilized outside of semiconducting, including in ceramic plates, thin filament pyrometry, foundry crucibles, bulletproof jackets, and auto clutches. Because it may operate at greater temperatures, higher current densities, and higher blocking voltages because of the wider bandgap, higher thermal conductivity, and larger critical electric field [10–13], one of its earliest uses in electrical applications was as a lightning arrester in a high-voltage power system. Schottky diodes, MOSFETs, IGBTs, and power electronics are examples of SiC's recent use in electronics.

Compared to SiC technology, GaN has a greater bandgap energy and higher electron mobility [14]. It has also progressed in the low- to medium-voltage, high-frequency sector. For high-frequency applications based on lateral transistors, GaN is more effective. Both materials can offer better performances than the Si devices on the market [15,16], but the many technological processes for transistor manufacture must be properly integrated. GaN-based Field Effect Transistors (FETs), also known as, GaN High Electron Mobility Transistor (HEMTs) can switch faster than Si power transistors. GaN HEMTs have a tiny physical growth, which enables the devices to be more energy-efficient and high voltage application while providing extra space for external components. The properties of Si, SiC, and GaN have been demonstrated in Figure 1.

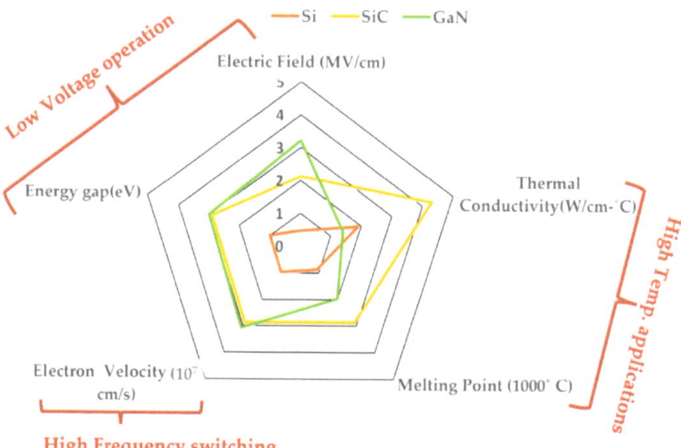

Figure 1. Summary of Si's, SiC's, and GaN's relevant material properties [1].

SiC is the third hardest substance on earth and is known to be as hard as a single crystal. The SiC energy gap ranges from 2 to 3.3 eV depending on the polytype crystal structure. SiCs are the ideal choice among commercially available WBG devices due to their greater power rating, quicker switching frequency, much lower switching losses, and capacity to handle higher junction temperatures than Si-based devices. Another promising wide bandgap semiconductor material GaN [16–19] is an example of the third generation of semiconductor materials, with a wider bandgap (bandgap width of more than 3.4 eV), high critical breakdown electric field, high anti-radiation ability, and rapid electron saturation velocity. GaN material has a wide range of applications and is one of the most efficient ways to save energy and reduce consumption worldwide. It summarizes recent research on GaN technology, demonstrating the slow but steady development of a local GaN supply chain.

Wide bandgap power electronic semiconductors are now in the state of active research and development. Researchers are constantly investigating novel device architectures, packaging methods, and heat management strategies to improve further the performance, reliability, and efficiency of wide bandgap devices. These initiatives tackle issues including enhancing device dependability, cutting manufacturing costs, and boosting system-level integration. Wide bandgap power electronic semiconductor applications are also being expanded into new markets, including 5G wireless communications, the Internet of Things (IoT), and sophisticated medical equipment. Wide bandgap devices thrive in these applications because they need excellent power density, quick switching times, and high-frequency functioning.

Ultrawide bandgap (UWBG) semiconductors like diamond enable incredibly promising electrical gadgets. Diamond has a band gap of 5.5 eV broad, more than five times that of Si [17]. Diamond transistor devices can theoretically switch at frequencies over 100 GHz and function at temperatures higher than 600 °C. High-power diamond Schottky diodes for power electronics, diamond UV detectors for monitoring flames, and diamond radiation detectors for physics research are a few unusual applications. Diamond's excellent heat conductivity and breakdown voltage enable incredibly small, effective power devices. The difficulties in doping diamonds to create trustworthy ohmic connections and the constraints in manufacturing large single-crystal diamond wafers are obstacles to developing diamond electronics. However, in situ-doped polycrystalline diamond films and diamond-on-Si techniques are advancing the field [17]. Diamond has the potential for orders of magnitude of improvements in power density, operating temperature, radiation hardness, and switching speed compared to traditional electronics. Deep space missions to power grid electronics could benefit from revolutionary applications if the diamond's full potential is realized. Due to their high-efficiency power conversion, electric vehicles have longer driven ranges and require less time to charge. Utilizing wide bandgap technology, renewable energy systems like solar and wind power may maximize energy collecting and grid integration [18]. Moreover, their high-power density and enhanced thermal management capabilities are advantageous for aerospace and defense applications.

This manuscript first provides the background on power electronics and wide bandgap materials. After that, in separate sections for each material, it describes the properties, historical development, devices, applications, and difficulties of the major semiconductors: Si, SiC, GaN, and diamond. It then assesses recent advancements in IGBTs and MOSFETs and compares performance characteristics like bandgap, breakdown voltage, and switching frequency across the materials. Adoption trends and the current commercial environment are also covered. The conclusion summarizes the results and provides a forecast for future developments. The review is organized to present a thorough technology overview, evaluate the state of the art, and give strategic recommendations to researchers and industry professionals working on the future generation of high-efficiency power semiconductors.

The focus of this article is to provide an extensive review of wide and ultrawide bandgap power semiconductor devices. From the earliest device inventions to the most recent market offerings, it describes the historical development in research and commercialization for each material. The article examines important properties of several devices, including power diodes, MOSFETs, IGBTs, bandgap, critical electric field, voltage/current ratings, switching frequency, packaging, and dependability. SiC and GaN are given much attention because they compete directly with Si devices due to their bigger bandgaps. Challenges like material availability, thermal management, cost reduction, and gate drive complexity are examined. The current commercial environment is evaluated, following adoption patterns and technological advancements resulting in gains in aerospace, renewable energy, electric vehicles, and other applications needing high-efficiency power conversion [4,5]. The assessment integrates advancements across the wide bandgap power semiconductor spectrum to inform and direct future innovation in this promising field.

2. Si

The preceding discussion demonstrates that Si power devices remain the workhorse technology in power electronics applications despite rising competition from WBG power devices. Si is a Group 14 (IVA) member in the periodic table of elements. Si is also part of the carbon family. These family elements include C, Ge, Sn, and Pb. Si is a metalloid, one of only a few elements with metal and nonmetal properties. Aside from oxygen, Si is the second most abundant element on Earth's crust. Si was established in 1960 by the 11th General Conference on Weights and Measures, CGPM, Conférence Générale des Poids ET Mesures [19]. The CGPM is the international authority that ensures the wide dissemination of Si and modifies it as necessary to reflect the latest advances in science and technology. It has a diamond cubic crystal structure with a lattice parameter of 0.543 nm [20]. The historical overview of Si wafer diameter and crystal weight increase goes beyond the scope of this article. In Table 1, the historical timetable of Si evaluation is presented.

Being a semiconductor, the element, ceramics, and bricks are used for making transistors. It is a vital component of Portland cement. Si materials are used in components of electronic devices. It also makes solar cells [21–25] and parts for computer circuits [26]. A solar cell is a device that converts sunlight into electrical energy [27–30]. A rectifier is an electrical device that converts alternating current to direct current. The most important Si alloys are those made with Fe, Al, and Cu. When Si is produced, scrap iron and metal are sometimes added to the furnace [31,32]. Several waterproofing systems employ Si as a component for water purification. Si is used in many mold release agents and molding compounds. It is also a component of ferroSi—an alloy widely used in the steel industry.

Table 1. Evaluation of power electronic Si-based semiconductor device.

Year	Device	Specifications	Milestone	Features
1950s	Thyristor/SCR	Up to 1000 V, 100 A	First thyristor invention	Low on-state loss
1956	Power diode	Up to 300 V, 10 A	First commercial Si power diode	Higher switching speed than selenium diodes
1958	Power transistor	Up to 60 V, 10 A	First Si power transistor	Higher gain and frequency than germanium transistors
1960s	SCR	Up to 1000 V, 100 A	Widespread SCR adoption	Phase control for AC power control
1970s	Power MOSFET	Up to 500 V, 10 A	First commercial power MOSFET	Higher switching speed than BJTs
1978	IGBT	600 V, 10 A	Invention of IGBT	MOSFET speed with BJT bidirectional capability
1980s	BJT	Up to 1200 V, 10 A	Improved high-voltage BJTs	Optimized for high-voltage applications
1980s	MOSFET	Up to 900 V, 100 A	Trench gate and VDMOS	Reduced on-resistance
1990s	MOSFET	Up to 900 V, 100 A	Double-diffused MOSFET	Further on-resistance reduction
1990s	IGBT	1200 V, 50 A	Trench gate IGBT	Lower losses than planar IGBTs
2000s	IGBT	6500 V, 1200 A	3rd-gen trench gate IGBT	Near ideal switching behavior
2000s	MOSFET	900 V, 150 A	Super junction MOSFET	Low on-resistance, fast switching
2010s	IGBT	6500 V, 1500 A	4th-gen field stop IGBTs	Minimized tail current
2020s	IGBT, MOSFET	Improvement	SiC and GaN emerging	Improved, but Si is phased out for WBG

The diamond cubic crystal structure of Si has a face-centered cubic (FCC) lattice with a basis of two Si atoms. Table 2 overviews Ribbon and Multi-Si technology improvements within the next 5–8 years [33]. However, an analysis of Mono-Si technology was not carried out because the current Mono-Si technology still has too much uncertainty.

Table 2. Future of Multi-Si, Mono-Si, and Ribbon-Si technology.

Wafer Technology	Multi-Si	Mono-Si	Ribbon-Si
Si feedstock production (MWp)	160	-	-
Crystallization and wafer (mm)	150 × 150	125 × 125	150 × 150
Cell processing (cells)	72	-	-
Module assembly	Frameless	Framed	Frameless
Wafer thickness (μm)	285–>150	270–300	300–>200
Module efficiency (%)	13.2–>16	14–>15	11.5–>15

2.1. Si Diode

For many years, Si diode power semiconductors have been crucial in several applications because of their dependable and effective rectification capabilities. Low forward voltage drops, high current-carrying capacity, and exceptional thermal stability are all positive traits of Si diodes [34]. Their extensive use can be ascribed to their proven dependability records, cost efficiency, and sophisticated production methods. Numerous devices, such as power supplies, inverters, rectifiers, and voltage regulators, use Si diodes. The Si Power Rapid Diode family bridges the gap between SiC diodes. Examples of such Si diodes are previously released emitter-controlled diodes and Infineon's existing high-power 600 V/650 V diode. Moreover, the TRENCHSTOP™ 5 and high-speed 3 IGBT (Insulated Gate Bipolar Transistor) and CoolMOSTM are good partners for the Rapid 1 and Rapid 2 diodes [35]. For usage in automotive, industrial power control, power management, sensor solutions, and security in Internet of Things applications, Infineon Technologies provides a comprehensive selection of ready-to-use semiconductor design solutions and reference schematics. Si diodes are ultra- and hyper-fast and have outstanding performance with a voltage range of 600–1200 V [36].

As mentioned earlier, the gap between SiC diodes and emitter-controlled diodes is filled by the Rapid 1 and Rapid 2 power Si diodes, which are a complement to the current high-power 600 V/650 V diodes. The new families of hyper- and ultra-fast diodes provide exceptional efficiency and dependability while striking the ideal balance between price and performance. The additional 50 V provides higher reliability.

The 650 V Rapid 1 Diode: The Rapid 1 diode series has the lowest conduction losses, and the smooth recovery minimizes EMI emissions with a 1.35 V temperature-stable forward voltage (FV). The equipment is ideal for power factor correction (PFC) topologies, frequently used in large home appliances like air conditioners and washers.

The 650 V Rapid 2 Diode: The family of Rapid 2 diodes is designed for applications switching between 40 kHz and 100 kHz by providing a low reverse recovery charge (Q_{rr}) and time (t_{rr}) to reduce the reverse conduction times associated with the power switch turn-on losses and to provide maximum efficiency [37].

2.2. Si MOSFET

MOSFETs are extensively utilized power semiconductors that have completely changed the power electronics industry. FETs) have become the most significant device in the semiconductor industry due to Lilienfeld's 1930 [38] patent on the idea and Kahng and Atalla's 1960 [39] practical implementation of Si/Si dioxide. The development of this industry has been characterized by an exponential pattern known as Moore's law over the past seven decades [40]. Today's metal-oxide-semiconductor field-effect transistor (MOSFET) has undergone several modifications, evolving from a single-gate planar MOSFET to a multiple-gate non-planar MOSFET. Nevertheless, it has been and will continue to be the mainstay of the semiconductor industry for the foreseeable future.

MOSFETs made of Si rely on modulating the conductive channel that forms between a semiconductor layer's source and drain terminals. The device's bulk is a Si substrate that has been extensively doped; the gate insulator is a thin Si dioxide layer. A voltage applied to the gate terminal generates an electric field that regulates the channel's conductivity. A positive voltage repels the majority carriers (for an N-channel MOSFET, electrons) from the channel, resulting in a depletion area and decreasing the conductivity of the channel [41]. Applying a negative voltage draws in most of the carriers and improves the conductivity of the channel. The MOSFET may change between the ON state (conducting) and the OFF state (non-conducting) by varying the gate voltage. This idea makes it possible to effectively manage the MOSFET's ability to handle power and current flow. These components have low on-resistance and can handle large voltages and currents, making efficient power conversion and control possible [6]. The conductivity of Si MOSFETs may be precisely controlled by using a thin Si dioxide layer as the gate insulator. They offer quick switching times, little gate drive needs, and superior thermal performance.

The main electrical specs for five commercial Si power MOSFETs with current ratings ranging from 2.5 A to 40 A are provided in Table 3. Input capacitance C_{iss} and on-state resistance $R_{ds(on)}$ grow along with the current rating, whereas gate-drain capacitance C_{gd} remains mostly constant. A 2.5 A MOSFET, for example, has a C_{iss} of 1800 pF and $R_{ds(on)}$ of 0.3, whereas a 40 A device has a C_{iss} of 10,000 pF and $R_{ds(on)}$ of 0.05. Across different power MOSFETs, a trade-off exists between a higher current capability and electrical characteristics such as input capacitance and on-resistance. Moreover, this table illustrates how Si power MOSFET specs and performance scale across various current ratings.

Table 3. Characteristics of selected Si MOSFETs [42].

Part Number	Current (A)	C_{ISS} (pF)	C_{GD} (pF)	$R_{ds(on)}$ (25) (Ω)
IXFP20N50P3M	2.5	1800	7	0.3
IXFP20N50P3M	5	1800	7	0.3
IXFH16N50P3	10	1515	7	0.3
IXFR64N50P	20	9700	30	0.095
IXFR80N50Q3	40	10,000	115	0.05

Several sectors, including the automobile, renewable energy, industrial automation, and telecommunications, use Si MOSFETs extensively. The subject of power electronics has been profoundly influenced by Si MOSFETs, which are incredibly adaptable and dependable power semiconductors. Si MOSFETs have evolved into crucial components in various applications, from automotive and renewable energy to industrial automation and telecommunications, because of their low on-resistance, high voltage and current handling capacity, and quick switching rates. They are a popular option for power electronic systems due to their superior thermal performance and compatibility with well-known production techniques, ensuring effective power conversion and control. The widespread use of Si MOSFETs demonstrates how important a role they have played in boosting the overall performance and dependability of power electronic systems and devices. They are the favored option for power electronic systems due to their dependability, high efficiency, and compatibility with established production methods.

2.3. Si SuperJunction MOSFET

To break the Si 1-D constraint, the super-junction (SJ) concept for vertical power devices was established in the mid-1990s [43–45]. When the device is turned off, a vertical P layer or P column is introduced to compensate for the charges in the N drift layer. This approach is highly similar to the RESURF concept [46], which has been used in various lateral power devices. The drift area of these devices has a special design that alternates P- and N-type regions, allowing for a more even dispersion of the electric field. For high-power applications, the SuperJunction design lowers on-resistance and boosts efficiency. SJ MOSFETs are designed to reduce the electric field concentration and provide improved voltage-blocking capabilities by forming a depletion area with several tiny cells [47].

Moreover, lower conduction losses, quicker switching times, and enhanced thermal properties are all improved by this design. As a result, the electric field in an SJ device has changed from a triangle to a blue rectangular form, and the N drift layer doping has increased. Forming the vertical P column is the most difficult part of making SJ MOSFET. There are two popular methods, both of which are commercially employed. Si SJ MOSFETs have gained significant attention in power electronics, particularly in applications such as power supplies, LED lighting, and motor drives. Their advanced design and improved efficiency contribute to higher power density and system performance. As a result, Si SJ MOSFETs are rarely used in applications that need a third-quadrant operation, such as voltage source inverters.

2.4. Si IGBT

Si IGBTs are critical power semiconductors that have transformed the field of power electronics. These power devices have revolutionized by combining the advantages of MOSFETs and bipolar junction transistors (BJTs) [48]. They offer high voltage and current handling capabilities while maintaining low on-state voltage drop and fast switching speeds. The structure of a Si IGBT comprises three layers: an N-type collector, a P-type base, and an N-type emitter. By regulating the conductivity of the base region through the voltage applied to the gate terminal, the IGBT enables efficient power switching. The performance and efficiency of Si IGBTs have recently been improved for various applications. One key development is integrating cutting-edge trench gate architectures and novel cell designs. These developments have improved switching speeds, decreased on-state losses, and reduced conduction losses. Furthermore, recent research has concentrated on improving the thermal management of the IGBT, enabling better power densities and increased reliability [49]. Due to improvements in power conversion efficiency, increased power density, and system performance, they are now more appropriate for various applications, including electric cars, renewable energy systems, and industrial automation.

Additionally, current Si IGBT advances have concentrated on reaching greater voltage ratings while lowering power losses. One significant development is using cutting-edge gate-driving methods and cell structure optimization in IGBTs. These developments have increased voltage ratings, decreased conduction and switching losses, and enhanced efficiency [50]. Advanced production methods and materials have also made it possible for better thermal management, which has increased power density and enhanced dependability. The latest advancements in Si IGBT technology have opened the door for creating smaller and more effective power electronic systems in various sectors.

The main features and specifications of popular Si power semiconductor devices, such as BJTs, diodes, MOSFETs, IGBTs, and thyristors, are compared in Table 4. For instance, Si IGBTs have medium conduction, high switching, and overall high-power losses while operating up to 1.2 kV blocking voltage and 50 A current rating. They work well in situations requiring medium voltage and medium frequency. In contrast, despite their slow switching speed, Si thyristors can handle high-voltage, low-frequency applications with up to 4 kV voltage and 3000 A current capacity. Si MOSFETs, which have a blocking voltage of 600 V and a current rating of 100 A, as well as minimal conduction and switching losses, fill the low-voltage, high-frequency market niche. Each Si device has defined features based on its advantages; however, it is constrained by the characteristics.

Table 4. Specifications of Si-based semiconductor device.

Parameter	Unit	Si BJT	Si Diode	Si MOSFET	Si IGBT	Si Thyristor/GTO
Voltage rating	kV	0.8	1.2	0.6	1.2	4
Current rating	A	15	40	100	50	3000
Switching frequency	kHz	10	20	20	20	<1
Channel resistance	Ω-cm^2	-	-	30	-	-
Off-state breakdown voltage	kV	0.8	0.6	0.6	1.2	4
Maximum junction temperature	°C	150	150	150	150	150
Conduction losses		High	Low	Medium	Medium	Low
Switching losses		Medium	Low	Low	High	High
Power losses		High	Low	Medium	High	Medium
Applications		Low voltage, low frequency	Rectification, voltage clamping	Low voltage, high frequency	Medium voltage, medium frequency	High voltage, low frequency

Table 5 compares four common Si-based power semiconductor devices—diodes, MOSFETs, super junction MOSFETs, and IGBTs—on features like scalability, cost, failure modes, and typical applications, showing the tradeoffs between different devices for use in power electronics and motor drives. MOSFETs and IGBTs are more scalable and suitable for high-power applications like motor drives. At the same time, diodes tend to be lower cost but limited to simpler rectification and voltage clamping circuits.

Table 5. Comparison of Si-based semiconductor devices.

Features	Si Diode	Si MOSFET	Si-SJ MOSFET	Si IGBT
Scalability	Scalable to high power levels	Scalable to medium power levels	Scalable to high power levels with parallel modules	Scalable to high power levels with parallel modules
Cost	Low to moderate cost compared to other power semiconductors	Low to moderate cost compared to other power semiconductors	Moderate to high cost compared to other power semiconductors	Moderate cost compared to other power semiconductors

Table 5. *Cont.*

Features	Si Diode	Si MOSFET	Si-SJ MOSFET	Si IGBT
Type of failure	Typically fails due to reverse breakdown, overcurrent, or excessive temperature.	Typically fails due to overvoltage, overcurrent, or overheating	Typically fails due to overvoltage, overcurrent, or overheating	Typically fails due to overvoltage, overcurrent, or overheating
Applications	Power supplies, rectifiers, freewheeling diodes, flyback diodes, voltage clamping, snubber circuits, battery charging	Switched mode power supplies, lighting, audio amplifiers, consumer electronics, automotive applications	Power supplies, solar inverters, server power supplies, industrial applications	Motor drives, industrial applications, renewable energy systems, UPSs, electric vehicles

3. Silicon Carbide (SiC)

SiC is a semiconducting material with outstanding physical, chemical, and electrical properties, making it very suitable for fabricating high-power, low-loss semiconductor devices. Moreover, commercially available SiC devices have lower switching/conduction loss, superior thermal stability, and greater temperature tolerance. SiC devices are, therefore, a very promising alternative to converters designed for high-temperature applications [51]. SiC power electronic devices have a theoretically allowed junction temperature of 600 °C due to the semiconductor's wide bandgap, around three times that of Si material [52]. On the other hand, the fabrication of these devices is rather intricate owing to the same properties of SiC, like its chemical inertness and hardness. It took over a hundred years to develop SiC electronics up to its modern state when power SiC devices possessing higher efficiency than their Si counterparts became commercially available and widely used in numerous applications. The resistance of the SiC material to an electric field is ten times greater than the resistance of the Si. As a result, SiC devices could be considered capable of withstanding the same blocking voltage with a 10-times-thinner material [53].

A tetrahedral crystalline structure is formed when each Si atom shares its electrons with four carbon atoms. Different SiC poly varieties can be made from this fundamental structure. Shown as Figure 2, SiC is the only chemical compound of group IV elements. It has a strictly stoichiometric concentration ratio of Si and carbon (C) atoms. It should not be mixed with solid solutions, which may be formed by other group IV elements and may have variable component concentration ratios (e.g., SixGe1-x) [54].

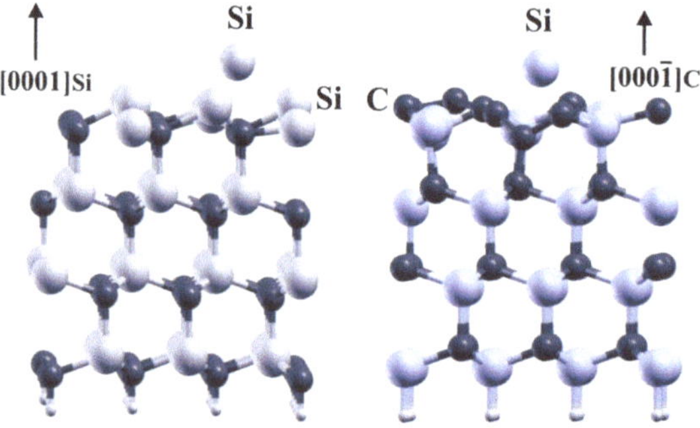

Figure 2. Side view of SiC lattice [54].

SiC devices can operate at higher switching dynamics. The thermal energy required for an electron to pass from the valence band to the conduction band (forming an electron-hole pair) is considerable in WBG materials. As a result, even at high junction temperatures, the WBG device's electric characteristics are retained within defined bounds. This enables

SiC semiconductor devices to function at high temperatures. Currently, the operating temperature range for SiC devices on the market is between 200 °C and 300 °C [55]. Therefore, SiCs are the ideal choice among commercially available WBD devices due to their greater power rating, quicker switching frequency, much lower switching losses, and capacity to handle higher junction temperatures than Si-based devices [54].

3.1. Discovery of SiC

As previously mentioned, SiC is an excellent material for high-power electronics and high-temperature applications because it has a wide bandgap and good thermal stability [56]. However, 2D SiC offers incredible new features missing from bulk SiC materials due to quantum confinement and surface effects [57–59]. SiC has arguably had the longest history of all the semiconducting materials used in electronics. SiC was found as a manufactured material, which makes its discovery exceptional in and of itself. Jöns Jacob Berzelius (1779–1848), in his 1824 report, most likely made the first observation of a chemical molecule bearing Si–C bonds [60]. He could conduct experiments in peculiar settings because of his acknowledged expertise in experimental methods. He also discovered several new chemical elements, including Si, and many other accomplishments. Berzelius made the extremely cautious claim that he had identified a substance that, when burned, created an equal number of Si and carbon atoms [61].

Seventy years later, the first validated SiC synthesis took place by chance [62]. American engineer Edward Goodrich Acheson (1856–1931) experimented with a newly developed electrical furnace in 1891 to create synthetic diamonds (highly demanded by the industry as an abrasive material) [24]. B. I. Ozernikova and A. P. Bobrievich discovered the first naturally occurring SiC of terrestrial origin in sediments of the Tyung River (Siberia) in 1956 and diamond-bearing kimberlite pipes in Yakutia, USSR, respectively, in 1957 [63]. From Schottky diodes in the 1990s to high voltage SiC MOSFETs and IGBTs in the 2020s, Table 6 traces the evolution of SiC devices, highlighting key developments like the first commercial devices, normally off JFETs, and trench gate MOSFETs that enabled higher voltage, current, and frequency capabilities compared to Si devices for uses like EVs and PV inverters. SiC device advancements and the emergence of GaN/SiC hybrids are boosting adoption and enabling performance above traditional Si power electronics.

Table 6. Evaluation of power electronic SiC-based semiconductor device.

Year	Device	Specifications	Milestone	Features
1990s	Schottky diode	600 V, 1 A	First commercial SiC diode	Higher voltage capability than Si
2001	JFET	1200 V, 5 A	First SiC transistor	Higher bandgap than Si devices
2006	MOSFET	1200 V, 10 A	First commercial SiC MOSFET	Higher frequency capability than Si IGBTs
	MOSFET	1700 V, 100 A	Trench gate SiC MOSFETs	Reduced on-resistance
2010s	JFET	1700 V, 50 A	Normally-off SiC JFETs	Simpler gate drive than depletion mode
	BJT	1200 V, 15 A	Higher current SiC BJTs	Improved SOA over Si BJTs
	SBD	1700 V, 20 A	Low loss SiC Schottky diodes	Faster switching than Si PiN diodes
2015	MOSFET	3300 V, 24 A	Higher voltage SiC MOSFETs	Expanding adoption in EV/PV markets
2018	IGBT	3300 V, 100 A	First commercial SiC IGBTs	Entering higher power applications
2020s	MOSFET	>10 kV, >100 A	Voltage and current increase	Replacing Si IGBTs and thyristors
Future	IGBT	>10 kV, >100 A	SiC IGBT refinement	Competing with Si IGBTs
	GaN	GaN/SiC	Combining GaN and SiC	Performance greater than either alone

3.2. Material Growth

Wide bandgap semiconductor SiC has outstanding features that make it highly sought-after for various applications. Epitaxy, the process through which SiC crystals grow, is essential for creating high-performance electronics. Chemical vapor deposition (CVD) and physical vapor transfer (PVT) are two growth techniques that have been used [64]. In CVD, SiC is deposited on a substrate due to the high-temperature decomposition of a precursor gas that contains Si and carbon. SiC source material in PVT is sublimated and recrystallizes onto a colder substrate. To produce high-quality single-crystal SiC, both processes require exact temperature control and a suitable growing environment. Growth rate, shape, and crystal quality are influenced by temperature gradients, gas flow rates, and crystal orientation [65]. Larger, higher-quality SiC crystals with fewer flaws have been made possible by improvements in growth methods and equipment, allowing for the manufacture of power electronic devices, high-frequency devices, and sensors that provide greater performance and efficiency compared to conventional materials [66]. The widespread use of SiC in several technological applications will be facilitated by further research and development efforts in SiC growing processes, which promise future improvements in crystal quality, scalability, and cost-effectiveness.

3.3. SiC Diode

Due to the greater SiC dielectric critical field than its Si counterparts, a blocking voltage rise of 10 times above that of Si is achievable with the same thickness of the SiC drift layer. Compared to Si diodes, SiC's high thermal conductivity has several benefits, including the ability to operate at higher current density ratings and reduce the size of cooling systems. SiC SBDs have been commercially available since 2001 and have continuously increased in the blocking voltage and conduction current ratings. There are essentially three types of SiC power diodes [67]. The PN junction and the Schottky Barrier Diode (SBD) junction are two semiconductor mechanisms that can create a diode, and there is no conductivity modulation in the SBD. From 600 V to 10 kV, Si PN junction diodes dominate the market. The scenario is fully reversed in the case of SBD on WBG material, such as SiC. The high critical electric field E_c in SiC reduces the resistance of an SBD significantly. As a result, conductivity modulation is neither necessary nor desirable. Because SBD does not store any charge, it can achieve near-zero reverse recovery loss. As a result, a WBG diode based on the Schottky mechanism is nearly optimal. The junction barrier Schottky (JBS) diode structure, which provides an area to shield the Schottky region in a reverse blocking state, can be used to reduce leakage current.

The PN junction of the JBS diode can become conductive at a high forward bias, giving it a stronger surge capability than the SBD [68]. On the other hand, the SiC PN junction diode will have to overcome a forward voltage drop of roughly 3 V across the PN junction, making it exceedingly undesirable from the standpoint of conduction loss, even if the drift area resistance can be decreased via conductivity modulation. The JBS structure is typically used in SiC diodes above 600 V. Because the off-state leakage current in JBS is decreased, devices can be rated at temperatures as high as 175 °C. Majority of the carrier electrons still are used for device conduction, and advantages of the SiC PIN diode over JBS or SBD is its substantially lower leakage current, which makes it an ideal option for high-voltage and high-temperature operation.

3.4. SiC MOSFET

SiC power MOSFETs are quickly gaining widespread application. Reliability issues like bias temperature instability and gate oxide cracking are mostly under control [69]. It is the chosen SiC three-terminal switch due to the well-established gate-driving technique and user base in Si MOSFET and IGBT compared to SiC JFET and BJT. It is commercially available in voltages ranging from 650 V to 1700 V, with higher current (5–600 A) modules [70]. SiC MOSFETs provide clear prospects for improving operating frequency, efficiency, and power density, but their employment is complicated by several unwanted side effects brought on by their rapid switching speed [71]. The characteristics of five commercial SiC MOSFETs are listed in Table 7, along with their component numbers, maximum current ratings in Amps, input capacitances (C_{ISS}) in pF, gate-drain capacitances (C_{GD}) in pF, and drain-source on-state resistances ($R_{ds(on)}$) in Ohms. SiC MOSFETs can achieve relatively low on-resistances—0.013 Ohms for an 80 A device, which enable high-frequency switching, but at the expense of greater input and gate capacitances than Si MOSFETs.

Table 7. Characteristics of selected SiC MOSFETs [28].

Part Number	Current (A)	C_{ISS} (pF)	C_{GD} (pF)	$R_{ds(on)}$ (25) (Ω)
IMW120R350M1H	2.5	180	1	0.35
C3M0280090J	5	150	2	0.28
IMW120R220M1H	10	289	2	0.22
C3M0120100K	20	350	3	0.12
SCT4026DE	40	2320	9	0.026
SCT4013DR	80	4580	10	0.013

When compared to the IGBT system, the operation frequency of SiC MOSFET-based converters has increased by one or two orders of magnitude, as demonstrated in Figure 3. SiC MOSFETs can also achieve zero switching loss under specific situations. A 1.2 kV SiC MOSFET module was recently proven to operate at 3.38 MHz [72,73]. The P regions shelter the gate oxide in the planar structure, so the peak electric field near the oxide is decreased. The gate-oxide stability problem in planar MOSFETs has been overcome, and high-reliability performance has been attained [74].

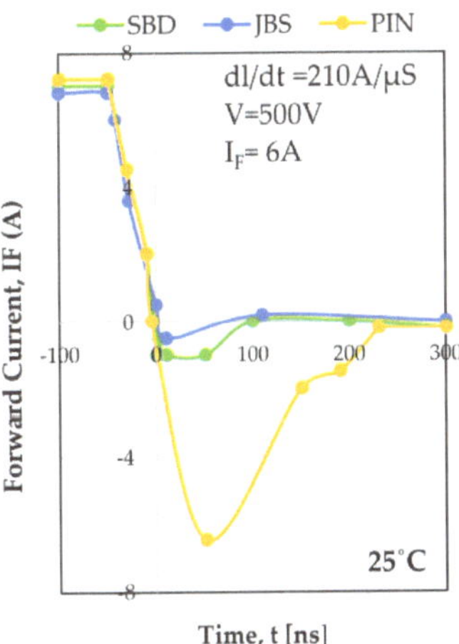

Figure 3. SiC diodes' turn-off current waveforms at 25 °C inductive load [73].

General Electric (GE) also exhibited the industry's first dependable SiC MOSFET with a 200 °C junction temperature [75,76]. This is more difficult in the trench device. Many different trench structures exist to protect the trench gate's bottom [77–82]. Another key motivator for SiC MOSFET innovation, in addition to enhancing electrical performance, is reliability. The three key criteria for evaluating dependability are high-temperature gate bias, high-temperature reverse bias, and high-humidity, high-temperature reverse bias. One of the most important challenges in the fabrication of SiC-MOSFETs was the lack of a reliable insulator for the gate terminal. A proper insulator is needed to achieve stable forward I–V characteristics and a stable gate threshold voltage. For this reason, the SiC-MOSFET was commercialized later than the SiC-JFET. Most MOSFETs contain a PIN diode inherent to their structure. This diode has a forward bias voltage around 2.5 V. During the conduction of the body diode, if the MOSFET is turned on, the forward characteristic of the body diode can be virtually improved, and, thus, the conduction losses are reduced. Figure 4 shows the calculated converter efficiency for a 1200-V SiC MOSFET system versus that is based on an IGBT/SiC diode hybrid power module. These graphs are critical for designing power converters for various applications.

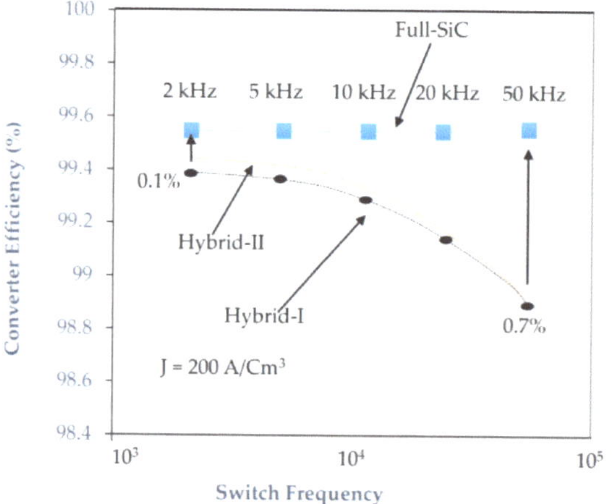

Figure 4. Calculated converter efficiency for a 1200-V SiC MOSFET system versus that based on an IGBT/SiC diode hybrid power module [83].

Switches, solenoids, encoders, generators, and electric motors are the primary electromechanical devices that connect the digital and physical worlds. The capacity of each of these gadgets to translate electrical impulses into mechanical motions is what gives them their enchantment. The need for more control, efficiency, and capabilities from these electromechanical devices rises as fields like automated manufacturing, electronic vehicles, sophisticated building systems, and smart appliances develop [84]. It investigates how improvements in SiC MOSFETs are expanding the possibilities of electric motors, which hitherto relied on Si IGBTs for power inversion. Similar to the new power electronic converters, SiC devices could be utilized for motor driver applications for novel electric machines for various applications [85–90]. This development increases the potential of motor drive applications across all industries.

The cabling between the drive inverters and the motor driver may be significantly reduced by bringing the motor driver assembly to the motor's local position, resulting in considerable cost savings. Figure 5 shows seven motors of a robotic arm, that is required to be powered by 21 different cables, which might require hundreds of meters of costly and intricate cabling infrastructure, in a typical Si IGBT power cabinet. Two lengthy cables that link to each motor's motor drive within the local motor assembly can be used in a SiC MOSFET motor drive system to decrease the number of cables [84]. There are specific applications where IGBTs may still be better suited, as is true for all types of components; however, SiC MOSFET inverters offer several distinct advantages over Si IGBTs, making them very attractive solutions for motor drive and a wide range of other applications.

3.5. SiC IGBT

SiC IGBT has received a lot of interest in the domains of high voltage transmission, smart grid, and pulse power since it is the highest voltage switch. SiC IGBTs have not been commercialized because of their inherent flaws and crude manufacturing methods. The stated SiC IGBT devices' exceptional static and dynamic performance and significant dv/dt during hard switching challenge the power conversion system. SiC IGBT has a lot of potential solutions; however, comparisons with Si IGBT and SiC MOSFET reveal significant discrepancies. The potential SiC IGBT appears to have a chance to displace Si devices in the future based on early experiments in high-voltage fields.

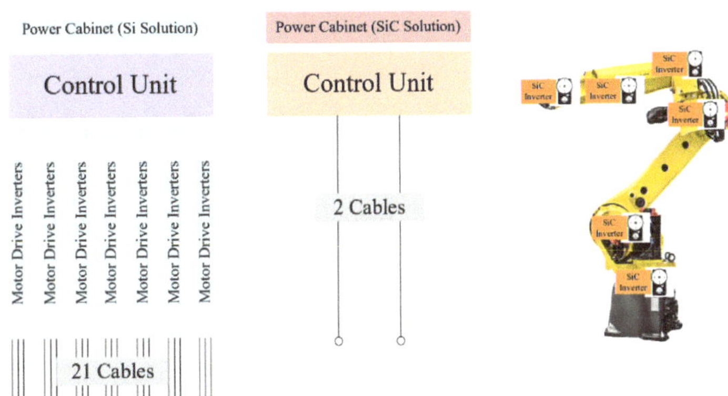

Figure 5. Comparison of a Si IGBT vs. SiC MOSFET system control of a robot arm [84].

High voltage direct current transmission (HVDC), industrial applications, traction systems, and new pulsed power applications are the key markets for Si IGBT with high voltage and current. Ultra-high voltage in these domains is intended to decrease the number of devices connected in series and simplify converter topologies [91]. According to one source, Si IGBTs can withstand a maximum voltage of 8.4 kV [92], which is about as high as Si devices can go. The continued advancement of Si IGBT in these sectors is also severely constrained by frequency and operating temperature. SiC exhibits greater breakdown field strength, inherent temperature, thermal conductivity, and carrier saturation drift velocity as a wide bandgap material [93]. As can be seen in Table 8, the SiC IGBT device is more competitive in high-voltage, high-temperature, and high-power areas. The SiC IGBT with 4H-polytype and n-channel is chosen among SiC polytypes and channel types because it has a low forward voltage (V_f), quick switching, and a large safe working region. Wide-base PNP transistors in 4H-SiC N-IGBTs have excellent bulk mobility and low current gain, contributing to a favorable trade-off between V_f and switching loss.

Table 8. Electrical characteristics of SiC-IGBT and Si-IGBT devices [94].

Manufacturer	Advanced Power Technology	Semikron	Mitsubishi
Device type	SiC-IGBT	SiC-IGBT	Si-IGBT
Part number	APT60GF120JRDQ3	SK25GH063	CM150DY-24A
Used experiments	SPT	Three-phase inverter	AGPU, SPT, Three-phase inverter
IC (A)	2.1	2.1	2.1
V (V)	3.0	2.3	2.4
R_{ce}(on) (mΩ)	33	33	356
Turn-on energy ET, on (mJ)	14.6	1.1	4
Turn-off energy ET, off (mJ)	6.5	0.8	16

SiC IGBTs have a lower V_f than Si IGBTs under the same blocking capability but have lower bulk mobility than Si IGBTs. Even though SiC IGBTs have not yet been commercialized, significant advancements have been made over the previous 30 years, as seen in Figure 1. Because of the readily accessible n+ substrate with low resistivity and defect density, the SiC p-channel IGBT has been widely investigated from 1996 when the first SiC IGBT was produced [95] through 2010. Continuous improvements have been made to the P-IGBT's performance, notably since the charge storage layer (CSL) was introduced [96]. Due to immature technology and a P-type substrate with a high resistivity and defect density during this time, the constructed N-channel IGBT performs poorly [97].

Because free-standing technology offers a way to develop a P+ collector on an N+ substrate, the research focus of SiC IGBT shifted to SiC N-channel IGBT before Cooper's freestanding technology proposal in 2010 [98]. After that, SiC N-IGBT displays ever-improving static and dynamic properties. The SiC IGBT has a blocking voltage that exceeds 27 kV [99], making it a promising device in high-voltage areas. In comparison to the SiC MOSFET of the same rated voltage, the SiC IGBT might cut the differential specific on-resistances ($R_{on,sp,diff}$) by more than one order of magnitude. Therefore, even if the switching loss (E_{sw}) of the SiC IGBT is larger than that of SiC MOSFET, it is promising for power conversion systems with more than 100 kW transmission power. Recently, the first solid-state transformer prototypes and the first Marx generators have also been made using the 12–15 kV SiC IGBT modules that were developed [100–102].

SiC-IGBTs are more effective than Si-IGBTs because of their switching speed. The gate-to-emitter voltage rise quickly to allow for the SiC-IGBT to switch quickly. Therefore, to charge the input capacitance C_{iss}, a greater gate current is needed [103]. Generally, to switch the IGBT off, the gate current capability is increased by reducing the external turn-on and turn-off gate resistors even though the same current capability is needed. In order for the gate current to rise as quickly as required, the gate driver's stray inductance must also be kept to a minimum [94]. SiC-IGBTs need a negative gate-to-emitter voltage like that found in Si-IGBTs to obtain a quick and secure turn-off transient. A SiC-IGBT driver typically supplies a gate-to-emitter voltage of +20 V positive and 5 V negative [104].

SiC BJTs, JFETs, Schottky diodes, and MOSFETs are four popular SiC power semiconductor devices described and shown in Table 9. SiC devices outperform Si in high-voltage, high-frequency applications like PV inverters, EV systems, and power supplies. It compares parameters like bandgap, electron mobility, voltage/current ratings, switching frequency, power losses, and figures of merit. The table shows the compromises made by various SiC devices for application-specific performance optimization.

Table 9. Specification of the SiC-based semiconductor device.

Parameter	Unit	3C-SiC BJT	6H-SiC JFET	4H-SiC SBD	4H-SiC MOSFET
Bandgap	eV	2.2	3.0	3.26	3.26
Critical electric field	Mv/cm	1.5	3	3	3
Electron mobility	cm^2/V-s	1000	370	800	800
Saturated electron drift velocity	cm/s	2e6	2e6	2e6	2e6
Thermal conductivity	W/cm-K	4.9	4.9	4.9	4.9
Lattice mismatch	%	3.5	3.5	3.5	3.5
Wafer size	mm	100	150	150	150
Voltage rating	kV	1.2	1.7	1.7	1.7
Current rating	A	15	10	24	24
Switching frequency	Hz	10	50	-	100
Resistivity	Ω-cm	10^4	10^4	10^4	10^4
Channel resistance	Ω-cm^2	120	35	-	80
Stress levels of voltage	V	5000	10,000	10,000	10,000
Stress levels of current	A	15	10	24	24
Off-state breakdown voltage	kV	1.2	1.7	1.7	1.7
Maximum junction temperature	°C	500	500	600	600
Temperature range	°C	−55 to 250	−55 to 250	−55 to 300	−55 to 300

Table 9. *Cont.*

Parameter	Unit	3C-SiC BJT	6H-SiC JFET	4H-SiC SBD	4H-SiC MOSFET
Temperature stability	°C	1	1	1	1
Conduction losses		Medium	Low	Low	Low
Switching losses		Medium	Medium	-	High
Power losses		Medium	Medium	Low	Medium
Baliga's figure of merit		51	204	408	408
Johnson's figure of merit		8	31	62	62
Applications		Medium voltage, medium frequency	High voltage, high frequency	High voltage rectifier	High voltage, high frequency

3.6. Applications and Emergence of SiC Power Electronics

SiC is employed in semiconducting and other items such as armored vehicles, ceramic plates, thin filament pyrometry, foundry crucibles, and auto clutches. SiC was initially used in electrical applications as a lightning arrester in a high-voltage power system because engineers and scientists realized SiC works well even in the presence of high volts and high temperatures. SiC devices are appropriate for a wide range of applications in aerospace and space missions [105,106], despite the necessity for high-temperature dependable device packaging to be developed [107,108]. Schottky diodes, MOSFETs, and power electronics are some of the most recent electrical devices that use SiC.

Applications for SiC include sandblasting injectors, automobile water pump seals, bearings, pump parts, and extrusion dies. These applications use SiC's exceptional hardness, abrasion resistance, and corrosion resistance [109]. SiC is undoubtedly durable and versatile, with applications ranging from semiconductors for Schottky diodes to use as an abrasive polishing material. Its exceptional qualities include sublimation, great chemical inertness and corrosion resistance, excellent thermal characteristics, and the capacity to develop as a single-crystal structure.

The introduction of the first mass-produced electrical vehicles (EVs) to the market in 2008—Tesla Motors' debut of its first all-electric vehicle—significantly impacted the development of SiC power electronics. Two components of the electrical power train significantly impact the performance of these cars. They are an inverter and a battery charger that transform the DC power from a battery pack into AC power for a motor. Since the battery capacitance in EVs restricts the amount of on-board stored energy, the efficiency of power conversion by these units is crucial. The bulk of modern electric vehicles (EVs) and the first electric vehicles (EVs) used inverters with Si IGBTs and conversion efficiencies ranging from 80% to 95%. Even at 95% efficiency, these inverters waste too much energy and need liquid cooling. Compared to the electrical motors, they are bigger and heavier. The conversion efficiency of the inverter might be increased to 99% [110] by swapping Si IGBTs for SiC MOSFETs while being significantly lighter and smaller. The first commercial SiC power MOSFET was introduced by Cree, Inc. in 2011 [111]. This potential use of SiC power devices in the high-volume automotive sector sparked increased research into the design and technology of SiC devices.

Tesla introduced the Model 3 in 2017, the first electric vehicle to have inverters based on SiC MOSFETs. By the time this article was published, each week's manufacturing of automobiles used 48 SiC MOSFETs with a 650 V/100 A rating made at the STMicroelectronics fab in Catania, Italy [112]. At the same time, Tesla Model 3's traction motor spins at 17,900 revolutions per minute, and China's NEV Technology Roadmap 2.0 aims to reach a motor speed of 25,000 rotations per minute by 2035 [113]. DENSO, a global automotive manufacturer, built its first inverter using SiC semiconductors. The new Lexus RZ, a Toyota luxury brand's first specifically designed battery electric vehicle (BEV) model, will employ this inverter, which is part of the eAxle, an electric drive module designed by BluE Nexus Corporation [114]. It appears that Toyota is producing its first BEV in large quantities through the Lexus Division. SiC and GaN are competing to see which is superior for power electronics. They both outperform Si in either situation.

The market for SiC power devices is expanding rapidly, and the SiC industrial sector today exhibits significant diversity, with successful businesses operating in various ways. This predicts that SiC power electronics will continue to flourish as an industrial technology during the coming several decades. The enormous potential of SiC as a material for high-temperature and high-frequency electronics, which is still not realized and is awaiting convincing demonstration of SiC's superiority over conventional semiconductors for these applications, is another factor driving the further development of SiC technology.

Over the past 13 years, GE Aviation has committed over $150 million to developing SiC technology, solidifying its position as the market leader. Because of characteristics like high-temperature tolerance, low losses, and higher frequencies, SiC enables lighter, more effective, and higher-performance power electronics. To reach extremely high-power density and efficiency goals, GE is utilizing SiC technology in aerospace applications, including hybrid electric aircraft propulsion. GE can create lighter and more potent systems for electric ground vehicles and other applications due to SiC. GE Aviation is well-positioned to promote SiC power device adoption moving ahead due to its extensive aerospace knowledge and experience.

The SiC-based generator controller developed by GE Aerospace shows a significant advancement in power electronics when compared to industrial Si-based generator controllers, as shown in Table 10. Its benefits are glaringly obvious in many important areas. First off, the SiC controller is extraordinarily effective at harvesting and managing electrical power within a small footprint due to an amazing fourfold improvement in power density relative to its size. With a twofold increase in power density per unit weight, its power-to-weight ratio is also improved, which results in both increased efficiency and a decrease in the overall weight of the controller. This drop is accompanied by a 50% reduction in physical size, highlighting SiC's promise for applications that conserve space. Notably, the SiC-based controller excels at increasing conversion efficiency. It increases DC/AC conversion efficiency from 94% to an astounding 99%, guaranteeing little energy is lost in the process. Similarly, it improves efficiency in DC/DC conversions from 84% to 95%, maximizing power transfer. SiC technology improves efficiency in AC/DC conversions from 85% to 92%, leading to more effective power conversion and energy use.

Table 10. Comparison between Si-based and SiC-based generator controller [115,116].

Parameters	Unit	Si-Based Generator Controller	SiC-Based Generator Controller
Output power	kW	100	200
Aux SSPC channels	kW	No (Excitation Controller)	120 (2ea 600 V × 100 A)
Bi-directional	-	Yes	Yes
Space claim (volume)	L	33.6	17.4
Size	mm	444 × 400 × 189	350 × 350 × 142
Power density (size)	kW/L	3	12
Weight	Kg	27.7	25
Power density (weight)	kW/kg	3.6	8
Coolant temp, max	°C	85	105
Ambient temp, max	°C	71	121
Communications	-	IEEE 1394B Bus	CANBus

Figure 6 shows the evaluation of SiC modules and SiC technology utilized in aerospace applications by GE Aerospace in conjunction with Power America, the Department of Energy, and the National Renewable Energy Laboratory. Specifications for various SiC power module variants from GE Aerospace are listed in Table 11. The modules have voltage ratings of 1200 V and 1700 V and come in half-bridge, dual-bridge, six-switch, and six-pack versions. Ratings range from 425 A to 1425 A. On-state resistance, thermal resistance, size, and maximum junction temperature are important factors mentioned. With small footprints up to 100 mm × 140 mm and operational temperatures up to 175 °C, the table demonstrates great power density and temperature capability of SiC modules. SiC modules can be configured in a wide range of ways to accommodate various power electronic circuit topologies and the current/voltage ratings needed for aviation applications.

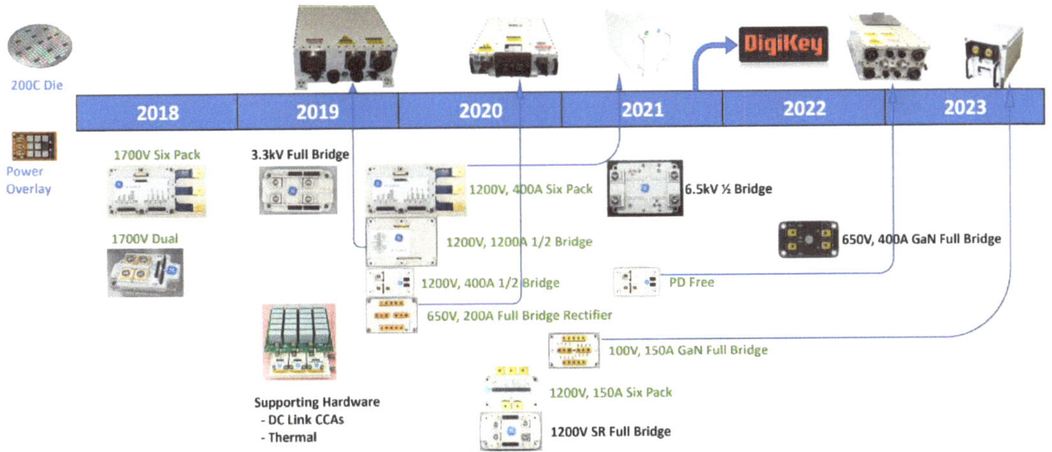

Figure 6. Evaluation of wideband power semiconductor modules designed and manufactured by GE Aerospace [115–117].

Table 11. SiC power module by GE Aerospace [117].

Type of SiC Module	Part Number	Voltage Rating (V)	Current Rating (A)	R_{DS}(on) @25 °C (mΩ)	Thermal Cooling System ($R_{th(j-c)}$) (K/W)	Size (Width × Length) (mm)	Maximum Junction Temp (°C)
Half-bridge	GE12047CCA3	1200	475	3.1	0.1	48 × 86	175
	GE17042CCA3	1700	425	3.8			
	GE12090CDA3	1200	875	1.6	0.03	100 × 140	
	GE17080CDA3	1700	765	1.9			
	GE12160CEA3	1200	1425	1.0	0.03	90 × 134	
	GE17140CEA3	1700	1275	1.2			
Dual-bridge	GE12047BCA3	1200	475	3.1	0.1	48 × 86	
	GE17042BCA3	1700	425	3.8			
6-switch	GE12050HEA3	1200	6 × 475	3.1	0.1	90 × 134	
	GE17045HEA3	1700	6 × 425	4.8			
6-pack	GE12050EEA3	1200	3 × 475	3.1	0.1		
	GE17045EEA3	1700	3 × 425	3.8			

3.7. Challenges of SiC Power Device Development

Although high voltage (HV) SiC devices perform better than their Si counterparts, additional considerations must be made to use them effectively. The HV SiC device application must overcome obstacles in a variety of time frames. The majority of difficulties are caused by relatively brief time intervals (between 0.01 and 1 μs), such as electromagnetic interference (EMI), packaging, and gate drive design. The output of a SiC-based converter can have advantages in terms of output with a high-frequency range, low total harmonic distortion (THD), and high control bandwidth, to name a few. Newly developed converter topologies and PWM schemes could be utilized with SiC devices to compare the viability of the device [118–121]. HV SiC device application technology is still in the early stages of development [122]. The material's properties present a significant problem for SiC manufacturing. Generally, SiC takes more energy, longer, and higher temperatures for crystal growth and processing due to its extreme hardness, which is nearly diamond-like [123].

3.7.1. Packaging

Building power modules with small and fast device dies requires addressing electrical, thermal, and mechanical challenges. Figure 7 depicts a typical power module packing.

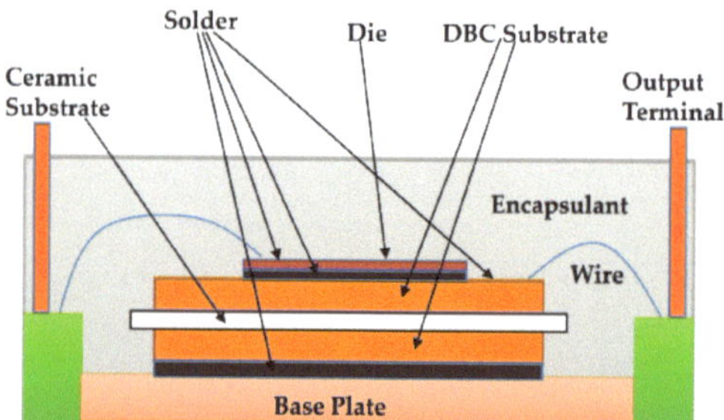

Figure 7. SiC power module packaging structure.

3.7.2. Electrical Insulation

The packaging of HV SiC power modules presents a significant problem regarding electrical insulation design. Special attention should be given to vulnerable places that are easily compromised, such as the die, substrate, and output terminal. SiC devices have substantially greater voltage ratings than Si devices, although SiC devices have thinner dies. The electric field around the die from the anode to the trace that is soldered with the cathode gets noticeably stronger with the higher voltage rating and thinner die. A recent generation 10 kV SiC MOSFET, for instance, has a die thickness of 100 μm and an average electric field of 100 kV/mm [124], but a 1.7 kV IGBT has a die thickness of 209 μm and an electric field of only 8.1 kV/mm [123]. As a result, HV SiC device packaging surrounding the die has an electrical field concentration that is ten times greater than Si's.

3.7.3. Parasitics

The internal connecting wires inside the module and the external power stage contain the stray inductance. The combination of these stray inductance components affects how well the gadget switches. Using a decoupling capacitor [122,123] can minimize the influence of stray inductance in the external power stage. The stray inductance inside the module is the main subject of this subsection. They primarily have three effects on device performance: voltage overshoot, ringing, and switching speed restriction.

3.7.4. Power Module with Multi-Dies

Developing the multi-die in parallel connection for HV SiC-based power modules is important due to the single die's restricted current rating of present HV SiC devices, typically approximately 20 A per die. For instance, the 10 kV/240 A SiC MOSFET power module is made up of 18 dies connected in parallel [125].

3.7.5. Gate Drive

The gate drive connects the power semiconductor device and control. Power semiconductors are turned on and off by transferring the control signal from the gate drive to the drive signal. Gate drive design is crucial for SiC devices to achieve their maximum potential. Two crucial factors, efficiency and dependability, should be considered when designing the gate drive. The HV SiC devices do not have commercially accessible gate drivers because they are unique power devices. Its gate drive needs to be the subject of research [122].

3.7.6. Electromagnetic Interference

The converters using HV SiC devices may experience significant EMI due to the high dv/dt caused by the quick switching speed and the huge parasitic capacitance due to the compact size. Both EMI filters and special EMI reduction techniques are desirable. The EMI and dv/dt filters in motor drives are essential for preventing the voltage doubling effect that results from high dv/dt [126]. High dv/dt will significantly increase the grid-side conduction EMI in applications involving the grid. EMI filters are required to ensure that the power conversion system satisfies the criteria for grid-connected converters [127].

3.8. Future Trends of SiC

SiC devices have several advantages and many difficulties from both the device and application viewpoints. These serve as a guide for upcoming developments in research:

3.8.1. Voltage-Derating Guidelines

A voltage-derating design guideline must be devised and SiC devices' field dependability for various applications must be shown. This is particularly crucial for applications like aviation systems [127], where reliability is essential. On the one hand, a SiC device in a power conversion system may have a greater current rating than a Si device, resulting in a larger thermal ripple. On the other hand, the SiC device's ability to operate at greater temperatures suggests stricter requirements for the packaging materials.

3.8.2. Improving Manufacturing Techniques for Affordability

To make SiC devices more affordable for system applications, manufacturing techniques must be improved for higher yields [128]. The majority of significant SiC producers are switching to 150 mm SiC epitaxial wafers. There are also debates over switching to a fabless manufacturing method, which would allow SiC power devices to be produced in Si fabs. Because the quality would be ensured by the already existing facilities and established production infrastructure, such a shift would result in lower prices. It is difficult, nevertheless, due to the rigidity of SiC equipment and processes.

3.8.3. Considerations for SiC Device Gate Driver Design

SiC devices switch significantly faster than Si devices, which presents difficulties for the design of the gate driver [129,130]. First, the SiC device's larger dv/dt injects more common mode current through a miller capacitor into the gate loop, creating a positive spurious gate voltage. An improved gate driver with active dv/dt and di/dt control is a growing trend. Additionally, a greater dv/dt injects a higher common mode current through the isolation barrier to the primary side of the gate driver, limiting the coupling capacitance. Second, SiC devices have a smaller die size and a faster current rise under the fault. Thus, the gate driver's short circuit protection response requirement is larger, and the SiC device's parallel or series connection is more susceptible to timing errors.

3.8.4. Novel EMI Filter Design

It is necessary to look at the novel EMI filter design [108]. The EMI noise is 10 to 100 times higher due to the faster switching transient and higher switching frequency. When used in high-voltage and high-power applications, this becomes very difficult. Firstly, SiC power converters require the modeling and prediction methods of EMI noise. On the other side, novel filter and shielding designs are required.

3.8.5. Reducing Commutation Loop for Enhanced SiC Device Performance

Using novel system designs with a reduced commutation loop is crucial, such as the SiC device package and system busbar structure [72]. Although it is known that a SiC power converter may supply more current, this is only true when the breakdown voltage of the device is lower than the device voltage stress during the transition. In other words, the voltage overshoot may restrict the system's current rating rather than the thermal. This issue is extremely significant for some new applications, such as the 1.5 kV DC photovoltaic system [131]. Up until now, laminated and multi-layer laminated busbar structures have been the preferred option. Investigating a low-inductance capacitor is also necessary.

3.8.6. Exploring Cooling Techniques for SiC Device Reliability

SiC chips' lower die size and, thus, increased loss density present new design difficulties for heat management techniques [132]. Additionally, the SiC chips' lower thermal capacitance could lead to a larger temperature ripple, which could present problems from a reliability standpoint. Phase change cooling, liquid jet impingement cooling, and other effective cooling techniques are some prospective, promising alternatives [133]. The advancing 3D printing technologies may also be leveraged to support original ideas.

3.8.7. Advancements in Ancillary Components for High Temperature

SiC power devices' capacity to operate at high temperatures necessitates advancements in ancillary components such as high-temperature capacitors, packaging, control electronics, gate drivers, and sensors [134]. The bus bars that connect the high-temperature devices to the capacitors make them particularly difficult since they raise the temperature of the capacitors even further. To make the capacitors usable with high-temperature switches, cooling may be required. In addition, Si-on-insulator (SOI) or SiC-based high-temperature gate drivers are required. The typical temperature range for SOI technology is 225 °C or lower. Since SiC has a lower channel mobility than other semiconductors and is, therefore, not appropriate for very low-voltage applications, SiC-based integrated circuit design is complex.

3.8.8. Lowering Engineering Effort and Costs

Due to their lack of familiarity with these novel devices, many potential SiC users are also concerned about the expensive non-recurring engineering expense. The conventional SiC PEBB, like the Si PEBB, can dramatically lower engineering effort and development costs, opening the door for commercializing SiC applications [135]. The heat sink, the robust gate driver, the high-bandwidth and noise-free controller, and the low-inductance busbar may all be incorporated. The entire system might then be constructed entirely around this PEBB for various purposes.

4. GaN

Due to its capacity to provide much better performance across a variety of applications while using less energy and physical space to do so compared to current Si technologies, GaN is gaining significance. GaN technologies are becoming crucial in some applications where Si as a power conversion platform has reached its physical limits, while in other applications the advantages of efficiency, switching speed, compactness, and high-temperature operation combine to make GaN increasingly alluring.

Table 12 depicts the evolution of GaN power semiconductor devices from 1990s HEMTs to current MISFETs (Metal Insulator Semiconductor Field Effect Transistor) and IGBTs, highlighting significant developments such as the introduction of the first enhancement mode transistors, the dependability of the GaN MISFET gate oxide, and the extension into high-voltage EV applications. With growing vertical GaN, MISFETs, and GaN-on-Si, which overcome the limits of early HEMTs for better voltage ratings and reliability to replace and exceed Si MOSFETs and IGBTs, GaN enables higher-frequency switching and current density compared to Si devices.

Table 12. Evaluation of power electronic GaN-based semiconductor device.

Year	Device	Specifications	Milestone	Features
1990s	HEMT	50 V, 1 A	First GaN transistor	Higher frequency than Si and GaAs
2000s	HEMT	200 V, 1 A	Enhancement mode GaN	Normally off operation
2010s	HEMT	600 V, 30 A	High-voltage GaN transistors	Replacing Si MOSFETs in adapters
	MISFET	200 V, 1 A	GaN MISFETs	Gate oxide reliability improvements
2015	HEMT	1200 V, 15 A	>1 kV rating achieved	Entering high-voltage applications
2019	MISFET	650 V, 20 A	Commercial GaN MISFETs	Reduced gate leakage over HEMTs
2020s	HEMT	3.3 kV+, 100 A+	High current density	Targeting EV traction inverters
	MISFET	3.3 kV+, high current	Further MISFET refinement	Improved reliability over HEMTs
Future	IGBT	1.2 kV+, high current	GaN on Si IGBTs	Improving Si substrates and vertical GaN
	Thyristor	Medium voltage, high current	GaN thyristors	High-current applications

GaN-based power devices first appeared in 2000, with the GaN FET being manufactured on a SiC substrate utilizing radio frequency standards. Following that, with the advancement of material growth techniques, GaN power devices have made a quantum jump in quality [136]. GaN has a strong atomic junction because it is made up of nitrogen, a light element. Their net parameter, which refers to the distance between atomic unit cells and their crystalline structure within the same material, is less than that of other semiconductors from the III–V groups of the periodic table [137]. As a result, GaN has better electrical properties than Si dioxide.

A switch to GaN technology will assist in fulfilling demand while reducing carbon emissions as the world's energy needs grow. It has been demonstrated that the design and integration of GaN may provide next-generation power semiconductors with a carbon footprint 10 times less than that of slower, older Si chips, as shown in Figure 8. In order to strengthen the argument for GaN, it is predicted that switching all data centers from Si to GaN will cut energy loss by 30–40%, resulting in savings of over 100 TWhr and 125 Mtons of CO_2 by 2030 [138].

GaN has appeal for reasons other than only operational performance and system-level efficiency gains. GaN offers a strong "green" edge over older, slower Si because a GaN power IC chip could save 80% in manufacturing and process chemicals and energy and more than 50% in packaging.

Currently, GaN wafer diameters of 2″ have been obtained at a cost that is 10 times that of SiC and up to 100 times that of Si [139]. However, in contrast to SiC, GaN has been widely used in optoelectronics and radio frequency applications due to its broad energy band and potential high-frequency properties [140]. A lot of electronics, such as radio transmitters, plasma generators, MRI scanners, power converters, and wireless power transfer (WPT), among many others, depend on radio frequency (RF) power [141]. GaN's wide bandgap, huge critical electric field, strong electron mobility, and reasonably good thermal conductivity make it appealing for high-voltage, high-frequency, and high-temperature applications. GaN-based materials can emit light in various visible wavelengths (violet, blue, and green), as well as for high frequency and high-power applications, due to their wide range of energy bandgap [142–144].

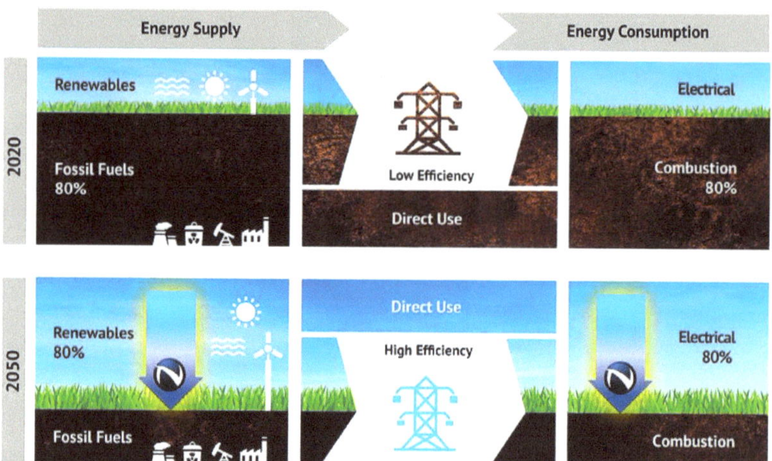

Figure 8. GaN technology to significantly reduce carbon emissions and energy consumption [138].

GaN and its alloys, such as AlGaN and InGaN, were the subject of much research. Other options are used, including Si, sapphire, and SiC substrates, which are less expensive. These substrate materials match the lattice properly and are thermally compatible with GaN. GaN-based devices are currently commercialized in the photonics field, although this semiconductor material is still in the early stages of development for power applications. Due to their cheaper cost, GaN rectifiers on Si or sapphire substrates, and their remarkable trade-off between on-resistance and breakdown voltage, these devices have gained much attention in recent years. For 600 V and lower voltage applications, the GaN heterojunction field-effect transistor (HFET) has been commercially introduced [145,146]. The source and drain contacts are interconnected using many metal layers. Another notable WBG material is GaN, which has features ideal for power device applications. Early GaN HFETs are depletion mode devices; to achieve an enhancement device, a cascade setup with a Si MOSFET will be required [146].

Over the past ten years, GaN has risen to the top of the materials used to make power devices. Next-generation power devices use the compound semiconductor material GaN. Due to its advantages over Si-based devices, such as outstanding high-frequency characteristics, it is starting to be adopted [147]. GaN is a fantastic material for fabricating high-speed/high-voltage components since it has the widest energy gap, critical field, and saturation velocity among semiconductors for which power devices are already on the market [148]. Without doping, a two-dimensional electron plasma with high mobility and a high channel density is created using AlGaN/GaN heterostructures. The presence of spontaneous and piezoelectric polarization makes this possible.

Additionally, the ability to build these devices on large-size, cheaper Si substrates offers a financial benefit that enables one to use Si CMOS equipment and affordable facilities [149]. Subsequently, due to their low switching charges and parasitic capacitances, GaN transistors have lower switching and resistive losses [148]. Device scaling and monolithic integration, which have advantages in terms of downsizing, enable a high-frequency operation. Since GaN is a more recent material than Si, it is crucial to fully understand and characterize the trapping and deteriorating processes in order to increase device stability and dependability [148]. GaN-based devices are currently commercialized in the photonics field, although this semiconductor material is still in the early stages of development for power applications.

4.1. GaN Diode

The advancement of the GaN substrate in recent years has been the driving force behind the development of vertical GaN-based devices. The vertical GaN diode, on the other hand, has, at this early stage, become a hot research area due to the relatively immature technology for the vertical triode [150]. Vertical GaN SBDs are similar to AlGaN/GaN SBDs in that they both exhibit low conduction loss and high switching speeds with little reverse recovery time in frequency fields [151,152]. However, the latter has a higher current density and fewer leakage paths.

The bulk of GaN Schottky power diodes disclosed up to this point feature either lateral or quasi-vertical layouts because there are not many GaN substrates with electrical conductivity [153]. Although the forward voltage drop is still significant, lateral GaN rectifiers have shown breakdown voltages as high as 9.7 kV on sapphire substrates [154]. GaN rectifiers constructed on Si or sapphire substrates are quite popular because of their lower cost. Due to the recent availability of high-temperature HVPE (hydride vapor phase epitaxy) free-standing GaN substrates, 600 V operating voltage GaN Schottky diodes are soon to be made available on the market to compete with SiC Schottky rectifiers [155]. In addition, commercial GaN Schottky diodes operating in the 600 V to 1.2 kV voltage range will be made available by the industry very soon. JBS GaN diodes, however, could enhance the performance of GaN-based power rectifiers in the 600 V to 3.3 kV range; however, contact resistance to implanted p-type GaN still needs to be reduced [156].

There are two types of GaN power diodes: the GaN Schottky barrier diode (SBD) and the GaN power diode (PN). When a typical GaN power rectifier is turned on, electrons must pass over the Schottky barrier of the SBD, resulting in a high turn-on voltage that is incompatible with lowering device losses [151]. Rapid progress in the growth and fabrication of vertical GaN (v-GaN) diodes has been made in the past few years, with the device unipolar figures of merit (UFOMs) exceeding those of SiC [157–160]. As the peak electric field is present primarily in the bulk rather than along the surface as in lateral devices, v-GaN diodes can operate in systems requiring higher-voltage hold-off. This enables competition with SiC and Si diodes in voltage regimes above 600 V [161].

The vertical structure is extensively employed in general-purpose power electronic devices that can provide a higher current. This structure includes the benefits of both lateral and vertical structures, but it also has the downsides of both. Its advantage is that it can be used with existing processes and can be created in large sizes [162]. Dang et al. reported Au/Pt-GaN Schottky diodes with up to 550 V breakdown voltage in 2000 [163]. In 2009, Arslan et al. used the metal organic chemical vapor deposition (MOCVD) technology to make a Ni/Au-AlGaN/GaN heterojunction Schottky diode and examined its current transport under various temperature settings [164]. The first GaN commercial integrated power devices, the iP2010 and iP2011, were developed in 2010 by US International Rectifier, based on the GaN SBD technology platform—GaNpowIR. Micro GaN, a German business, launched the 600 V line of products for high-power, high-voltage applications in 2010, which included the Schottky diode MGG1TO617. Its turn-on voltage, turn-on resistance, and drain-source voltages are all less than 0.3 V, 329 m, and 600 V, respectively, and the leakage current is only 1 mA, reducing switching losses significantly. It has been used in aerospace and defense, power conversion, and traction application [165].

In 2011, EPC Corporation released its GaN line of products, with a maximum voltage of 300 V and a low on-resistance of only 150 m. Sanken Electric, a Japanese company, uses GaN-based SBD and HEMT solutions in DC/DC converters and plans to introduce 600 V diodes in 2012. Panasonic and Sharp have since released 600 V GaN-based SBDs. With the help of ARPA-E (American Energy Advanced Research Projects Agency) and the military, Avogy, now Nexgen Power Systems, is rapidly expanding its product line to include not only 600 V GaN SBD commercial products but also 1700 V PN-type diodes, which are used in solar and wind energy inverters, electric vehicles, power conversion, and aerospace applications [166]. Avogy's commercial GaN diode products are shown in Table 13. Because the technique for forming PN junctions on GaN materials is still in its beginning, in order to improve device performance and get over the limits of basic GaN SBDs and p-n diodes, several high-tech vertical GaN power rectifiers were created. Basic p-n diodes have a high forward voltage drop and a big reverse recovery current, whereas basic SBDs have significant reverse leakage and low breakdown voltage as downsides [167].

Table 13. GaN rectifier diode specification by Avogy, now known as Nexgen Power Systems [166].

Model	Type	URRM/V	If/A	IR/μA	QC/nC
AVDO2A600A	SBD	600	2	150	4
AVDO5A120A	PN	1200	5	0.1	7
AVDO5A170A	PN	1700	5	0.1	14

4.2. GaN MOSFET

Five commercial GaN MOSFETs' characteristics are included in Table 14, along with their component numbers, maximum current ratings in Amps, input capacitances (C_{ISS}) in pF, gate-drain capacitances (C_{GD}) in pF, and drain-source on-state resistances ($R_{ds(on)}$) in Ohms. The high electron mobility and current density of GaN enable GaN MOSFETs to achieve extremely low on-resistances, such as 0.035 Ohms for a 40 A device, as shown in the table. However, compared to the equivalent Si MOSFETs, input and gate capacitances are larger, which can reduce the performance of high-frequency switching.

Table 14. Characteristics of commercial GaN MOSFETs [28].

Part Number	Current (A)	C_{ISS} (pF)	C_{GD} (pF)	R_{ds}(on) @25 °C (Ω)
TP65H150G4PS	2.5	307	1	0.15
TP65H150G4PS	5	307	1	0.15
TP65H150G4PS	10	307	1	0.15
TP65H070L	20	600	4	0.072
TP65H035WSQA	40	1500	14	0.035

For the vertical GaN MOSFETs, the cell pitch of our GaN MOSFETs is around 3–5 times larger than that of the SiC MOSFETs [168]. At the same time, lateral GaN MOSFETs benefit from consistent and broad conduction band migration in high-voltage power switching, making them less vulnerable to hot electron injection and a better replacement for SiC MOSFETs and GaN HEMTs. Although our GaN MOSFETs' cell pitch is too wide, the performance of the vertical GaN trench MOSFETs is in line with the top SiC MOSFET performance. However, compared to SiC devices, the doping concentrations of GaN FETs are relatively lower [169].

4.3. GaN Heterojunction Field-Effect Transistor (HEFT)

GaN heterostructure field-effect transistors (HFETs), a kind of wide bandgap semiconductor electronic components, are popular in high-frequency and high-power applications due to their advantages of having a high breakdown voltage and high electron mobility [134,135]. Vertical power devices based on GaN material are still at a very early research stage [136–148], and there are currently no commercial vertical power devices accessible due to the difficulty in producing low-cost GaN epitaxial wafers, which is required to construct vertical power devices.

The market adoption of normally off GaN-based power electronics solutions will also be influenced by the reduction of material costs and the enhancement of material quality, both of which impact the dependability of the devices. Accordingly, significant research efforts by the scientific community would be needed over the following years to achieve a thorough knowledge of the physics of GaN-based materials and devices [149]. Nitride-based electrical devices show great promise due to their high electron mobility and saturation velocity, high sheet carrier concentration at heterojunction interfaces, strong breakdown field, and low thermal impedance of GaN-based films produced over SiC or bulk aluminum nitride (AlN) substrates [150]. It is predicted that the specific on-state resistance (R_{on}) of FETs will be lower than that of Si or gallium arsenide. FETs with very low on-state resistance are very effective for low-loss power-switching devices like inverters.

4.4. GaN High Electron Mobility Transistor (HEMT)

GaN HEMTs are aggressively used in high-performance compact power supplies for fast chargers, data centers, light detection and ranging (LiDAR), and other applications because they exhibit exceptional performance as power-switching devices, including ultra-high switching frequencies and high conversion efficiency [16,170]. It is important to note that the on-state losses and switching losses can greatly decrease while maintaining the desired normally off feature in a cascade GaN HEMT built from a high-voltage D-mode GaN HEMT and a high-speed, low-voltage, and Si MOSFET [170]. It is rather relatively simple since controlling a cascade GaN HEMT is identical to driving a Si-MOSFET. On the other hand, due to E-mode GaN HEMT's high voltage and current slew rates, low threshold voltage, and low allowed gate voltages, operating an E-mode GaN HEMT requires considering several complex parameters. A negative driving voltage can be utilized to assure successful turn-off operations, although it may increase reverse conduction loss and require additional power supply units. A decent alternative is a voltage clamp [170].

Due to their rapid switching speed and low conduction losses, GaN HEMT are particularly appealing for high-frequency applications. Because there are no impurities (dopants) in the 2DEG area, electron mobility is high, allowing for low resistance and quick switching. The majority of GaN-HEMT transistors are lateral structures. In the heterojunction formed by GaN, the polarization electric field significantly modulates the distribution of energy bands and charges; GaN heterojunction field effect transistors dominate the GaN transistor. The device structure is also called HEMT [171].

4.4.1. Enhanced GaN HEMT

Due to the polarization characteristics of the conventional GaN HEMT, in the most used voltage-type power converter, power switches are required to be in a normally off state from the perspective of safety and energy saving, so a lot of research work is now focused on implementing enhanced GaN HEMT devices. When the enhanced GaN HEMT is in a normally off state, short circuit elimination techniques are widely used for protection [172–174]. At present, large international semiconductor companies, such as America's MicroGaN, Transphorm, EPC, Germany's Infineon, Japan's Panasonic, and Canada's GaN systems, have introduced GaN HEMT devices, the highest voltage reaching 1200 V.

4.4.2. High-Voltage Cascade GaN HEMT

As mentioned, GaN devices now play a role in high-voltage applications due to the introduction of high-voltage cascade GaN HEMTs. As illustrated in Figure 9, the cascade structure combines a high-voltage normally-on GaN HFET and a low-voltage normally-off type Si MOSFET into a new mixing tube, resulting in a normally off state [175]. It is a voltage-controlled device in which the gate is activated, 2DEG is produced, and the transistor is turned on when the negative voltage between the gate and the source is greater than the threshold voltage. The transistor switches off when the voltage between the gate and the source is less than the threshold voltage. Because the on-state and switching losses of Cascode GaN HEMTs are relatively low, and the diode has stronger reverse recovery properties than Si MOSFETs, the power system's efficiency can be greatly improved [175].

Table 15 lists the parameters for GaN HEMTs, GaN MISFETs, and GaN Schottky diodes. Bandgap, electron mobility, voltage/current ratings, switching frequency, and power losses are just a few comparisons made. High critical electric field, high current density, and high frequency switching are some of GaN devices' key benefits over Si, while thermal conductivity and lattice mismatch still present problems. GaN HEMTs and diodes enable high-frequency performance, and MISFETs, which aim to replace Si MOSFETs and IGBTs, offer greater dependability.

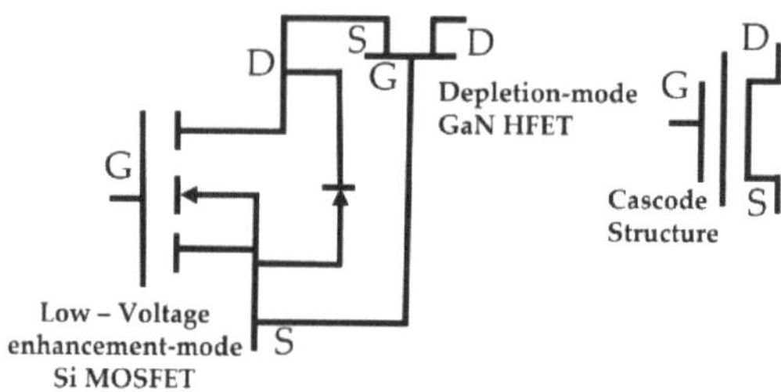

Figure 9. Device structure of cascade GaN HEMT.

Table 15. Specification of the different GaN-based semiconductor devices.

Parameter	Unit	GaN HEMT	GaN MISFET	GaN SBD
Bandgap	eV	3.4	3.4	3.4
Critical electric field	Mv/cm	3.3	3.3	3.3
Electron mobility	cm^2/V-s	2000	1500	2000
Saturated electron drift velocity	cm/s	1×10^5	1×10^5	1×10^5
Thermal conductivity	W/cm-K	1.3	1.3	1.3
Lattice mismatch	%	15	15	15
Wafer size	mm	150	150	150
Voltage rating	kV	1.2	1	1.2
Current rating	A	30	20	30
Switching frequency	Hz	1000	500	-
Resistivity	Ω-cm	106	106	106
Channel resistance	Ω-cm^2	8	10	-
Stress levels of voltage	V	600	500	600
Stress levels of current	A	30	20	30
Off-state breakdown voltage	kV	1.2	1	1.2
Maximum junction temperature	°C	250	250	250
Temperature range	°C	−55 to 250	−55 to 250	−55 to 250
Temperature stability	°C	-	-	-
Conduction losses		Medium	Medium	Low
Switching losses		Low	Medium	-
Power losses		Medium	Medium	Low
Baliga's figure of merit		26	26	26
Johnson's figure of merit		46	23	46
Applications		High frequency	Medium frequency	Rectifier

4.5. Applications of GaN

Long-employed in the manufacture of RF and LED components, GaN is now becoming more widely accepted in various power-switching and conversion applications. GaN-based ICs may meet this requirement by delivering reliable operation at greater temperatures, saving space, and enhancing system performance and efficiency. The innovative semiconductor material GaN has received much attention and is used extensively in many different sectors. GaN outperforms conventional Si-based devices in terms of power efficiency, speed, and durability due to its special characteristics, including high electron mobility, broad bandgap, and high breakdown voltage.

The power GaN development market is predicted to increase from $126 million in 2021 to $2 billion in 2027, at a compound annual growth rate (CAGR) of 59%, according to market research firm Yole Développement [176], which is shown in Figure 10. The consumer sector, which includes fast chargers, Class-D audio, power banks, and time-of-flight sensors in smartphones and tablets, will contribute significantly to this expansion. GaN offers greater power density, improved thermal efficiency, and more compact and lightweight solutions in various applications. The convergence is driving growth in the telecom/datacom and automotive/mobility sectors to a 48 V power supply in both high-power density computers and vehicle DC–DC converters. Data centers need less power when using 48 V systems, and GaN performs better under these conditions than Si does. Manufacturers of solar microinverters, optimizers, and energy storage systems are increasingly developing using GaN for improved efficiency, higher power density, and increased dependability as the adoption of renewable energy sources quickens. A number of firms, including BRC Solar and Solarnative, have introduced GaN-based solutions.

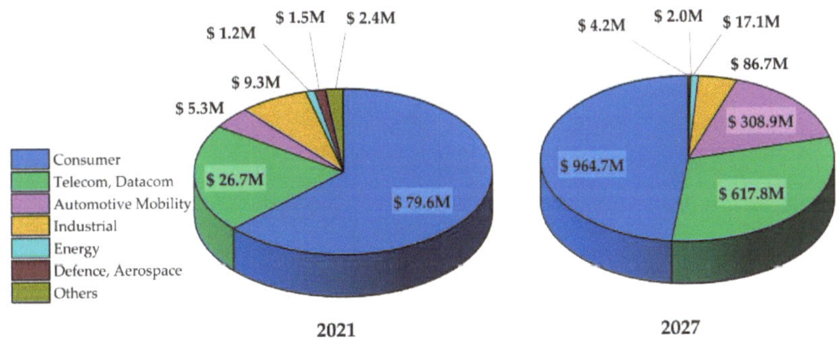

Figure 10. 2021–2027 power GaN device market projected revenue [176].

There is a critical need for energy-efficient solutions due to the expansion of the cloud and rising demand for data centers. Using GaN technology, which enables a single-stage power conversion process and eliminates the intermediary step of converting power from 48 V to 12 V, is one option to deal with the problem. Significant energy savings are provided by this direct conversion from 48 V to the necessary 1 V at the point of load. GaN technology is also used in autonomous cars, especially in LIDAR systems. GaN technology accelerates the transmission of laser beams, improving lidar's resolution and mapping capabilities and paving the way for autonomous vehicles and augmented reality applications. Devices made of GaN are widely used in a variety of industries. GaN is useful for ion thrusters, solar power conversion, robotics, and lidar in space due to its built-in radiation tolerance. GaN is also transforming motor drives, making it possible for eMobility, personal robots, and drones to have smaller, lighter, and more effective systems [176]. GaN-based power solutions improve the efficiency and dependability of solar micro-inverters and energy storage devices in renewable energy. GaN's wireless power sources aid medical technology by enhancing implanted device charging and allowing portable imaging equipment for procedures like colonoscopies and MRI scans. The development of wireless power, which enables wireless device charging and is revolutionizing our way of life for everything from smartphones to home appliances, is also being fueled by GaN.

GaN RF components are used in phones and laptops to send and receive GSM and WiFi signals, and GaN is increasingly being used in the chargers and adapters that power these devices. The mobile fast-charging industry is the biggest market for power GaN at the moment. GaN power ICs can enable three times quicker charging in adapters that are half as big and heavy as sluggish, Si-based solutions. Additionally, the retail launch price of GaN for single-output chargers is around half that of earlier best-in-class Si chargers and up to three times cheaper for multi-output chargers.

Servers in data centers are also using GaN power semiconductors. Si's capacity to handle electricity effectively and efficiently encounters 'physical material' barriers as data center traffic increases. Consequently, high-speed GaN integrated circuits (ICs) replace the outdated, sluggish Si chip. Major gains in efficiency are made possible by the consolidation of data center technology, a novel HVDC design strategy, and the well-proven dependability of highly integrated, mass-produced GaN power ICs. Global Si-to-GaN data center upgrades are predicted to cut energy loss by 30–40%, resulting in over 100 TWhr and 125 Mtons of CO_2 emissions saved by 2030, shown in Figure 11 [138]. Therefore, using GaN marks another step toward the data center industry's carbon net-zero aspirations.

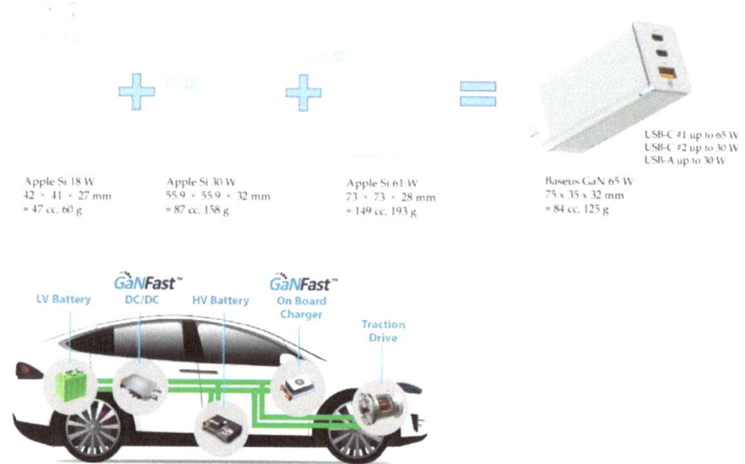

Figure 11. GaN technology in various power sectors [138].

GaN is becoming the preferred technology in the automotive sector for power conversion and battery charging in hybrid and electric cars. Inverters used in solar power installations, power conversion plans for motor drives, and other industrial applications increasingly use GaN-based power products. There has long been speculation that GaN will eventually replace Si power transistors. The application areas of GaN are electric vehicles, high-speed railways, household appliances, industrial motors, aerospace, smart grid, solar energy generation, wind power generation, and large capacity [141]. It is helpful for various RF applications, including satellite communications and 5G, 6G, and mobile communications, since it can operate at high frequencies, control high levels of power, and sustain a high operating voltage [177].

Moreover, GaN faced increased competition as a result of recent developments in Si SJ technology and the introduction of SiC power MOSFETs. In light of this, a reappraisal is necessary, especially in the 600 V domains involving all three [16]. The zero reverse recovery charge of lateral GaN HEMTs is one of its distinguishing characteristics. When the drain voltage drops below the total of the gate potential and the threshold voltage, a reverse channel is formed since there are no p-n junctions and current flows in a polarization-induced 2DEG [16]. Because of this property, GaN HEMTs are the best option for applications requiring continuous switching on a reverse-biased device, such as in half-bridge or full-bridge topologies. The suitability of GaN HEMTs for dual-gate structures to achieve bidirectional blocking and conducting devices [178] or to integrate multiple power devices on a common die, which makes it easier to achieve higher integration levels than with discrete components, is another intriguing feature of these devices [179].

4.6. Challenges of GaN Power Device

There are still various difficulties to overcome in order to replace Si technology and become mainstream. By raising the switching frequencies, GaN transistors can produce power-switching systems that are extremely compact and highly efficient [180]. GaN devices are now making excellent strides and making their way into the market, but several issues must be addressed. Vertical GaN power devices with breakdown voltages greater than 5 kV are possible as a result of advancements in substrate technology and field engineering optimization. It is also necessary to examine how the breakdown mechanism in these devices came to be [181]. Due to its potential to revolutionize power electronics with increased efficiency, better power densities, and quicker switching rates compared to conventional Si devices, GaN power devices have attracted a lot of interest in recent years. However, the development of GaN power devices faces the following major obstacles.

4.6.1. Material Growth

GaN-based power devices are built on high-quality epitaxial material. SiC and sapphire have a smaller lattice misfit and greater thermal conductivity than Si, critical benefits for high-power devices [181]. Although the lattice mismatch between Si and GaN is substantial, its cost is modest, and the lattice mismatch can be mitigated by introducing a buffer layer for stress management, which restricts its applicability. The relationship between the thermal expansion coefficient of GaN and common substrates is shown in Figure 12 as a function of the lattice constant.

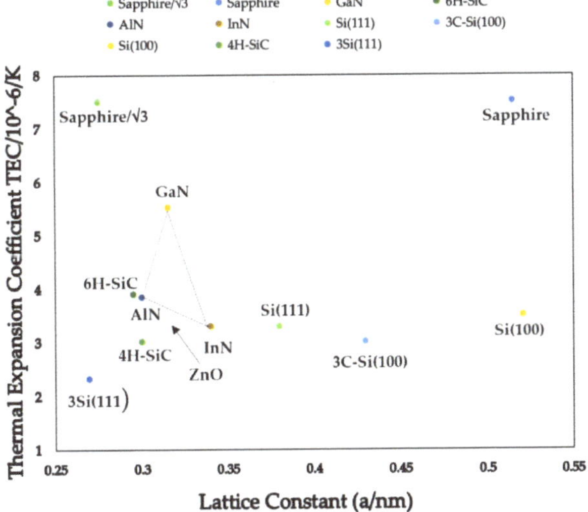

Figure 12. The thermal expansion coefficient of GaN and common substrates as a function of lattice constant.

4.6.2. Suppression of Current Collapse Effect

The current collapse effect of AlGaN/GaN HEMT devices poses a severe threat to GaN power device success. It is also one of the biggest issues with today's GaN power devices. As illustrated in Figure 13, when a significant bias is given to the drain, the leakage current degrades [182,183]. At present, the mechanism of the current collapse effect of GaN devices largely includes the following types:

1. The current collapse is triggered by carrier traps caused by deep-level centers in the material.
2. The 2DEG concentration in the AlGaN/GaN conductive channel decreases due to the change in polarization charge generated by the surface state and the surface effect, resulting in current collapse.
3. Because the material structure and the energy band structure boundary are so essential, even a minor disruption will cause the 2DEG to collapse [184,185].

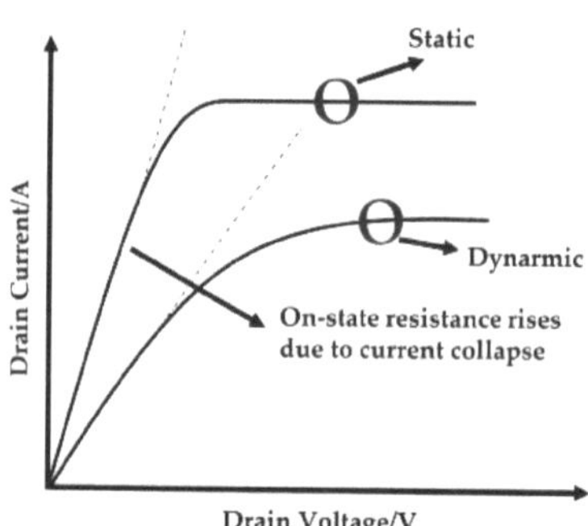

Figure 13. Current collapse effect.

4.6.3. Packaging

GaN power devices have garnered considerable attention in recent years due to their exceptional performance compared to traditional Si-based devices. These devices offer superior power density, rapid switching speeds, reduced on-resistance, and enhanced thermal conductivity. Appropriate packaging solutions are essential for handling their distinctive properties and optimize their performance to ensure their efficient and dependable operation. Packaging plays a vital role in safeguarding GaN power devices against external factors like moisture, temperature fluctuations, mechanical strain, and electrical interference. It also facilitates efficient heat dissipation and establishes electrical connections with the device. A common approach for packaging GaN power devices involves utilizing surface-mount technology (SMT) packages [186]. These packages typically employ ceramic or plastic materials with a metal lead frame or solder balls for electrical connections. The package design comprises a die attach area where the GaN power device chip is mounted, bond wires or flip-chip connections to establish electrical connections, and thermal vias or heat sinks to enhance heat dissipation.

The package encompasses a ceramic or plastic body with metal leads or solder balls for electrical connections. The GaN device chip is affixed to the die attach pad at the package's center. Electrical connections between the chip and the leads or solder balls are established using bond wires or flip-chip connections. The package may also incorporate heat sink structure to augment heat dissipation. Package designs can vary depending on the specific application and power requirements. For high-power applications, packages with larger thermal vias, enhanced heat sink structures, and improved thermal interface materials may be employed to manage increased power dissipation effectively.

Furthermore, the ongoing exploration of packaging technologies for GaN power devices aims to further enhance their performance. This involves the development of advanced materials with superior thermal conductivity, innovative interconnection like copper clip bonding or direct copper bonding, and novel package designs to minimize parasitic inductance and capacitance. Moreover, the packaging of GaN power devices is crucial for their development, ensuring reliability, efficient thermal management, and optimal electrical performance. Continued innovation in packaging techniques aim to overcome unique challenges and unlock the full potential of GaN power devices in various applications, including power electronics, automotive systems, and renewable energy.

4.6.4. Gate Driver

The gate drive function assumes a critical role in the operation and performance of GaN power devices, exerting control over device switching by supplying the appropriate voltage and current signals to the gate terminal of the GaN transistor. It is imperative to meticulously design the gate drive circuitry to ensure the efficient and dependable operation of GaN power devices. Considerations related to the gate drive for GaN power devices are outlined as follows:

1. Voltage and Current Levels: GaN power devices typically necessitate higher gate voltage levels in comparison to conventional Si-based devices. Operating with gate voltages ranging from 6 V to 10 V or even higher, the gate drive circuitry must be capable of generating and sustaining these elevated voltage levels. GaN power devices exhibit low gate capacitance, enabling faster switching but demanding careful attention to the gate driver's current capability.

2. Gate Driver ICs: GaN power devices often employ specialized gate driver integrated circuits (ICs). These ICs are specifically designed to deliver the required voltage and current levels, incorporate protection features, and enhance the overall performance of GaN transistors. Protection features may include under-voltage lockout (UVLO), over-current protection, and short-circuit protection to ensure device safety during operation.

3. High-Speed Switching: GaN power devices are renowned for their rapid switching speeds, which can present challenges in gate drive design. To achieve optimal performance, the gate drive circuitry must be capable of providing high-speed rise and fall times to minimize switching losses. This necessitates meticulous consideration of the gate driver's output impedance, gate trace layout, and the impact of parasitic elements in the gate circuit.

4. Gate Resistance: The appropriate selection of gate resistance is crucial for GaN power devices. A suitable gate resistor aids in dampening ringing effects and mitigating the risk of oscillations in the gate voltage waveform. It also limits current during switching transitions to prevent excessive power dissipation. The value of the gate resistor should be carefully optimized based on the specific GaN device and application requirements.

5. Gate Layout and Layout Considerations: A well-designed gate layout is essential in order to minimize parasitic inductances and capacitances that can negatively impact device performance. This entails keeping gate traces as short as possible, reducing loop areas, and employing techniques like guard rings and vias to manage parasitic effects. An optimized gate layout contributes to faster switching speeds, lower losses, and overall improvement in device performance.

Table 16 compares features, including split outputs, bootstrap voltage control, and target applications for five common GaN driver ICs from top manufacturers that drive GaN FETs in half-bridge designs. With configurations ranging from general purpose to automotive-qualified, the drivers offer crucial gate drive isolation and enhanced GaN switching performance as GaN transistors become more prevalent in power supply and upcoming electric vehicle applications [187,188]. The advantages of GaN FETs' high-frequency capability and efficiency can be maximized with the right choice of GaN driver.

Table 16. Most popular GaN driver solutions [31].

Manufacturer	Model	Split Outputs	Bootstrap Voltage Management	Configuration	Features
pSemi	PE29101	Yes	Yes	Half-bridge	Frequency < 33 MHz
	PE29102	Yes	No	Half-bridge	Frequency < 33 MHz

Table 16. *Cont.*

Manufacturer	Model	Split Outputs	Bootstrap Voltage Management	Configuration	Features
Texas Instruments	LMG1205	Yes	Yes	Half-bridge	Automotive-qualified
	LM5113-Q1	Yes	Yes	Half-bridge	General purpose
μi	uP1966A	Yes	Yes	Half-bridge	General purpose

4.6.5. Electrical Insulation

In order to maintain adequate electrical isolation between various components, avoid electrical leakage, and improve the device's overall dependability and safety, electrical insulation is a crucial component in the development of GaN power devices. Figure 14 illustrates challenges associated with electrical isolation for GaN devices. Barriers that withstand high voltages and prevent accidental electrical hookups are made using insulation materials and procedures. As it guarantees dependable operation and provides protection from electrical failure, electrical insulation is essential for the development of GaN power devices. Dielectric materials having sufficient dielectric strength, thermal stability, and compatibility are used to create insulation between conductive components. Insulation layers, edge isolation techniques, and the proper packaging insulation are utilized to prevent short circuits, lessen parasitic effects, and maintain electrical integrity. By giving effective electrical insulation first importance, GaN power devices may increase performance and operational reliability in many applications.

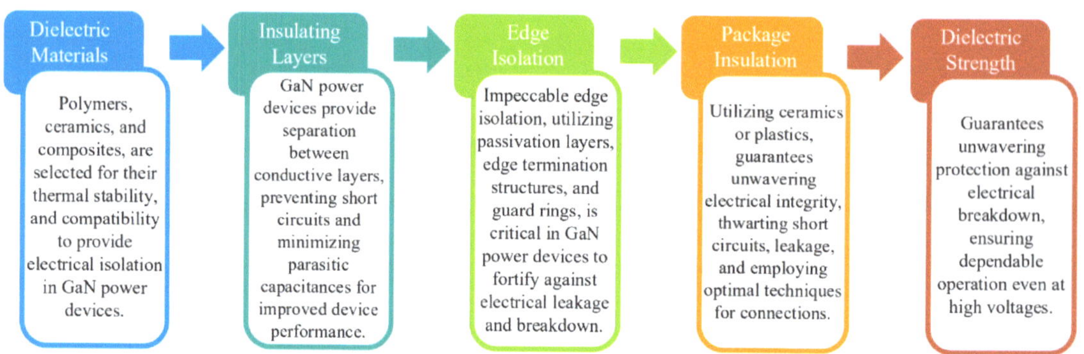

Figure 14. Challenges associated with electrical isolation for GaN devices.

4.6.6. Electromagnetic Interference

GaN power device development is heavily concerned with electromagnetic interference (EMI), as these devices can produce high-frequency switching signals that could travel as undesired electromagnetic emissions and interact with other electronic systems. To ensure the proper operation of GaN power devices and avoid interference with nearby devices or delicate electronic equipment, EMI management is essential.

1. EMI Sources: Fast switching transitions caused by GaN power devices result in high-frequency harmonics and transient currents. These may produce electromagnetic emissions that spread through the design of the device, the circuit traces, and the connections to the outside world. During switching events, voltage peaks, ringing, and current loops are the main EMI causes in GaN power devices.
2. EMI Mitigation Techniques: Several techniques are employed to mitigate EMI in GaN power devices. These include:

- Filtering: The use of passive components such as capacitors, inductors, and ferrite beads to suppress high-frequency noise and attenuate unwanted harmonics.
- Shielding: Incorporating shields, conductive enclosures, or grounded metal layers around sensitive components to contain electromagnetic fields and prevent their propagation.
- Layout Optimization: Carefully consider trace routing, component placement, and grounding techniques to minimize loop areas, reduce parasitic inductance, and control impedance.
- Grounding and Bonding: Establishing proper grounding and bonding practices to minimize ground loops, reduce voltage differentials, and provide an effective return path for high-frequency currents.

3. Compliance with EMI Standards: GaN power devices must adhere to particular electromagnetic compatibility (EMC) requirements to ensure their functioning is below acceptable EMI limits. Evaluations of radiated emissions conducted emissions, and vulnerability to outside electromagnetic fields are all part of compliance testing.
4. EMI Filtering Components: Choosing the right EMI filtering components is essential when creating GaN power devices. These parts must be tuned for the device's working frequency range, have high-frequency filtering capabilities and low parasitic elements, and be parasite-free.
5. EMI Simulation and Modeling: EMI behavior in GaN power devices may be predicted and examined using simulation and modeling techniques throughout the design process. Using these technologies, engineers may reduce EMI problems by optimizing circuit design, filtering tactics, and grounding procedures.

4.7. Future Trend of GaN

Due to the relatively recent industrial introduction of GaN, future developments are a crucial topic of discussion when examining the potential uses of this technology in various applications [189]. The high cost of these devices, the current GaN devices' limited voltage rating, the complex gate driver design and control complexity, the area-specific thermal resistance in GaN-based IC development, and packaging issues to ensure these devices' long-term reliability present the biggest challenges and areas for improvement. Below, each of these elements is explored to determine where GaN's future lies [190]. However, GaN will ultimately replace old Si in data centers, home solar energy systems, and other consumer applications like fast chargers and other consumer applications. GaN technology will be used more frequently in electric vehicles' onboard chargers [191].

4.7.1. Cost Reduction

GaN devices can be manufactured on Si substrates, a standard industry procedure today, to address the cost issue connected with GaN. GaN's attractive material features and the development of fabrication facilities that are compatible with complementary metal-oxide semiconductors (CMOS) work together to provide power devices that perform better and are more affordable. Increased demand for power converter applications may result from the development of high-power ICs using GaN on Si wafers, which can further lower these costs [192].

As semiconductor technology has advanced since the commercialization of GaN technology, the unit cost of these transistors has dropped considerably. For instance, manufacturers like GaN Systems now sell the 650 V, 15 A e-mode GaN for around $12, whereas, formerly, a standalone GaN MOSFET had a unit price of about $75. This demonstrates that, as more power electronics applications adopt GaN technology, the cost is anticipated to drop further during the ensuing years due to economies of scale. GaN HEMTs can be widely used in power electronics due to their low cost, which greatly improves the performance and efficiency of electrified transportation mediums, lowering the entire system cost. Therefore, GaN HEMTs are more advantageous than they are expensive, effectively resulting in cost savings for production and operation [190].

4.7.2. Thermal Management

The vertical GaN devices feature higher breakdown voltage and current than the lateral GaN designs without growing the chip's size. Vertical structures are more dependable. In addition, vertical GaN devices have easier thermal management than lateral ones [190]. Vertical GaN structures have many advantages, and a lot of research and development is being carried out to make these structures commercially viable. When working with GaN devices, thermal management is a crucial concern. In the development of GaN-based ICs, minimizing area-specific thermal resistance is crucial. Future GaN device technologies are anticipated to use diamond and SiC substrates with strong heat conductivity [191]. The terminals of lateral GaN on Si substrate devices are on the same side of the die.

As a result, these chips have bumps added to them for mounting purposes. These bumps only cover a small percentage of the die surface area and have a low thermal conductivity. Either through these bumps, known as topside cooling, or through their Si substrate material, known as backside cooling, the GaN transistors are cooled. The top side of the die has the greatest thermal performance for thermal dissipation [191]. Future designs of highly competitive power electronic converters might include a phase-leg power module based on GaN devices that incorporates the power stage, the gate driver control circuitry, and the cooling system into a single container based on the present integrated modules [191]. The combined module will have enhanced thermal management and high-power capacity. These modules might address some of the problems that electric transportation is now experiencing, as previously mentioned [189].

4.7.3. Gate Driver Design

The first GaN power IC Process Design Kit (PDK), which is known as All GaNTM, enables the monolithic integration of 650 V GaN IC circuits with GaN MOSFETs. The integration of the GaN driver with the GaN MOSFET is crucial for high-frequency operation. Due to its fast-switching transitions, the discrete GaN's gate is susceptible to noise and voltage spikes, which can lead to damage. Noise can be reduced to some extent by integrating the GaN MOSFET with the driver. However, the GaN MOSFET's inclusion in a multi-chip module is not without its difficulties, as larger losses are caused by the impedance between the GaN MOSFET's gate and the Si driver output. The best possible efficiency, speed, and robustness can be attained with monolithic integration [189]. Since the driver is integrated, one module can contain the logic circuitry, startup protection, dv/dt control, and dv/dt robustness. Two MOSFETs are combined with the driving and protection circuitry in half-bridge power ICs. The level-shifter losses are 10 times lower with the 650 V GaN power IC than Si [190]. The development of more efficient, more powerful, and less expensive power systems will be made possible by next-generation monolithic integration, which includes enhanced I/O features, over-current, and over-temperature protection [190].

In the near future, it is anticipated that gate drivers with inbuilt short circuit protection will be developed. This is particularly helpful in applications that adopt a modular approach, in which case the size of the parallelized packages is crucial. For instance, aircraft DC/DC converters are frequently built using the modular technique. The size of the power electronic components designed for industry continues to shrink as a result of advances in semiconductor technology. More power-dense DC/DC converter designs are possible with integrated gate drivers. GaN HEMT gate drive design calls for meticulous design considerations. For turn-on and turn-off, a separate gate resistor is usually advised. This is significant because the RGON has control over the dv/dt slew rate. The turn-on gate resistor's value must be carefully chosen because, if it is too high, switching will be slowed down and result in larger switching losses, and, if it is too low, gate oscillation will cause more switching losses. It is suggested that the turn-off gate resistor, RGOFF, for GaN System's GS66508 be between 10 and 20ω. RGOFF enables quick pull-down for a powerful gate drive and starts from 1 to 2ω. Figure 15 depicts a general layout of the gate driver schematic [193].

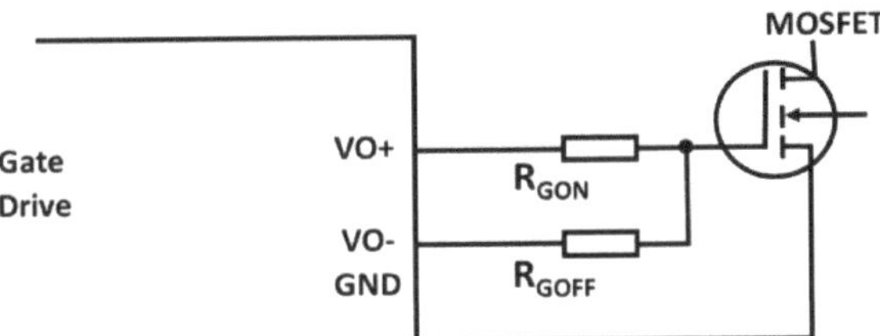

Figure 15. GaN E-HEMT gate driver.

4.7.4. Motor Drive

Because of its exceptional features, GaN has been suggested as a legitimate substitute for conventional Si-based MOSFETs and IGBTs in the motor control field. GaN technology offers effective, light, and low-footprint solutions with up to 1000 times the switching frequency of Si and lower conduction and switching losses. GaN power transistors have a switching speed that can reach 100 V/ns, which allows engineers to employ inductors and capacitors with lower values and smaller sizes [194]. Low R_{DS}(on) improves energy efficiency and enables more compact dimensions by reducing the amount of heat produced. GaN-based devices require capacitors with greater working voltages, higher dV/dt transient tolerance, and lower equivalent series resistance [194] than Si-based devices do. GaN has a high breakdown voltage (50–100 V, as opposed to the typical 5 to 15 V values attainable with conventional semiconductors), enabling power devices to run at greater input energies and voltages without harm. This is an additional benefit that GaN offers. It may reach a wider bandwidth with higher switching frequencies, which enables the implementation of motor control algorithms with greater precision. Additionally, it is possible to attain a degree of efficiency with variable frequency drive (VFD) motor control that is not possible with ordinary Si MOSFETs and IGBTs. Subsequently, because the motor speed can be ramped up and down, the load may be maintained at the desired speed, and the VFD achieves an incredibly accurate speed control. Because GaN transistors switch significantly more quickly than their Si counterparts, parasitic inductances and losses are reduced, switching performance is improved (less than 2-ns rise and fall time), and the heat sink can be reduced in size or even eliminated. Very low switching losses in the GaN power stage enable greater pulse width modulation (PWM) switching frequencies with a peak efficiency of up to 98.5% at 100 kHz PWM [194].

4.7.5. 5G

GaN can amplify high-frequency signals (even those of the order of a few gigahertzes) very effectively, which opens up several real and exciting opportunities in the RF industry. Thus, it is feasible to develop high-frequency amplifiers and transmitters that can travel great distances and have uses in base stations, satellite communications, radar, early warning systems, and other technologies [194]. In terms of increased capacity and efficiency, reduced latency, and all-around connectivity, 5G offers substantial advantages as the next-generation mobile technology. It is necessary to employ materials like GaN that can give high bandwidth, high power density, and excellent efficiency values for the use of various frequency bands, including the sub-6-GHz band and the millimeter-wave (mmWave) (above-24-GHz) band [194].

Due to its physical characteristics and crystalline structure, GaN can allow higher switching frequencies than comparable laterally diffused MOSFET devices at the same applied voltage, resulting in a substantially smaller footprint. RF front-end chipsets are necessary for emerging 5G technologies like massive multiple-input multiple-output (MIMO), and mmWave. The greatest option for high-power 5G and RF applications is GaN-on-SiC, which combines the high power density of GaN with the excellent thermal conductivity and low RF losses of SiC. There are currently a number of GaN-based products appropriate for 5G applications on the market, including multiple channel switches and low-noise amplifiers for massive 5G MIMO applications [194].

4.7.6. Data Centers

In the data center industry, where high performance and cost reduction are fundamental concerns, the marriage of GaN with Si also presents significant prospects. Voltage converters are frequently used in data centers, where cloud servers run continuously, with typical values of 48 V, 12 V, and even lower voltages for powering the multiprocessor system cores [194]. Power-conversion efficiency has emerged as a crucial consideration for businesses looking to achieve net-zero carbon emission, especially those running data centers and providing cloud computing services. This is due to the constantly rising worldwide electricity generation. GaN technology may significantly address the need for more power in less space for data centers by attaining improved efficiency in converters and power supply, size reduction, and better thermal control, which lowers costs for the provider [194]. AC/DC converters are quite prevalent in data centers, which first regulate the bus voltage to a DC value using a power factor correction (PFC) front-end stage. Next, a DC/DC stage steps down the bus voltage and produces a galvanically isolated and controlled DC output (48 V, 12 V, and more). The PFC stage maximizes real power [194] by keeping the power supply's input current synced with the main voltage.

5. Ultrawide Bandgap Semiconductor

AlGaN/AlN, diamond, and Ga_2O_3 are examples of ultrawide bandgap (UWBG) semiconductors, which have bandgaps that are substantially wider than those of traditional wide bandgap materials like SiC and GaN [195]. Due to their extraordinarily wide bandgaps, which allow for properties like high breakdown voltages, high operating temperatures, and high-power densities, these materials have the potential to lead to significant advancements in electrical and optoelectronic devices. In contrast to more established semiconductors like gallium arsenide and Si, UWBG materials are still in a very early stage. In areas like substrate availability, doping, comprehending carrier transport mechanics, and constructing useful devices, considerable obstacles need to be addressed.

As the bandgap increases, the figures of merit for devices like power switches scale positively, indicating that UWBG materials could permit improved performance. However, UWBG materials are still in their infancy and face formidable obstacles in areas including material production, doping, substrate accessibility, and fundamental physics comprehension. This article examines the current state-of-the-art and points out important areas for future research in material growth, physics, devices, and applications.

Creating large-diameter native substrates, comprehending growth dynamics, and attaining controlled doping, particularly p-type doping, are important issues in the field of materials. The creation of innovative UWBG materials such as BN alloys offers opportunities. To describe high-field carrier dynamics in UWBG materials, new transport physics models are required. It is also difficult to confine carriers using UWBG heterostructures. Opportunities exist in both TCAD modeling and thermal transfer. The need for advancements in vertical device topologies, contacts, packaging, and other areas makes applications like high-voltage power electronics, RF and microwave devices, deep UV optoelectronics, and severe environment electronics intriguing.

An overview of the advancement of semiconductors based on diamond from the 1980s to the present is shown in Table 17. It follows the development of early research devices, such as Schottky diodes built of single-crystal diamond, through more sophisticated polycrystalline diamond devices, such as vertical MOSFETs and lateral MOSFETs. The table lists each device's significant accomplishments, technical details, and features.

Diamond bipolar transistors, diamond Schottky diodes, and diamond MISFETs are three different categories of diamond-based semiconductor devices that are compared in Table 18. The advantages of diamond's high bandgap, high critical electric field, high thermal conductivity, and high breakdown voltage are shared by all three devices. Moreover,

Table 18 highlights diamond devices' high-voltage, high-power, high-temperature, and high-frequency working capabilities.

Table 17. Evaluation of diamond-based semiconductor device.

Year	Device	Specifications	Milestone	Features
1980s	Schottky diode	Single-crystal research only	First diamond electronics	Extremely high bandgap
1990s	Schottky diode	Single-crystal, <100 V	Small single-crystal diodes	High-temperature operation
2000s	Schottky diode	Polycrystalline, 400 V	First polycrystalline devices	Manufacturable on poly diamond films
2010s	Vertical JFET	Polycrystalline, 50 V	First diamond vertical transistors	Blocking voltage and operating temperature increase
2018	Lateral MOSFET	Polycrystalline, 200 V	Diamond lateral MOSFET	High current density demonstrated
2020s	Vertical MOSFET	Polycrystalline, >1 kV	Diamond vertical power MOSFETs	Targeting commercial viability
Future	Bipolar transistor	Polycrystalline, >1 kV	High-voltage diamond bipolar transistors	Complementing MOSFETs
	Thyristor	>10 kV blocking voltage	Ultra-high voltage rectifiers and switches	Surpassing limitations of existing technologies
	IGBT	>10 kV, high current density	Diamond IGBTs	Theoretical capabilities unmatched by any material

Table 18. Specification of diamond-based semiconductor device.

Parameter	Unit	Diamond Bipolar Transistor	Diamond Schottky Diode	Diamond MISFET
Bandgap	eV	5.45	5.45	5.45
Critical electric field	Mv/cm	10	10	10
Electron mobility	cm^2/V-s	1800	2200	1800
Saturated electron drift velocity	cm/s	2.7×10^7	2.7×10^7	2.7×10^7
Thermal conductivity	W/cm-K	22	22	22
Voltage rating	kV	>10	>10	>10
Current rating	A	5	10	5
Switching frequency	Hz	>1000	-	>1000
Resistivity	Ω-cm	>1011	>1011	>1011
Channel resistance	Ω-cm^2	<1	-	<1
Stress levels of voltage	V	>10,000	>10,000	>10,000
Stress levels of current	A	5	10	5
Off-state breakdown voltage	kV	>10	>10	>10
Maximum junction temperature	°C	>600	>600	>600
Temperature range	°C	−55 to 250	−55 to 250	−55 to 250
Conduction losses		Very Low	Very Low	Very low
Switching losses		Low	-	Low
Power losses		Very Low	Very Low	Very low

Table 18. *Cont.*

Parameter	Unit	Diamond Bipolar Transistor	Diamond Schottky Diode	Diamond MISFET
Baliga's figure of merit		1650	1650	1650
Johnson's figure of merit		288	288	288
Applications		High power, high frequency	Rectifier	High power, high frequency

6. Current Innovation and Comparison

WBG and future UWBG semiconductors will be used to replace conventional Si power devices in the new generation of power devices for power converters. The commercial Si IGBT dominant power switch's current maximum breakdown voltage capacity is 6.5 kV. In any case, a Si-based gadget could not function above 200 °C. These inescapable physical restrictions severely reduce the efficiency of modern power converters, necessitating, among other things, complicated and expensive cooling systems [196]. These novel power semiconductor materials can make electric energy transformations more efficient, leading to a more judicious use of electricity and a smaller carbon impact.

SiC and GaN are the most desirable candidates among the WBG semiconductors because they already offer a good compromise between their theoretical properties (blocking voltage capability, operation temperature, and switching frequency) and commercial presence [15]. Both are ideal candidates to replace Si in the following wave of high-power and high-frequency electronics due to their wide bandgaps, resulting in higher breakdown voltage and operation temperature than Si [197].

Electrical quantities like voltage, frequency, and operating temperature define the application of the power system in power electronics. The physical characteristics of Si, SiC, GaN, and diamond materials are listed in Table 19 [198–200]. Furthermore, Table 19 compares key physical properties and device parameters of Si, SiC, GaN, and diamond semiconductors. It shows that wide bandgap semiconductors like SiC, GaN, and especially diamond have superior attributes like higher breakdown fields, thermal conductivity, and switching frequencies compared to Si, enabling high-power and high-frequency applications. Diamond has the greatest combination of properties overall such as a very high breakdown field and thermal conductivity.

Table 19. Comparisons of the specifications of Si, SiC, GaN, and diamond.

Parameter	Unit	Si	SiC	GaN	Diamond
Bandgap	eV	1.1	3.0–3.4	3.4–3.6	5.45
Critical electric field	Mv/cm	0.3	3	3.3	10
Electron mobility	cm^2/V-s	1500 to 2000	100 to 600	1000 to 2000	2200
Saturated electron drift velocity	cm/s	10^5	2×10^6	10^5	2.7×10^7
Thermal conductivity	W/cm-K	1.5 to 2.0	3.0 to 4.9	1.0 to 1.5	22
Lattice mismatch	%	-	3.5	15	-
Wafer size	mm	300	150	150	-
Voltage rating	kV	<1	10	1.2	>10
Current rating	A	100	20	30	10
Switching frequency	Hz	20 k	400 k	1 M	>1 M
Power density (size-wise)	W/cm^2	~5	~10	~30	>100
Power density (weight-wise)	W/g	~5	~10	~20	>100

Table 19. *Cont.*

Parameter	Unit	Si	SiC	GaN	Diamond
Resistivity	Ω-cm	10	104	106	>1011
Channel resistance	Ω-cm^2	30	5	8	<1
Stress levels of voltage	V	600	10,000	600	>10,000
Stress levels of current	A	100	100	30	10
Off-state breakdown voltage	kV	0.6	10	1.2	>10
Maximum junction temperature	°C	150	600	250	>600
Temperature range	°C	−55 to 200	−55 to 300	−55 to 250	−55 to 250
Temperature stability	°C	-	1	-	-
Conduction losses		High	Low	Medium	Very Low
Switching losses		Low	High	Medium	Low
Power losses		High	Medium	Medium	Very Low
Baliga's figure of merit		1	408	26	1650
Johnson's figure of merit		1	62	46	288
Applications		Low voltage low power	High voltage, high power	High frequency	High power, high frequency

The bandgaps of WBG semiconductors are roughly three times, and the electric field values of SiC polytypic and GaN are comparable and much greater than those of Si. Higher radiation hardening and high-temperature operation are advantages for WBG semiconductors. For Si, this occurs at a temperature of about 150 °C [201], where around 900 °C is the inherent temperature of 4H-SiC, and the temperature range between 300 °C and 800 °C has been used to test manufactured AlGaN/GaN HEMTs [202]. They can be utilized in harsh environments where Si-based devices are ineffective [203].

In comparison to Si, SiC, and GaN, diamond has an extraordinarily wide bandgap of 5.45 eV and the highest electrical breakdown field, thermal conductivity, electron velocity, and current density. This makes diamond an exceptional material for creating semiconductor devices with an extremely high power density, high frequency, and high temperature. To construct useful diamond electronic devices, however, significant obstacles still exist in creating large, high-quality diamond substrates and accomplishing doping and metallization. With certain manufacturing challenges, SiC and GaN also have very wide bandgaps when compared to Si, enabling high-voltage, high-power, and high-speed devices. Overall, these ultra-wide bandgap semiconductors promise to significantly outperform Si devices in terms of performance.

In the context of electronics, certain constraints that apply to optoelectronics are relaxed. For instance, the bandgaps of these materials do not need to be direct, making SiC a viable option. Additionally, light emission efficiency is not as critical for electronics, making GaN and aluminum GaN (AlGaN) suitable materials, not just indium GaN (InGaN) [17]. Among these materials, SiC has a longer history and has seen sustained investment since the late 1970s, particularly through the U.S. Department of Defense Office of Naval Research. This investment has led to significant advancements in SiC material synthesis and quality, as well as progress in various device technologies.

Over the past few years, 4H-SiC has drawn more attention as a suitable material for high-voltage power applications. The first Schottky barrier diode (SBD) was promoted in 2001, and this marked the realization of the SiC-based power electronics goal [204,205]. Photovoltaic (PV) inverters, power supply, and power factor correction circuits (PFC) are the main applications for SiC diodes. Figure 16 shows the variation of V_{th} for a 4H-SiC MOSFET with a nitridated SiO2 layer as a function of stress time and temperature.

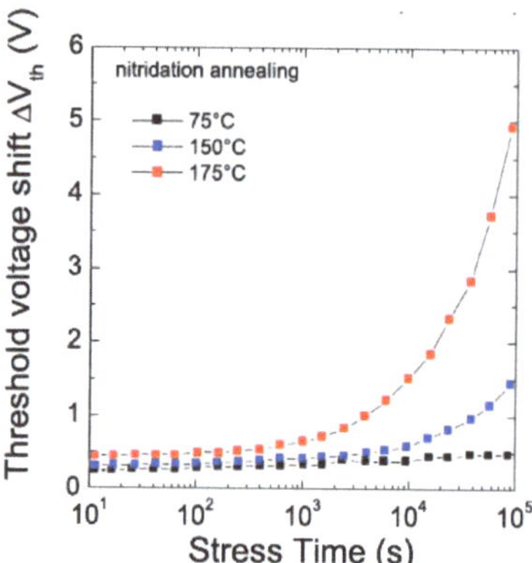

Figure 16. Threshold voltage shift V_{th} with relation to stress time at various temperatures [204].

Due to its sensitive gate dielectrics, the SiC MOSFET offers questionable reliability; the major issue is the comparably poor channel mobility. It is a suitable choice for the fabrication of power electronics devices with high break-down voltage, low specific R_{ON}, and high-frequency switching operations due to physical properties like a high saturation velocity and a high critical electric field [206]. GaNs' stronger critical electric field and higher electron mobility should assure substantially superior efficiency compared to SiC [15]. The GaN HEMT is inherently a normally on device because of the existence of the 2DEG, which can be seen in Figure 17a,b [17].

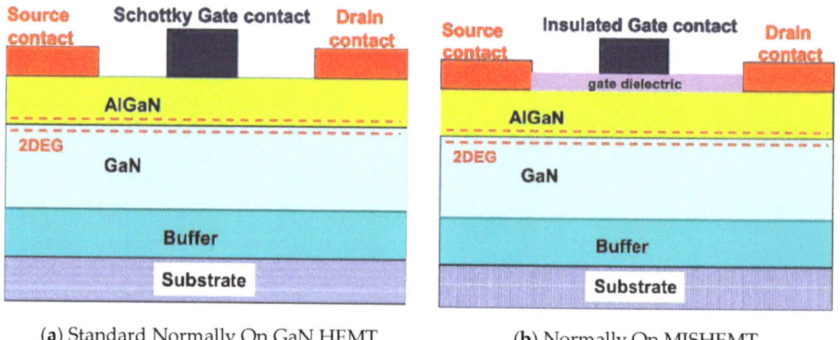

(**a**) Standard Normally On GaN HEMT　　　　(**b**) Normally On MISHEMT

Figure 17. Schematic of configurations for normally on AlGaN/GaN HEMTs with (**a**) Schottky gate; and (**b**) and insulated gate [17].

In general, the AlGaN/GaN heterojunction typically generates a two-dimensional electron gas (2DEG) and a high sheet carrier density (of around 10 cm) as a result of piezo-electric polarization [207]. Moreover, HEMTs are devices whose functioning is predicated on the presence of 2DEG [15,204]. Developing well-engineered normally off GaN HEMT technologies, improving insulator/GaN interfaces, enhancing efficient metallization plans for GaN-on-Si frameworks, and other issues are a few of the difficulties.

Ultrawide bandgap (UWBG) semiconductors represent an interesting and difficult new field of study in semiconductor materials, physics, devices, and applications, with bandgaps considerably larger than the 3.4 eV of GaN [17]. Furthermore, these UWBG materials have the potential for far higher performance than standard WBG materials because many figures of merit for device performance scale with the increasing bandgap in a very non-linear way. The commonly used Baliga figure of merit (BFOM) [208] is defined as $V^2_{BR}/R_{ON\text{-}SP}$ in the straightforward situation of a low-frequency unipolar vertical power switch, for instance, where $R_{ON\text{-}SP}$ stands for the specific on-resistance, which is the inverse of the conductance per unit area while the switch is on [208,209]. The device's capacity to block electricity while turned off and/or its conductivity per unit area when turned on increases with increasing BFOM. The BFOM scales as about the sixth power of the semiconductor bandgap because the critical electric field scales roughly as the square of the semiconductor bandgap.

To put this trio of materials in perspective, Table 20 lists some of their physical characteristics as well as the state-of-the-art values for three metrics crucial for device applications, for instance, the quality of their substrates as determined roughly by dislocations per square centimeter and substrate diameter, their capacity for p-type and n-type doping, and their capacity for n-type doping.

Table 20. Material characteristics of WBG and UWBG semiconductors [17].

Parameter	Unit	WBG			UWBG	
		GaN	4H-SiC	AlGaN/AlN	β-Ga$_2$O$_3$	Diamond
Bandgap	eV	3.4	3.3	Up to 6.0	4.9	5.5
Thermal conductivity (ε)	Wm^{-1}K^{-1}	253	370	253–319	11–27	2290–3450
Substrate quality (dislocations)	per cm^2	$\approx 10^4$	$\approx 10^2$	$\approx 10^4$	$\approx 10^4$	$\approx 10^5$
Substrate diameter	inch	8 (on Si)	8	2	4	1
Demonstrated p-type dopability	-	Good	Good	Poor	No	Good
Demonstrated p-type dopability	-	Good	Good	Moderate	Moderate	Moderate

Moreover, Table 20. lists some of the material characteristics of WBG and UWBG semiconductors, as well as the most recent developments in the following areas, such as substrate dislocation density, substrate diameter, and bulk p-type and n-type doping levels [17]. From an electronics perspective, they possess the following advantageous features, for example, a wide range of direct bandgaps, spanning from 3.4 eV to approximately 6.0 eV; high breakdown fields, with AlN exhibiting values exceeding 10 MVcm^{-1}; high electron mobility, reaching bulk mobilities of up to 1000 cm^{2}V^{-1}s^{-1}; high saturation velocities exceeding 10^7 cms^{-1}; and the relative ease of n-type doping with Si, which has a relatively small donor ionization energy, especially up to approximately 80–85% aluminum content [210,211].

From an optoelectronics perspective, AlGaN alloys enable the direct generation of emission wavelengths shorter than 365 nm, extending into the ultraviolet A, B, and C bands. Similarly, InGaN and AlGaN, are ternary alloys, which enables the use of heterostructures and bandgap engineering, a strategy successfully employed by other ternary alloys in the III–V materials family. UWBG applications, including high-power and high-frequency electronics, radiation detectors, electron emitters for ultra-high-voltage vacuum switches and traveling wave tube cathodes, and thermionic emitters for energy conversion, are all made possible by the exceptionally beneficial characteristics of diamond. As shown in Figure 18, lateral metal-oxide-semiconductor field-effect transistor (MOSFET) devices have now been created using atomic-layer-deposited (ALD) dielectric layers, even though these air-exposed surfaces are noticeably unstable.

Diamond also holds the highest known thermal conductivity of any material, which is particularly significant because heat removal is a major limiting factor in the performance of many power electronics and optoelectronics applications. Excellent electron emissivity on hydrogen-terminated surfaces, surface transfer doping made possible by these surfaces, [15] room-temperature UV exciton emission, and optical defect centers brought on by the nitrogen-vacancy (N-V) and Si-vacancy (Si-V) complexes are a few additional special qualities. For emerging quantum information systems, these defect centers have been proposed as a physical platform for qubits [148].

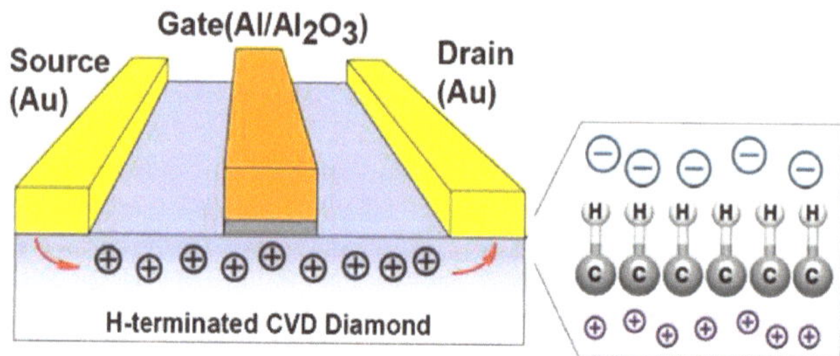

Figure 18. Diamond's hydrogen termination decreases its ionization energy and promotes electron transfer from the surface's valence band into other materials that adsorb the electrons [212].

Table 21 summarizes the state-of-the-art performance ranges for their respective technologies' evaluated powers and frequencies. It is possible to observe the trends in the applications of SiC technologies at higher powers, as the finest performance range of the technology gradually increases with the levels of the current in the transistor, and of the GaN technology at higher switching frequencies and lower power levels.

Table 21. State-of-the-art performance ranges for the evaluated powers and frequencies [28].

P (kW)	Si	SiC	GaN
1	-	Up to 14 kHz	14–500 kHz
2	-	Up to 28 kHz	28–500 kHz
4	-	Up to 55 kHz	55–500 kHz
8	-	Up to 110 kHz	110–500 kHz
16	-	1–500 kHz	-

A contemporary power semiconductor device switches rapidly between the ON and OFF states. A perfect switch may switch at any frequency and have no power losses in either the ON or OFF states. Losses do occur in practical devices, primarily in the ON state and during switching transitions. Three-terminal switches are required to produce the regulated ON and OFF transitions, with the third terminal controlling the transition by either supplying a voltage or a current signal. BJTs, thyristors, and gate turn-off (GTO) thyristors are current-controlled devices, whereas MOSFETs and IGBTs are voltage-controlled switches [213].

In contemporary power electronics converters, two terminal switches with unidirectional current flow capabilities, such as a diode, are also required. In general, the voltage and current ratings are any power device's most crucial parameters. Switching speed is another important specification that may be used to compare device capabilities. Devices with different breakdown voltage ratings are created for applications requiring a range of voltage levels [213]. Power devices with breakdown voltages above 600 V are often needed for important industrial and renewable energy applications, such as PV, wind, EV, and industry motor drives. Power devices with a voltage rating between 20 V and 600 V are commonly used for power supply applications in computers, mobile computing devices, and data centers [213]. Since the performance heavily depends on the voltage rating of the device, different devices can only be compared when they are made for the same breakdown voltage.

6.1. Voltage Rating

Modern power semiconductor devices are made by vertically stacking several P-type and N-type semiconductor layers on a substrate crystal wafer. The chip's main electrical terminals are on both sides. The switch function is accomplished by altering the device's conductivity from high in the ON state to low in the OFF state. When the maximum electric field reaches a critical breakdown field E_c, the voltage rating is commonly specified as the breakdown voltage. The semiconductor material in question determines E_c. A large OFF-state leakage current will be observed if the breakdown is reached electrically. The operation voltage is often chosen to be substantially lower than the breakdown voltage, due to the necessity of enduring transient overvoltage spikes, as well as the converter's long-term stability. The critical field E_c for Si is around 20 V/m, while E_c for broader bandgap materials like SiC and GaN is close to 300 V/m. For the three materials stated, this is graphically depicted in Figure 19.

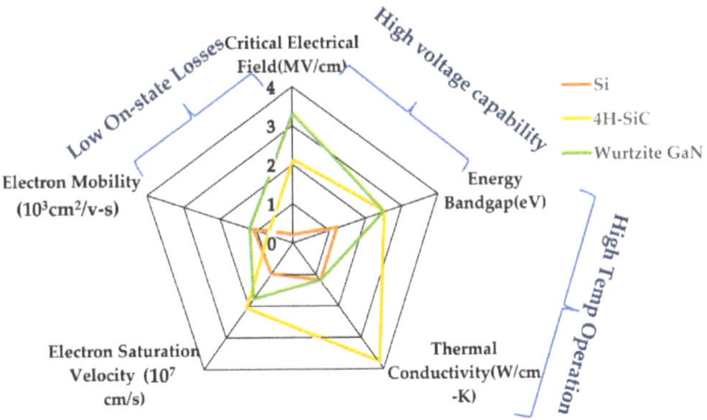

Figure 19. Critical material properties of Si, 4H-SiC, and Wurtzite GaN [214].

The use of wider bandgap power devices such as SiC and GaN has several advantages, including greater Ec and other desirable material features such as higher thermal conductivity. The same breakdown can be obtained in SiC material with an inner depletion thickness of less than 70 m and a surface termination region of roughly 200 m. As a result, achieving exceptionally high breakdown voltage in SiC power devices is significantly easier.

6.2. Current Rating

When the device conducts current in the ON state, the generated heat is manageable and does not cause the device to surpass its maximum operating temperature. Device innovation is motivated by increasing current density for a given breakdown voltage. Comparing absolute voltage and current ratings is one technique to assess state-of-the-art power devices, particularly their commercial readiness. This is depicted in Figures 20 and 21 for commercially available Si power devices and SiC and GaN power devices, respectively. Because of the excellent bipolar conduction mechanism in these two devices, the Si thyristor and Si diode have reached the highest voltage and current ratings. These two devices are also bundled in press-pack packaging and fabricated on a single wafer utilizing the level edge termination technique [215–217]. SiC and GaN power devices, first introduced to the market a decade ago, have made substantial progress in terms of commercially accessible voltage and current ratings. Clearly, there is still a significant disparity in the current ratings of Si power devices, as illustrated in Figure 20. As illustrated in Figure 21, hybrid devices built by Si IGBT and SiC diodes are being presented to fill this gap [218,219].

6.3. Switching Frequency

The power device in modern power converters must switch at high frequencies. Switching at higher frequencies has advantages such as improved dynamic response and smaller, lighter passive components. For high-density power electronics, reducing the size of passive components is crucial. As a result, switching frequency is another crucial metric to consider when comparing power devices. The device's switching losses during turn-on and turn-off limit the switching frequency's upper limit. As a result, the switching frequency is a tradeoff between conduction and switching loss, rather than a theoretical restriction on how fast the device can flip. Huang's thermal figure of merit (HTFOM) in

Table 22 can also be used as the switching frequency figure of merit (FOM). It indicates that hard-switching WBG switches will be limited in their frequency by poor thermal conductivity and/or small chip size [220]. Likewise, their chip size reductions have fallen short of the Huang chip area figure of merit (HCAFOM) prediction in Table 22.

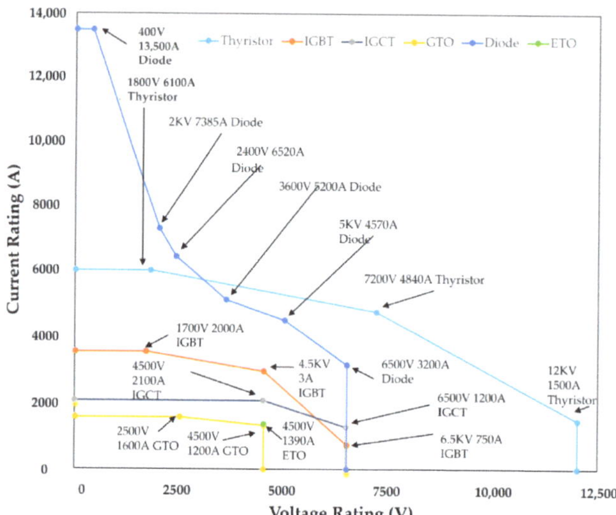

Figure 20. State-of-the-art commercial Si power devices in terms of the upper boundary of the voltage and current ratings achieved in a single-packaged device. The current rating shown is the DC rating at a case temperature of 85 °C.

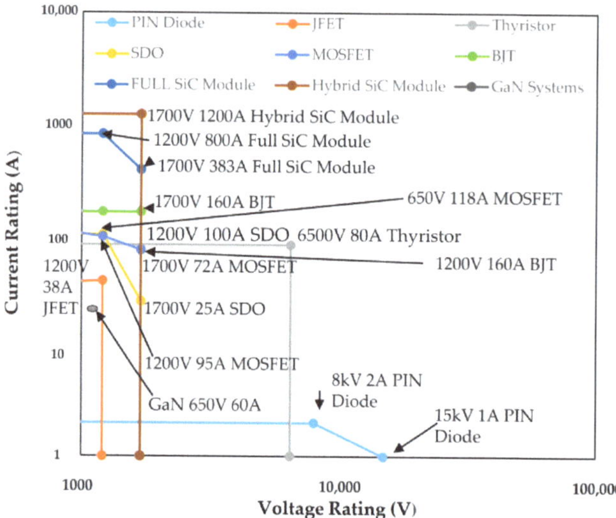

Figure 21. State-of-the-art commercial WBG power devices in terms of the upper boundary of the voltage and current ratings achieved in a single-packaged device. The current rating shown is the DC rating at a case temperature of 25 °C.

Table 22. Figure-of-merit comparison of Si, SIC, GaN, and diamond materials [214].

Parameter	Unit	Si	4H-SiC	Wurtzite GaN	Diamond
Electron mobility (μ)	cm^2/V-s	1360	700	1500	2200
Relative dielectric constant (ε)	-	11.7	9.7	8.9	5.7
Circuit electric field (Ec)	MV/cm	0.3	2.2	3.3	20
Thermal conductivity (σth)	W/cm-K	1.3	3.7	13.3	20
HMFOM	Ec $\sqrt{\mu}$	1	5.26	11.55	84.79
HCAFOM	$\varepsilon\,E^2c\,\sqrt{\mu}$	1	31.99	96.66	2753.9
HTFOM	σth/ε Ec	1	0.47	0.12	0.47

Table 23 indicates that die size decreases significantly when technology advances from standard Si MOSFETs to SJ MOSFETs to SiC. SiC MOSFETs are around 20 times smaller than Si MOSFETs. The size reduction in GaN is less significant than in SiC since the GaN device is a lateral power device rather than a vertical power device, and its $R_{ON\text{-}SP}$ decrease is lower. The greater die size, on the other hand, offers superior thermal performance. The lateral GaN has the added benefit of having a reduced capacitance/gate charge. The lateral structure is to blame for this. The SiC and GaN power devices are well-positioned to operate at higher frequencies in hard-switching or soft-switching converters due to significant reductions in DFOM1, DFOM2, and DFOM3.

Table 23. A device level comparison among three unipolar power transistors. Si MOSFET = IXTH30N60P, SJ MOSFET = IPD65R225C7, SJ MOSFET = C3M0280090D, GaN HFET = GS66504B estimated die size [220].

Parameter	Unit	Si MOSFET	Si-SJ MOSFET	SiC MOSFET	GaN HEFT
Breakdown voltage	V	600	650	900	600
Breakdown current	A	30	11	11.5	15
Wafer size	mm	-	200	100	150
Die area	mm^2	41	6.6	2.1	6.5
Current density	A/cm^2	-	170	540	230
R_{on}	Ω	0.24	0.22	0.28	0.11
DFOM1 ($R_{on} \times Q_g$)	Ω-nC	19.68	4.4	2.66	0.11
DFOM2 ($R_{on} \times Q_{gd}$)	Ω-nC	7.2	1.32	0.95	0.11
DFOM3 ($R_{on} \times Q_{oss}$)	Ω nC	18.9	83.7	6.93	3.08
DFOM4 ($R_{on} \times Q_{rr}$)	Ω-nC	960	1320	13.16	2.8
Rjc	°C/W	0.23	1.99	2.3	1
Normalized die cost	σth/ε Ec	-	1	4	3.6

6.4. Thin Wafer Field Stop IGBT (FS-IGBT)

IGBT technology became the essential notion virtually as soon as the planar power MOSFET was presented [221]. They are available in single switch and rotor configurations, with ratings ranging from 250 A to 1200 A. In motor control and drives, uninterruptible power supplies (UPS), transmission and distribution, commercial, construction, and agricultural vehicles (CAV), as well as traction uses, 4500 V and 6500 V IGBT modules are frequently used [222]. As a result, MOSFETs, GTOs, and BJTs have been rapidly phased out of medium- to high-power applications. In terms of technology, three decades of invention and industrialization have introduced various generations of IGBT technology. The implanted P collector is no longer reached by the depletion region. E_{off} may be modified due to the implanted collector, which allows for control of minority carrier injection. The non-punch through (NPT) IGBT, on the other hand, has a longer drift layer, which increases forward voltage (V_f) once more [220].

6.5. Reverse Conducting IGBT (RC-IGBT)

Because IGBTs lack a reverse conduction path, an externally packaged freewheeling diode (FWD) is required to let the current flow in the other direction. The reverse recovery loss should be reduced, and the recovery softness must be improved according to key design factors. A new generation of IGBTs with inbuilt FWD has been released [223–225]. The thin wafer manufacturing technology established for the FS-IGBT is used in the RC-IGBT. An N region is created by interrupting the backside P collector. The RC-IGBT chip can now take up the entire module footprint in an RC-IGBT power module. The MOS-controlled diode investigated this MOS control characteristic many years ago [226–228].

6.6. Reverse Blocking IGBT

To inhibit reverse voltage, certain significant renewable energy converters, such as the T-NPC three-level converter [229], need an IGBT in series with a diode. Changes in the collector junction doping concentration and edge termination must be performed to enhance the voltage. Reverse blocking IGBT (RB-IGBT) is one such RB-IGBT [230]. The N buffer layer must be removed to enhance the reverse breakdown voltage, converting the IGBT to an NPT-IGBT. To minimize the surface/edge electric field in the reverse direction, a new termination will be required. Deep diffusion or epitaxial regrowth following a deep etch can produce the latter.

6.7. Integrated Gate Commutated Thyristor (IGCT)

Because of the clear advantages of constructing a high-power device in a single device wafer and the extremely dependable press-pack packaging process, the Si thyristor, or SCR, has been and continues to be the most powerful semiconductor switch ever created. Until high-power IGBTs superseded megawatt gate turn-off (GTO) converters operating at a few hundred hertz, the high-power industry was dominated by megawatt GTO converters. The GTO's bad turnoff safe operation area (RBSOA) is one explanation for this. In the late 1990s, a substantial advancement was made to revitalize GTO technology. The GTO's gate drive circuit was the center of the innovation. The device begins to turn off when the current reaches about a third of the anode current. The thyristor action is still active because there is roughly a 2/3 current in the cathode/emitter junction currently. ABB has recently enhanced the IGCT's capabilities by incorporating an inbuilt freewheeling diode into the same wafer, resulting in a reverse conducting IGCT (RC-IGCT) [231]. The emitter turn-off (ETO) thyristor [232] aims to achieve unity gain turn-off. It is possible to accomplish a 5000 A snubber-less turnoff [233]. In case of emitter turn off thyristor (ETO), built-in current sensing is similarly simple to create, and the ETO, on the other hand, is currently not in commercial production [234].

6.8. Reliability and Application

Reliability and applications are critical factors to take into account in order to fully utilize wide bandgap power devices such as GaN and SiC in real-world systems. Significant improvements in power density, efficiency, and high-temperature operation are made possible by these devices. Robustness under dynamic switching conditions can be impacted by problems such as current collapse, threshold voltage instability, gate oxide breakdown, and electromigration. To reduce the negative effects on lifetime, adequate characterization and derating are required in addition to methods like gate drive optimization, sophisticated packaging, and layout strategies. For example, surface passivation and buffer layer modifications can reduce on-resistance degradation caused by current collapse [235]. By using field plates and optimizing dielectric thickness, gate reliability can be increased. GaN dies, substrates, and solders' acoustic mismatch cause stresses and defects during heat cycling, necessitating package co-design and modeling.

Utilizing wide bandgap capabilities in applications such as data center power supplies, naval electrical systems, EV charging, and renewable integration necessitates a comprehensive analysis covering device physics, packaging, thermal management, and system architectures. Wide bandgap devices are highly valuable when used in high-performance power electronics equipment because of their superior attributes such as faster switching, lower losses, and high-temperature capacity. However, their effective deployment in these devices requires a thorough understanding of degradation mechanisms, customized design strategies, and extensive qualification testing [16]. However, to do this, you need a multidisciplinary team with knowledge in materials science, accelerated testing, circuit design, application engineering, and device fabrication. Wide bandgap potential can be unlocked by holistic solutions, which also guarantee enough robustness to enable a seamless technology transfer.

Since the 1950s, Si has been the primary semiconductor substrate used in the production of power electronics equipment. On the other hand, the maximum theoretical efficiency of Si-based power-switching devices has been achieved [236]. High power losses, low switching frequencies, and decreased performance at high temperatures are some of the disadvantages of Si-based devices. As the need for distributed energy resource (DER) integration and urban electrification grows, a new class of advanced materials called wide bandgap (WBG) semiconductors has emerged. Among them are SiC, diamond, GaN, aluminum nitride (AlN), boron nitride (BN), zinc oxide (ZnO), and gallium oxide (Ga2O3). These materials have a great deal of potential for the upcoming power conversion technologies. Unipolar devices, such as MOSFETs, have limited rated voltage and current capabilities, but they can achieve high switching frequencies.

Bipolar Si semiconductors, on the other hand, enable the operation of high-power conversion devices at relatively low frequencies, but this requires larger and heavier passive components, which might not be appropriate for applications like power supplies, motor drives, automotive, and aerospace that have strict weight and volume limitations. Figure 22 illustrates how WBG materials show a compelling alternative to the limitations of Si by offering enhanced properties. These materials have higher electric breakdown fields, deeper doping concentrations, and thinner layers. These properties enhance their ability to block voltage and decrease drift resistance, which, in turn, reduces conduction losses. This implies that smaller WBG devices with the same on-resistance can have lower capacitance. A high saturation drift velocity allows for higher switching speeds with reduced capacitance because less energy is lost during each switching cycle [53]. Additionally, strong high-temperature performance and a reduction in leakage currents are guaranteed by the low intrinsic carrier concentration of WBG materials [237].

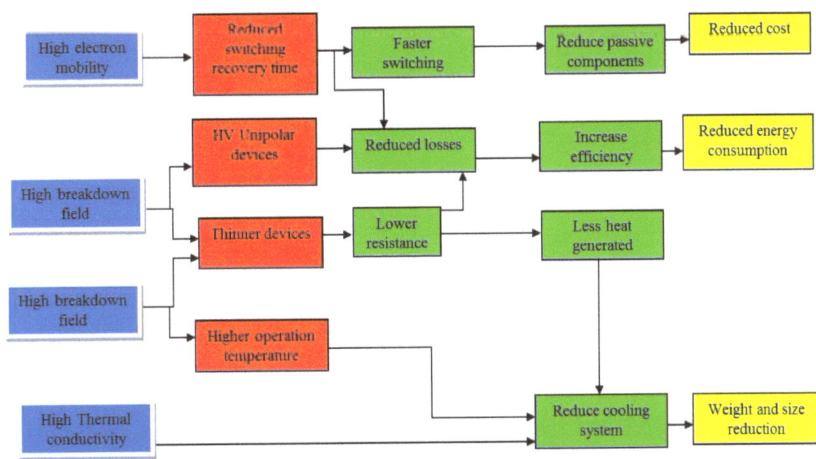

Figure 22. Flow of the characteristics of WBG semiconductor devices in terms of the parameter capabilities (blue); physical properties affected by WBG advantages (red); power electronics characteristics (green); and product benefits (yellow).

All the characteristics of WBG materials make them prominent semiconductor devices in high-output power equipment with increased efficiency, as well as smaller, lighter, and less expensive systems [238]. With small switching losses, WBG semiconductors can achieve 99% efficiency. When compared to Si devices, this indicates a reduction in energy losses of up to 75% [128]. Furthermore, it is possible to obtain higher switching frequencies. Due to Si limitations, frequencies higher than 20 kHz have not yet been achievable at power levels greater than tens of kilowatts; as a result, WBG materials provide better output quality and enable simpler circuit topologies by reducing the size and number of passive components [239].

6.9. Current Marketplace Scenario

A transformational phase is now taking place in the market environment for power electronic semiconductors, which includes Si, SiC, and GaN. Si has long dominated the industry because it is inexpensive and has reliable production methods, but SiC and GaN are emerging as disruptive technologies that are steadily capturing market share. SiC-based devices are appropriate for high-power applications like electric cars and renewable energy systems because they have higher power densities, faster switching rates, and reduced losses. On the other hand, GaN-based devices offer high-frequency operation and increased efficiency, finding use in data centers and small power converters. Si-based devices continue to rule the industry due to their maturity and widespread availability despite SiC and GaN's increasing acceptance. In the current market environment, SiC and GaN devices are gaining ground in high-power and high-frequency applications. In contrast, Si devices continue to predominate in low- to medium-power applications.

Power electronics are being used more often in various energy conversion end applications, propelling the considerable expansion of the worldwide power electronics market. The rising need for energy-efficient technology across a variety of end-user sectors is what drives this development. On 13 April 2023, in New York, GlobeNewswire shares [240] that, the market for power electronics will increase from USD 43.3 billion in 2022 to more than USD 94.21 billion by 2032, with a predicted compound annual growth rate (CAGR) of 8.3% from 2023 to 2032, which is shown in Figure 23.

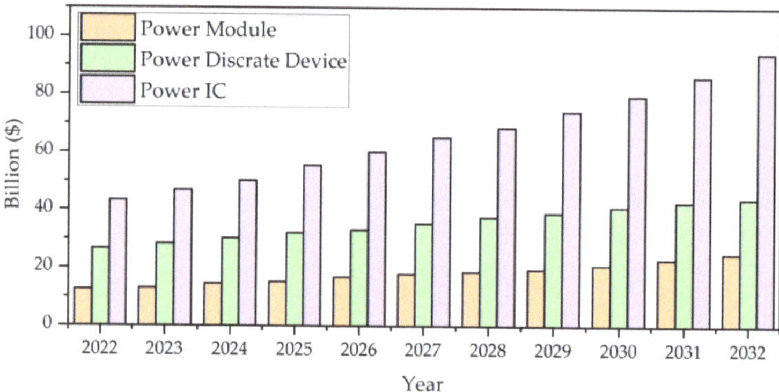

Figure 23. Power electronic market size projection: Will surpass USD 94.21 Billion revenue by 2032.

Power semiconductor devices comprising Si, SiC, GaN, and diamond are all compared in Table 24 for their features and uses. Voltage, current, frequency, applications, packaging, features, and manufacturers are given as important parameters for each type of device and material system. The chart displays the wide bandgap materials like SiC, GaN, and diamond for power electronics' high voltage and high-frequency capabilities.

Table 24. Features and values comparison among Si, SiC, GaN, and diamond.

Material	Device	Voltage (V)	Current (A)	Frequency (kHz)	Applications	Package	Features	Manufacturer
Si	IGBT	6500	1200	20	Motor drives, power supplies, converters	Discrete, Module	High power capability, easy to parallel	Infineon, STMicroelectronics
	MOSFET	900	150	100	Switching power supplies, motor drives	Discrete, Module	Fast switching, low losses	Infineon, ON Semiconductor
SiC	MOSFET	1700	100	100	EV drivetrains, PV inverters, power supplies	Discrete, Module	High efficiency, high frequency, high temperature	Wolfspeed, Rohm
	SBD	1700	20	-	HVDC, motor drives, battery charging	Discrete	Low loss rectification, fast recovery	Wolfspeed, Infineon
GaN	HEMT	1200	15	1000	Adapters, data center power, wireless power	Discrete, Module	High frequency, high efficiency	Efficient Power Conversion, Navitas
	MISFET	650	20	500	On-board chargers, power supplies, DC–DC converters	Discrete	Normally off, low losses	Transphorm, Panasonic
Diamond	Schottky Diode	400	0.1	-	High-temperature electronics	Research	Extremely high-temperature capability	Akhan Semiconductor
	JFET	50	0.1	-	High-temperature electronics	Research	High bandgap, temperature tolerance	Group4 Labs

Several commercially available power semiconductor devices from top producers like Infineon and STMicroelectronics are shown in Table 25. The table covers key characteristics of various Si, SiC, and GaN devices. These devices can be used for low, medium, and high voltage applications due to their wide voltage ratings of 20 V to 2 kV. The range of current ratings is 4.5 A to 400 A. For high-frequency GaN HEMTs, switching frequencies up to 2 MHz are mentioned. The wide bandgap SiC and GaN enable an operational temperature range of up to 200 °C. Standard packaging formats for discrete devices, such as TO-247 and TO-263, are displayed. For developing GaN technology, chip-scale packages are also listed. The chart shows how contemporary power semiconductors combine high voltage blocking capabilities, low loss switching, and high-temperature tolerance to allow performance advantages in various power electronics applications. For various applications, top manufacturers offer a wide range of Si, SiC, and GaN devices.

Table 25. Current commercial semiconductor devices and their ratings.

Manufacturing Company	Device Type	Part Number	Voltage Rating (V)	Current Rating (A)	Switching Frequency (Hz)	Operating Temp. (°C)
onsemi	Si Diode [241]	NXH80B120L2Q0SNG	1000 to 1200	30 to 400	10 k	−40 to 175
	SiC Diode [242]	FFSM1065A	650 to 1700	4 to 20	-	−55 to 175
	SiC MOSFET [243,244]	NVBG095N065SC1, NTH4L028N170M1	650 to 1700	30 to 81	1 M	14 to 175

Table 25. *Cont.*

Manufacturing Company	Device Type	Part Number	Voltage Rating (V)	Current Rating (A)	Switching Frequency (Hz)	Operating Temp. (°C)
Infineon	GaN Transistor [245,246]	IGOT60R042D1, IGLD60R070D1	400 to 600	60 to 12.2	100 k	−55 to 150
	Si Diode [247,248]	IDW100E60, IDP30E120XKSA1	600 to 1200	28 to 150	18 k to 100 k	−55 to 150
	SiC MOSFET [205,249]	IMBG65R022M1H, DF419MR20W3M1HFB11	650 to 2000	50 to 60	–	−55 to 150
	Automotive IGBT [250,251]	F450R07W1H3B11A, FS380R12A6T4B	650 to 1200	50 to 380	2 k to 50 k	−40 to 125
ROHM SEMICONDUCTOR	GaN HEMT [252]	GNE1040TB	150 to 150	10 to 20	-	−55 to 150
	SiC MOSFET [253,254]	SCT3017AL, SCT3017ALHR, SCT2H12NY	650 to 1700	21 to 04	1 M	−55 to 175
	Si MOSFET [255,256]	HP8JE5, HP8KC6, HP8MB5, UT6MA3	20 to 40	4.5 to 5.5	–	−55 to 150
	Ignition IGBT [257]	RGPR10BM40FH, RGPR20NL43HR	400–430	20 to 30	–	−55 to 175
ST Microelectronics	SiC Diode [258,259]	STPSC2006CW, STPSC40H12C	600 to 1200	10 to 40	1 M	−40 to 175
	SiC MOSFET [112,260]	SCT1000N170	650 to 1700	7 to 300	12 k to 25 k	−55 to 200
	GaN HEMT [261,262]	MASTERGAN1, MASTERGAN3	600 to 650	10 to 10.5	500 k to 2 M	−40 to 150
	IGBTs [263]	STG15M120F3D7, STG200G65FD8AG	300 to 1700	10 to 200	1 M	−55 to 150

7. Conclusions

Substantial advancements have been achieved in developing novel materials like SiC, GaN, and diamond, their commercialization, and the adoption of these wide and ultrawide bandgap power electronic semiconductors. Improvements in material quality, manufacturing processes, and device performance have boosted industrial use in many application fields. The capabilities and applications of these wide bandgap technologies are continuously being expanded by ongoing research and development. Wide bandgap power semiconductors, such as SiC, GaN, and ultrawide bandgap devices like diamond technology, have the potential to revolutionize power electronics due to their superiority over Si in terms of voltage blocking, switching speeds, efficiency, and thermal performance. SiC and GaN devices are progressively increasing in market share, particularly in electric vehicles, renewable energy, aerospace, and high-frequency applications, even if Si continues to rule the market now. Diamond's exceptionally wide bandgap could enable unprecedented power densities and high-temperature operation if manufacturing challenges can be overcome. Finally, these wide and ultra-wide bandgap semiconductors have the potential to eventually replace Si throughout the entire spectrum of power electronics due to their advantages over Si, influencing the development of next-generation, high-performance, energy-efficient devices. This article summarized the physical characteristics, difficulties, uses, and competitive environment of several prospective wide bandgap power electronic semiconductors, including the recently developed diamond technology.

Author Contributions: Conceptualization, S.M.S.H.R. and O.A.M.; methodology, S.M.S.H.R.; software, M.A.H. (Md. Asadul Haque) and M.A.H. (Md. Asikul Haque); validation, S.M.S.H.R. and O.A.M.; formal analysis, R.A. and M.K.H.; investigation, S.M.S.H.R., R.A., M.A.H. (Md. Asadul Haque), M.K.H., and M.A.H. (Md. Asikul Haque); resources, S.M.S.H.R., R.A., M.A.H. (Md. Asadul Haque), M.K.H., and M.A.H. (Md. Asikul Haque); data curation, S.M.S.H.R., R.A., M.A.H. (Md. Asadul Haque), M.K.H., and M.A.H. (Md. Asikul Haque); writing—original draft preparation, S.M.S.H.R., R.A., M.A.H. (Md. Asadul Haque), M.K.H., and M.A.H. (Md. Asikul Haque); writing—review and editing, S.M.S.H.R., R.A., M.A.H. (Md. Asikul Haque), and O.A.M.; supervision, O.A.M. All authors have read and agreed to the published version of the manuscript.

Funding: This research received no external funding.

Data Availability Statement: Not applicable.

Conflicts of Interest: The authors declare no conflict of interest.

References

1. Rafin, S.M.S.H.; Ahmed, R.; Mohammed, O.A. Wide Band Gap Semiconductor Devices for Power Electronic Converters. In Proceedings of the 2023 Fourth International Symposium on 3D Power Electronics Integration and Manufacturing (3D-PEIM), Miami, FL, USA, 1–3 February 2023; pp. 1–8.
2. Palmour, J.W. SiC power device development for industrial markets. In Proceeding of the IEEE International Electron Devices Meeting (IEDM) 2014, San Francisco, CA, USA, 15–17 December 2014; pp. 1.1.1–1.1.8.
3. Chow, T.P. Progress in high voltage SiC and GaN power switching devices. *Mater. Sci. Forum* **2014**, *778-780*, 1077–1082. [CrossRef]
4. Rafin, S.M.S.H.; Islam, R.; Mohammed, O.A. Power Electronic Converters for Wind Power Generation. In Proceedings of the 2023 Fourth International Symposium on 3D Power Electronics Integration and Manufacturing (3D-PEIM), Miami, FL, USA, 1–3 February 2023; pp. 1–8.
5. Rafin, S.M.S.H.; Haque, M.A.; Islam, R.; Mohammed, O.A. A Review of Power Electronic Converters for Electric Aircrafts. In Proceedings of the 2023 Fourth International Symposium on 3D Power Electronics Integration and Manufacturing (3D-PEIM), Miami, FL, USA, 1–3 February 2023; pp. 1–8.
6. Zerarka, M.; Austin, P.; Toulon, G.; Morancho, F.; Arbess, H.; Tasselli, J. Behavioral study of single-event burnout in power devices for natural radiation environment applications. *IEEE Trans. Electron Devices* **2012**, *59*, 3482–3488. [CrossRef]
7. Yu, C.-H.; Wang, Y.; Liu, J.; Sun, L.-L.J.I.T.o.E.D. Research of single-event burnout in floating field ring termination of power MOSFETs. *IEEE Trans. Electron. Devices* **2017**, *64*, 2906–2911. [CrossRef]
8. Carr, J.A.; Hotz, D.; Balda, J.C.; Mantooth, H.A.; Ong, A.; Agarwal, A. Assessing the impact of SiC MOSFETs on converter interfaces for distributed energy resources. *IEEE Trans. Power Electron.* **2009**, *24*, 260–270. [CrossRef]
9. Alam, M.; Kumar, K.; Dutta, V. Comparative efficiency analysis for Si, SiC MOSFETs and IGBT device for DC–DC boost converter. *SN Appl. Sci.* **2019**, *1*, 1700. [CrossRef]
10. Biela, J.; Schweizer, M.; Waffler, S.; Kolar, J.W. SiC versus Si—Evaluation of potentials for performance improvement of inverter and DC–DC converter systems by SiC power semiconductors. *IEEE Trans. Ind. Electron.* **2010**, *58*, 2872–2882. [CrossRef]
11. Chow, T.P. SiC and GaN high-voltage power switching devices. *Mater. Sci. Forum* **2000**, *338–342*, 1155–1160. [CrossRef]
12. Guo, Y.-B.; Bhat, K.P.; Aravamudhan, A.; Hopkins, D.C.; Hazelmyer, D.R. High current and thermal transient design of a SiC SSPC for aircraft application. In Proceedings of the 2011 Twenty-Sixth Annual IEEE Applied Power Electronics Conference and Exposition (APEC), Fort Worth, TX, USA, 6–11 March 2011; pp. 1290–1297.
13. Nakamura, T.; Sasagawa, M.; Nakano, Y.; Otsuka, T.; Miura, M. Large current SiC power devices for automobile applications. In Proceedings of the 2010 International Power Electronics Conference-ECCE ASIA-, Sapporo, Japan, 21–24 June 2010; pp. 1023–1026.
14. Ranjbar, A. Applications of Wide Bandgap (WBG) Devices in the Transportation Sector. Recent Advances in (WBG) Semiconductor Material (e.g. SiC and GaN) and Circuit Topologies. In *Transportation Electrification: Breakthroughs in Electrified Vehicles, Aircraft, Rolling Stock, and Watercraft*; IEEE: Piscataway, NJ, USA, 2022; pp. 47–72.
15. Roccaforte, F.; Fiorenza, P.; Greco, G.; Nigro, R.L.; Giannazzo, F.; Iucolano, F.; Saggio, M.J. Emerging trends in wide band gap semiconductors (SiC and GaN) technology for power devices. *Microelectron. Eng.* **2018**, *187*, 66–77. [CrossRef]
16. Chen, K.J.; Häberlen, O.; Lidow, A.; lin Tsai, C.; Ueda, T.; Uemoto, Y.; Wu, Y. GaN-on-Si power technology: Devices and applications. *IEEE Trans. Electron Devices* **2017**, *64*, 779–795. [CrossRef]
17. Tsao, J.; Chowdhury, S.; Hollis, M.; Jena, D.; Johnson, N.; Jones, K.; Kaplar, R.; Rajan, S.; Van de Walle, C.; Bellotti, E. Ultrawide-bandgap semiconductors: Research opportunities and challenges. *Adv. Electron. Mater.* **2018**, *4*, 1600501. [CrossRef]
18. Al Mamun, M.; Paudyal, S.; Kamalasadan, S. Efficient Dynamic Simulation of Unbalanced Distribution Grids with Distributed Generators. In Proceedings of the 2023 IEEE Industry Applications Society Annual Meeting, Nashville, TN, USA, 29 October–2 November 2023; pp. 1–6.
19. National Institute of Standards and Technology (NIST). Available online: https://www.nist.gov/pml/owm/metric-si/understanding-metric (accessed on 29 September 2023).

20. Zulehner, W. Historical overview of Si crystal pulling development. *Mater. Sci. Eng.* **2000**, *73*, 7–15. [CrossRef]
21. Zaidi, B. Introductory chapter: Introduction to photovoltaic effect. *Sol. Panels Photovolt. Mater.* **2018**, 1–8. [CrossRef]
22. Zaidi, B.; Hadjoudja, B.; Felfli, H.; Chouial, B.; Chibani, A. Effet des traitements thermiques sur le comportement électrique des couches de silicium polycristallin pour des applications photovoltaïques. *Rev. De Métallurgie* **2011**, *108*, 443–446. [CrossRef]
23. Goetzberger, A.; Knobloch, J.; Voss, B. *Crystalline Si Solar Cells*; Wiley Online Library: Hoboken, NJ, USA, 1998; Volume 1.
24. Zaidi, B.; Hadjoudja, B.; Felfli, H.; Chibani, A. Influence of doping and heat treatments on carriers mobility in polycrystalline Si thin films for photovoltaic application. *Turk. J. Phys.* **2011**, *35*, 185–188.
25. Mathieu, H.; Bretagnon, T.; Lefebvre, P. *Physique des Semiconducteurs et des Composants Electroniques-Problèmes Résolus*; Dunod: Malakoff, France, 2001.
26. Zaidi, B.; Saouane, I.; Shekhar, C. Electrical energy generated by amorphous Si solar panels. *Si* **2018**, *10*, 975–979.
27. Zaidi, B.; Saouane, I.; Shekhar, C. Simulation of single-diode equivalent model of polycrystalline Si solar cells. *Int. J. Mater. Sci. Appl.* **2018**, *7*. [CrossRef]
28. Conibeer, G. Third-generation photovoltaics. *Mater. Today* **2007**, *10*, 42–50. [CrossRef]
29. Zaidi, B.; Belghit, S.; Shekhar, C.; Mekhalfa, M.; Hadjoudja, B.; Chouial, B. Electrical performance of CuInSe2 solar panels using ant colony optimization algorithm. *J. Nano-Electron. Phys.* **2018**, *10*, 05045. [CrossRef]
30. Sigfússon, T.I.; Helgason, Ö. Rates of transformations in the ferroSi system. *Hyperfine Interact.* **1990**, *54*, 861–867. [CrossRef]
31. Waanders, F.; Mans, A. Characterisation of ferroSi dense medium separation material. *Hyperfine Interact.* **2003**, *148*, 325–329. [CrossRef]
32. Borysiuk, J.; Sołtys, J.; Bożek, R.; Piechota, J.; Krukowski, S.; Strupiński, W.; Baranowski, J.M.; Stępniewski, R. Role of structure of C-terminated 4 H-SiC (000 1⁻) surface in growth of graphene layers: Transmission electron microscopy and density functional theory studies. *Phys. Rev. B* **2012**, *85*, 045426. [CrossRef]
33. Alsema, E.A.; de Wild-Scholten, M.J. Environmental impacts of crystalline Si photovoltaic module production. *Mater. Res. Soc. Symp. Proc.* **2006**, *73*. [CrossRef]
34. Baliga, B.J. Power Semiconductor Devices. Brooks/Cole. 1995. Available online: https://www.abebooks.com/servlet/BookDetailsPL?bi=31307574601 (accessed on 29 September 2023).
35. Ultra- and Hyper-Fast Si Diodes: 650V Rapid 1 and Rapid 2. Available online: https://www.infineon.com/cms/en/product/power/diodes-thyristors/Si-diodes/650v-rapid-1-and-rapid-2/ (accessed on 29 September 2023).
36. Infenion. Available online: https://www.infineon.com/cms/en/product/power/diodes-thyristors/Si-diodes/600v-1200v-ultra-soft/ (accessed on 29 September 2023).
37. Moller, H. *Semiconductors for Solar Cells*; Artech House Inc.: Norwood, MA, USA, 1993.
38. Edgar, L.J. Method and Apparatus for Controlling Electric Currents. US Patent 1,745,175, 21 January 1930.
39. Kahng, D. Si-Si dioxide field field induced surface devices. In Proceedings of the IRE-AIEE Solid-State Device Research Conference, Pittsburgh, PA, USA, 1960; Available online: https://scholar.google.com/scholar_lookup?title=Silicon%E2%80%93Silicon%20Dioxide%20Field%20Induced%20Surface%20Devices&author=D.K.%20State&publication_year=1960 (accessed on 29 September 2023).
40. Moore, G.E. Cramming more components onto integrated circuits. *IEEE* **1998**, *86*, 82–85. [CrossRef]
41. Sze, S.M.; Ng, K.K. *Physics of Semiconductor Devices*; John Wiley & Sons: Hoboken, NJ, USA, 2007.
42. Prado, E.O.; Bolsi, P.C.; Sartori, H.C.; Pinheiro, J.R. An overview about Si, Superjunction, SiC and GaN power MOSFET technologies in power electronics applications. *Energies* **2022**, *15*, 5244. [CrossRef]
43. Deboy, G.; Marz, N.; Stengl, J.-P.; Strack, H.; Tihanyi, J.; Weber, H. A new generation of high voltage MOSFETs breaks the limit line of Si. In Proceedings of the International Electron Devices Meeting 1998. Technical Digest (Cat. No. 98CH36217), San Francisco, CA, USA, 6–9 December 1998; pp. 683–685.
44. Fujihira, T.; Miyasaka, Y. Simulated superior performances of semiconductor superjunction devices. In Proceedings of the 10th International Symposium on Power Semiconductor Devices and ICs. ISPSD'98 (IEEE Cat. No. 98CH36212), Kyoto, Japan, 3–6 June 1998; pp. 423–426.
45. Chen, X.B. Breakthrough to the" Si limit" of power devices. In Proceedings of the 1998 5th International Conference on Solid-State and Integrated Circuit Technology, Beijing, China, 23 October 1998; pp. 141–144.
46. Appels, J.; Vaes, H. High voltage thin layer devices (RESURF devices). In Proceedings of the 1979 International Electron Devices Meeting, Washington, DC, USA, 3–5 December 1979; pp. 238–241.
47. Udrea, F.; Deboy, G.; Fujihira, T. Superjunction Power Devices, History, Development, and Future Prospects. *IEEE Trans. Electron Device* **2017**, *64*, 713–727. [CrossRef]
48. Baliga, B.J. *Insulated Gate Bipolar Transistors (IGBTs): Theory and Design*; World Scientific Publishing: Singapore, 2003.
49. VolLaska, T. Progress in Si IGBT Technology – as an ongoing Competition with WBG Power Devices. In Proceeding of the 2019 IEEE International Electron Devices Meeting (IEDM), San Francisco, CA, USA, 7–11 December 2019; pp. 12.2.1–12.2.4. [CrossRef]
50. Vobecký, J. Design and technology of high-power Si devices. In Proceedings of the 18th International Conference Mixed Design of Integrated Circuits and Systems—MIXDES 2011, Gliwice, Poland, 16–19 June 2011; pp. 17–22.
51. Guo, X.; Xun, Q.; Li, Z.; Du, S. SiC converters and MEMS devices for high-temperature power electronics: A critical review. *Micromachines* **2019**, *10*, 406. [CrossRef]

52. Tian, Y.; Zetterling, C.-M. A fully integrated Si-carbide Sigma–Delta modulator operating up to 500 °C. *IEEE Trans. Electron Devices* **2017**, *64*, 2782–2788. [CrossRef]

53. Garrido-Diez, D.; Baraia, I. Review of wide bandgap materials and their impact in new power devices. In Proceedings of the 2017 IEEE International Workshop of Electronics, Control, Measurement, Signals and Their Application to Mechatronics (ECMSM), Donostia, Spain, 24–26 May 2017; pp. 1–6.

54. Li, K. Wide Bandgap (SiC/GaN) Power Devices Characterization and Modeling: Application to HF Power Converters. Ph.D. Thesis, École Doctorale Sciences pour L'ingénieur (Lille), Lille, France, 2014.

55. Chen, C.; Luo, F.; Kang, Y. A review of SiC power module packaging: Layout, material system and integration. *CPSS Trans. Power Electron. Appl.* **2017**, *2*, 170–186. [CrossRef]

56. Chabi, S.; Kadel, K. Two-dimensional SiC: Emerging direct band gap semiconductor. *Nanomaterials* **2020**, *10*, 2226. [CrossRef]

57. Susi, T.; Skákalová, V.; Mittelberger, A.; Kotrusz, P.; Hulman, M.; Pennycook, T.J.; Mangler, C.; Kotakoski, J.; Meyer, J.C. Computational insights and the observation of SiC nanograin assembly: Towards 2D SiC. *Sci. Rep.* **2017**, *7*, 4399. [CrossRef]

58. Ferdous, N.; Islam, M.S.; Park, J.; Hashimoto, A. Tunable electronic properties in stanene and two dimensional Si-carbide heterobilayer: A first principles investigation. *AIP Adv.* **2019**, *9*, 025120. [CrossRef]

59. Chowdhury, C.; Karmakar, S.; Datta, A. Monolayer group IV–VI monochalcogenides: Low-dimensional materials for photocatalytic water splitting. *J. Phys. Chem. C* **2017**, *121*, 7615–7624. [CrossRef]

60. Berzelius, J.J. Untersuchungen über die Flussspathsäure und deren merkwürdigsten Verbindungen. *Ann. Der Phys.* **1824**, *77*, 169–230. [CrossRef]

61. Pensl, G.; Ciobanu, F.; Frank, T.; Krieger, M.; Reshanov, S.; Schmid, F.; Weidner, M. Sic material properties. In *Sic Materials And Devices*; World Scientific: Singapore, 2006; Volume 1, pp. 1–41.

62. Acheson, E.G. Carborundum: Its history, manufacture and uses. *J. Frankl. Inst.* **1893**, *136*, 279–289. [CrossRef]

63. Kaminskiy, F.; Bukin, V.; Potapov, S.; Arkus, N.; Ivanova, V. Discoveries of SiC under natural conditions and their genetic importance. *Int. Geol. Rev.* **1969**, *11*, 561–569. [CrossRef]

64. Choyke, W.J.; Patrick, L.; Ziegler, J. (Eds.) *SiC 2: Materials, Processing, and Devices*; Springer: Berlin/Heidelberg, Germany, 2018.

65. Kimoto, T.; Cooper, J.A. *Fundamentals of SiC Technology: Growth, Characterization, Devices and Applications*; John Wiley & Sons: Hoboken, NJ, USA, 2014.

66. Zhang, Y.; Suzuki, T. Epitaxial Growth of SiC—Review on Growth Mechanism and Technology. *J. Cryst. Growth* **2019**, *512*, 1–22.

67. Singh, R.; Capell, D.C.; Hefner, A.R.; Lai, J.; Palmour, J.W. High-power 4H-SiC JBS rectifiers. *IEEE Trans. Electron. Devices* **2002**, *49*, 2054–2063. [CrossRef]

68. Huang, X.; Wang, G.; Lee, M.-C.; Huang, A.Q. Reliability of 4H-SiC SBD/JBS diodes under repetitive surge current stress. In Proceedings of the 2012 IEEE Energy Conversion Congress and Exposition (ECCE), Raleigh, NC, USA, 15–20 September 2012; pp. 2245–2248.

69. Cheung, K.P. SiC power MOSFET gate oxide breakdown reliability—Current status. In Proceedings of the 2018 IEEE International Reliability Physics Symposium (IRPS), Burlingame, CA, USA, 11–15 March 2018; pp. 2B. 3-1–2B. 3-5.

70. Wang, J.; Jiang, X. Review and analysis of SiC MOSFETs' ruggedness and reliability. *IET Power Electron.* **2020**, *13*, 445–455. [CrossRef]

71. She, X.; Huang, A.Q.; Lucia, O.; Ozpineci, B. Review of SiC power devices and their applications. *IEEE Trans. Ind. Electron.* **2017**, *64*, 8193–8205. [CrossRef]

72. Guo, S.; Zhang, L.; Lei, Y.; Li, X.; Xue, F.; Yu, W.; Huang, A.Q. 3.38 Mhz operation of 1.2 kV SiC MOSFET with integrated ultra-fast gate drive. In Proceedings of the 2015 IEEE 3rd Workshop on Wide Bandgap Power Devices and Applications (WiPDA), Blacksburg, VA, USA, 2–4 November 2015; pp. 390–395.

73. Pérez-Tomás, A.; Brosselard, P.; Hassan, J.; Jorda, X.; Godignon, P.; Placidi, M.; Constant, A.; Millán, J.; Bergman, J. Schottky versus bipolar 3.3 kV SiC diodes. *Semicond. Sci. Technol.* **2008**, *23*, 125004. [CrossRef]

74. Langpoklakpam, C.; Liu, A.-C.; Chu, K.-H.; Hsu, L.-H.; Lee, W.-C.; Chen, S.-C.; Sun, C.-W.; Shih, M.-H.; Lee, K.-Y.; Kuo, H.-C. Review of SiC processing for power MOSFET. *Crystals* **2022**, *12*, 245. [CrossRef]

75. Losee, P.; Bolotnikov, A.; Yu, L.; Beaupre, R.; Stum, Z.; Kennerly, S.; Dunne, G.; Sui, Y.; Kretchmer, J.; Johnson, A. 1.2 kV class SiC MOSFETs with improved performance over wide operating temperature. In Proceedings of the 2014 IEEE 26th International Symposium on Power Semiconductor Devices & IC's (ISPSD), Waikoloa, HI, USA, 15–149 June 2014; pp. 297–300.

76. DiMarino, C.M.; Burgos, R.; Dushan, B. High-temperature SiC: Characterization of state-of-the-art SiC power transistors. *IEEE Ind. Electron. Mag.* **2015**, *9*, 19–30. [CrossRef]

77. Qingwen, S.; Xiaoyan, T.; Yimeng, Z.; Yuming, Z.; Yimen, Z. Investigation of SiC trench MOSFET with floating islands. *IET Power Electron.* **2016**, *9*, 2492–2499. [CrossRef]

78. Jiang, H.; Wei, J.; Dai, X.; Ke, M.; Deviny, I.; Mawby, P. SiC trench MOSFET with shielded fin-shaped gate to reduce oxide field and switching loss. *IEEE Electron. Device Lett.* **2016**, *37*, 1324–1327. [CrossRef]

79. Hiyoshi, T.; Uchida, K.; Sakai, M.; Furumai, M.; Tsuno, T.; Mikamura, Y. Gate oxide reliability of 4H-SiC V-groove trench MOSFET under various stress conditions. In Proceedings of the 2016 28th international Symposium on Power Semiconductor devices and ics (iSPSd), Prague, Czech Republic, 12–16 June 2016; pp. 39–42.

80. Song, Q.; Yang, S.; Tang, G.; Han, C.; Zhang, Y.; Tang, X.; Zhang, Y.; Zhang, Y. 4H-SiC trench MOSFET with L-shaped gate. *IEEE Electron. Device Lett.* **2016**, *37*, 463–466. [CrossRef]

81. Wang, Y.; Tian, K.; Hao, Y.; Yu, C.-H.; Liu, Y.-J. An optimized structure of 4H-SiC U-shaped trench gate MOSFET. *IEEE Trans. Electron Devices* **2015**, *62*, 2774–2778. [CrossRef]

82. Takaya, H.; Morimoto, J.; Hamada, K.; Yamamoto, T.; Sakakibara, J.; Watanabe, Y.; Soejima, N. A 4H-SiC trench MOSFET with thick bottom oxide for improving characteristics. In Proceedings of the 2013 25th International Symposium on Power Semiconductor Devices & IC's (ISPSD), Kanazawa, Japan, 26–30 May 2013; pp. 43–46.

83. Nakajima, A.; Saito, W.; Nishizawa, S.-i.; Ohashi, H. Theoretical Loss Analysis of Power Converters with 1200 V Class Si-IGBT and SiC-MOSFET. In Proceedings of the Proceedings of Pcim europe 2015; International Exhibition and Conference for Power Electronics, Intelligent Motion, Renewable Energy and Energy Management, Nuremberg, Germany, 19–20 May 2015; pp. 1–6.

84. Arrow. Available online: https://www.arrow.com/en/research-and-events/articles/advantages-of-sic-mosfet-vs-si-igbt (accessed on 29 September 2023).

85. Rafin, S.M.S.H.; Mohammed, O.A. Novel Dual Inverter Sub-Harmonic Synchronous Machines. *IEEE Trans. Magn.* **2023**, *59*, 8202605. [CrossRef]

86. Rafin, S.M.S.H.; Ali, Q.; Mohammed, O.A. Hybrid Sub-Harmonic Synchronous Machines Using Series and Parallel Consequent Permanent Magnet. In Proceedings of the 2023 IEEE Energy Conversion Congress and Exposition (ECCE), Nashville, TN, USA, 29 October – 2 November 2023; pp. 1–5.

87. Rafin, S.M.S.H.; Ali, Q.; Mohammed, O.A. Novel PM-Assisted Model of the Two-Layer Sub-Harmonic Synchronous Machines. In Proceedings of the 2023 International Applied Computational Electromagnetics Society Symposium (ACES), Monterey/Seaside, CA, USA, 26–30 March 2023; pp. 1–2.

88. Rafin, S.M.S.H.; Mohammed, O.A. Sub-Harmonic Synchronous Machine Using a Dual Inverter and a Unique Three-Layer Stator Winding. In Proceedings of the 2023 IEEE International Magnetic Conference-Short Papers (INTERMAG Short Papers), Sendai, Japan, 15–19 May 2023; pp. 1–2.

89. Rafin, S.M.S.H.; Ali, Q.; Lipo, T.A. A novel sub-harmonic synchronous machine using three-layer winding topology. *World Electr. Veh. J.* **2022**, *13*, 16. [CrossRef]

90. Rafin, S.M.S.H.; Ali, Q.; Khan, S.; Lipo, T.A. A novel two-layer winding topology for sub-harmonic synchronous machines. *Electr. Eng.* **2022**, *104*, 3027–3035. [CrossRef]

91. Kadavelugu, A.; Mainali, K.; Patel, D.; Madhusoodhanan, S.; Tripathi, A.; Hatua, K.; Bhattacharya, S.; Ryu, S.-H.; Grider, D.; Leslie, S. Medium voltage power converter design and demonstration using 15 kV SiC N-IGBTs. In Proceedings of the 2015 IEEE Applied Power Electronics Conference and Exposition (APEC), Charlotte, NC, USA, 15–19 March 2015; pp. 1396–1403.

92. Rahimo, M.; Kopta, A.; Eicher, S.; Kaminski, N.; Bauer, F.; Schlapbach, U.; Linder, S. Extending the boundary limits of high voltage IGBTs and diodes to above 8 kV. In Proceedings of the 14th International Symposium on Power Semiconductor Devices and Ics, Sante Fe, NM, USA, 7 June 2002; pp. 41–44.

93. Wang, G.; Huang, A.Q.; Wang, F.; Song, X.; Ni, X.; Ryu, S.-H.; Grider, D.; Schupbach, M.; Palmour, J. Static and dynamic performance characterization and comparison of 15 kV SiC MOSFET and 15 kV SiC n-IGBTs. In Proceedings of the 2015 IEEE 27th International Symposium on Power Semiconductor Devices & IC's (ISPSD), Hong Kong, China, 10–14 May 2015; pp. 229–232.

94. Abdalgader, I.A.; Kivrak, S.; Özer, T. Power performance comparison of SiC-IGBT and Si-IGBT switches in a three-phase inverter for aircraft applications. *Micromachines* **2022**, *13*, 313. [CrossRef]

95. Ramungul, N.; Chow, T.; Ghezzo, M.; Kretchmer, J.; Hennessy, W. A fully planarized, 6H-SiC UMOS insulated-gate bipolar transistor. In Proceedings of the 1996 54th Annual Device Research Conference Digest, Santa Barbara, CA, USA, 26 June 1996; pp. 56–57.

96. Zhang, Q.; Jonas, C.; Callanan, R.; Sumakeris, J.; Das, M.; Agarwal, A.; Palmour, J.; Ryu, S.-H.; Wang, J.; Huang, A. New Improvement Results on 7.5 kV 4H-SiC p-IGBTs with R diff, on of 26 mΩ·cm 2 at 25 °C. In Proceedings of the 19th International Symposium on Power Semiconductor Devices and IC's, Jeju, Republic of Korea, 27–31 May 2007; pp. 281–284.

97. Avram, M.; Brezeanu, G.; Avram, A.; Neagoe, O.; Brezeanu, M.; Iliescu, C.; Codreanu, C.; Voitincu, C. Contributions to development of high power SiC-IGBT. In Proceedings of the CAS 2005 Proceedings. 2005 International Semiconductor Conference, Sinaia, Romania, 3–5 October 2005; pp. 365–368.

98. Wang, X.; Cooper, J.A. High-voltage n-channel IGBTs on free-standing 4H-SiC epilayers. *IEEE Trans. Electron. Devices* **2010**, *57*, 511–515. [CrossRef]

99. Van Brunt, E.; Cheng, L.; O'Loughlin, M.J.; Richmond, J.; Pala, V.; Palmour, J.W.; Tipton, C.W.; Scozzie, C. 27 kV, 20 A 4H-SiC n-IGBTs. *Mater. Sci. Forum* **2015**, *821–823*, 847–850. [CrossRef]

100. Madhusoodhanan, S.; Hatua, K.; Bhattacharya, S.; Leslie, S.; Ryu, S.-H.; Das, M.; Agarwal, A.; Grider, D. Comparison study of 12kV n-type SiC IGBT with 10kV SiC MOSFET and 6.5 kV Si IGBT based on 3L-NPC VSC applications. In Proceedings of the 2012 IEEE Energy Conversion Congress and Exposition (ECCE), Raleigh, NC, USA, 15–20 September 2012; pp. 310–317.

101. Kadavelugu, A.; Bhattacharya, S.; Ryu, S.-H.; Van Brunt, E.; Grider, D.; Agarwal, A.; Leslie, S. Characterization of 15 kV SiC n-IGBT and its application considerations for high power converters. In Proceedings of the 2013 IEEE Energy Conversion Congress and Exposition, Denver, CO, USA, 15–19 September 2013; pp. 2528–2535.

102. Tripathi, A.; Mainali, K.; Madhusoodhanan, S.; Patel, D.; Kadavelugu, A.; Hazra, S.; Bhattacharya, S.; Hatua, K. MVDC microgrids enabled by 15kV SiC IGBT based flexible three phase dual active bridge isolated DC-DC converter. In Proceedings of the 2015 IEEE Energy Conversion Congress and Exposition (ECCE), Montreal, QC, Canada, 20–24 September 2015; pp. 5708–5715.

103. Han, L.; Liang, L.; Kang, Y.; Qiu, Y. A Review of SiC IGBT: Models, Fabrications, Characteristics, and Applications. *IEEE Trans. Power Electron.* **2021**, *36*, 2080–2093. [CrossRef]
104. Ciuti, G.; Ricotti, L.; Menciassi, A.; Dario, P. MEMS sensor technologies for human centred applications in healthcare, physical activities, safety and environmental sensing: A review on research activities in Italy. *Sensors* **2015**, *15*, 6441–6468. [CrossRef]
105. Funaki, T.; Balda, J.C.; Junghans, J.; Kashyap, A.S.; Mantooth, H.A.; Barlow, F.; Kimoto, T.; Hikihara, T. Power conversion with SiC devices at extremely high ambient temperatures. *IEEE Trans. Power Electron.* **2007**, *22*, 1321–1329. [CrossRef]
106. Katoh, Y.; Snead, L.L. SiC and its composites for nuclear applications–Historical overview. *J. Nucl. Mater.* **2019**, *526*, 151849. [CrossRef]
107. Ning, P.; Wang, F.; Ngo, K.D. High-temperature SiC power module electrical evaluation procedure. *IEEE Trans. Power Electron.* **2011**, *26*, 3079–3083. [CrossRef]
108. Wang, R.; Boroyevich, D.; Ning, P.; Wang, Z.; Wang, F.; Mattavelli, P.; Ngo, K.D.; Rajashekara, K. A high-temperature sic three-phase ac-dc converter design for> 100/spl deg/c ambient temperature. *IEEE Trans. Power Electron.* **2012**, *28*, 555–572. [CrossRef]
109. Carbosystem. Available online: https://carbosystem.com/en/Si-carbide-properties-applications (accessed on 29 September 2023).
110. Zabihi, N.; Mumtaz, A.; Logan, T.; Daranagama, T.; McMahon, R.A. SiC power devices for applications in hybrid and electric vehicles. *Mater. Sci. Forum* **2019**, *963*, 869–872. [CrossRef]
111. Durham, N. Cree Launches Industry's First Commercial SiC Power Mosfet. 2011. Available online: https://compoundsemiconductor. net/article/87227/Cree_Launches_Industry%E2%80%99s_First_Commercial_SiC_Power_MOSFET (accessed on 29 September 2023).
112. STMicroelectronics: STPOWER SiC MOSFETs. Available online: https://www.st.com/en/power-transistors/stpower-sic-mosfets.html (accessed on 29 September 2023).
113. Cai, W.; Wu, X.; Zhou, M.; Liang, Y.; Wang, Y. Review and development of electric motor systems and electric powertrains for new energy vehicles. *Automot. Innov.* **2021**, *4*, 3–22. [CrossRef]
114. Greencarcongress. Available online: https://www.greencarcongress.com/2023/04/20230404-denso.html (accessed on 29 September 2023).
115. Hamdan, M. Keynote: The Development and Use of SiC Power Devices in Aerospace Applications. In Proceedings of the IEEE Design Methodologies Conference 2023, Miami, FL, USA, 24-26 September 2023.
116. 200kW Integrated Starter Generator Controller. Available online: https://www.geaerospace.com/sites/default/files/2022-03/20 0kW-Integrated-Starter-Generator-Controller%20%281%29.pdf (accessed on 29 September 2023).
117. SiC Power Modules. Available online: https://www.geaerospace.com/sites/default/files/Si-Carbide-Power-Modules-2020.12. pdf (accessed on 29 September 2023).
118. Rafin, S.M.S.H.; Lipo, T.A.; Kwon, B.-I. Performance analysis of the three transistor voltage source inverter using different PWM techniques. In Proceedings of the 2015 9th International Conference on Power Electronics and ECCE Asia (ICPE-ECCE Asia), Seoul, Republic of Korea, 1–5 June 2015; pp. 1428–1443.
119. Rafin, S.M.S.H.; Lipo, T.A. A novel cascaded two transistor H-bridge multilevel voltage source converter topology. In Proceedings of the 2015 Intl Aegean Conference on Electrical Machines & Power Electronics (ACEMP), 2015 Intl Conference on Optimization of Electrical & Electronic Equipment (OPTIM) & 2015 Intl Symposium on Advanced Electromechanical Motion Systems (ELECTROMOTION), Side, Turkey, 2–4 September 2015; pp. 40–45.
120. Rafin, S.M.S.H.; Lipo, T.A.; Kwon, B.-i. Novel matrix converter topologies with reduced transistor count. In Proceedings of the 2014 IEEE Energy Conversion Congress and Exposition (ECCE), Pittsburgh, PA, USA, 14-18 September 2014; pp. 1078–1085.
121. Rafin, S.M.S.H.; Lipo, T.A.; Kwon, B.-i. A novel topology for a voltage source inverter with reduced transistor count and utilizing naturally commutated thyristors with simple commutation. In Proceedings of the 2014 International Symposium on Power Electronics, Electrical Drives, Automation and Motion, Ischia, Italy, 18–20 June 2014; pp. 643–648.
122. Ji, S.; Zhang, Z.; Wang, F. Overview of high voltage SiC power semiconductor devices: Development and application. *CES Trans. Electr. Mach. Syst.* **2017**, *1*, 254–264. [CrossRef]
123. Lovati, S. SiC Technology: Challenges and Future Perspectives. Available online: https://www.powerelectronicsnews.com/sic-technology-challenges-and-future-perspectives/ (accessed on 29 September 2023).
124. Ji, S.; Zheng, S.; Wang, F.; Tolbert, L.M. Temperature-dependent characterization, modeling, and switching speed-limitation analysis of third-generation 10-kV SiC MOSFET. *IEEE Trans. Power Electron.* **2017**, *33*, 4317–4327. [CrossRef]
125. Passmore, B.; Cole, Z.; McGee, B.; Wells, M.; Stabach, J.; Bradshaw, J.; Shaw, R.; Martin, D.; McNutt, T.; VanBrunt, E. The next generation of high voltage (10 kV) SiC power modules. In Proceedings of the 2016 IEEE 4th Workshop on Wide Bandgap Power Devices and Applications (WiPDA), Fayetteville, AR, USA, 7–9 November 2016; pp. 1–4.
126. Jackson, D. Reliability through increased safety insulation systems: The effect of high-speed switching on the motor insulation system. *Power Eng. J.* **2000**, *14*, 174–181. [CrossRef]
127. Bolotnikov, A.; Losee, P.; Permuy, A.; Dunne, G.; Kennerly, S.; Rowden, B.; Nasadoski, J.; Harfman-Todorovic, M.; Raju, R.; Tao, F. Overview of 1.2 kV–2.2 kV SiC MOSFETs targeted for industrial power conversion applications. In Proceedings of the 2015 IEEE Applied Power Electronics Conference and Exposition (APEC), Charlotte, NC, USA, 15–19 March 2015; pp. 2445–2452.
128. Armstrong, K.O.; Das, S.; Cresko, J. Wide bandgap semiconductor opportunities in power electronics. In Proceedings of the 2016 IEEE 4th Workshop on Wide Bandgap Power Devices and Applications (WiPDA), Fayetteville, AR, USA, 7–9 November 2016; pp. 259–264.

129. Sadik, D.-P.; Colmenares, J.; Tolstoy, G.; Peftitsis, D.; Bakowski, M.; Rabkowski, J.; Nee, H.-P. Short-circuit protection circuits for Si-carbide power transistors. *IEEE Trans. Ind. Electron.* **2015**, *63*, 1995–2004. [CrossRef]

130. Hu, J.; Alatise, O.; Gonzalez, J.A.O.; Bonyadi, R.; Alexakis, P.; Ran, L.; Mawby, P. Robustness and balancing of parallel-connected power devices: SiC versus CoolMOS. *IEEE Trans. Ind. Electron.* **2015**, *63*, 2092–2102. [CrossRef]

131. Al Mamun, M.; Paudyal, S.; Kamalasadan, S. Dynamics of Photovoltaic System with Smart Inverter Functions using Phasor Domain Model. In Proceedings of the 2023 IEEE Industry Applications Society Annual Meeting, Nashville, TN, USA, 29 October–2 November 2023; pp. 1–6.

132. Ning, P.; Liang, Z.; Wang, F. Double-sided cooling design for novel planar module. In Proceedings of the 2013 Twenty-Eighth Annual IEEE Applied Power Electronics Conference and Exposition (APEC), Long Beach, CA, USA, 17–21 March 2013; pp. 616–621.

133. Laloya, E.; Lucia, O.; Sarnago, H.; Burdio, J.M. Heat management in power converters: From state of the art to future ultrahigh efficiency systems. *IEEE Trans. Power Electron.* **2015**, *31*, 7896–7908. [CrossRef]

134. Ning, P.; Zhang, D.; Lai, R.; Jiang, D.; Wang, F.; Boroyevich, D.; Burgos, R.; Karimi, K.; Immanuel, V.D.; Solodovnik, E.V. High-temperature hardware: Development of a 10-kW high-temperature, high-power-density three-phase ac-dc-ac SiC converter. *IEEE Ind. Electron. Mag.* **2013**, *7*, 6–17. [CrossRef]

135. She, X.; Datta, R.; Todorovich, M.H.; Mandrusiak, G.; Dai, J.; Frangieh, T.; Cioffi, P.; Rowden, B.; Mueller, F. High performance SiC power block for industry applications. In Proceedings of the 2016 IEEE Energy Conversion Congress and Exposition (ECCE), Milwaukee, WI, USA, 18–22 September 2016; pp. 1–6.

136. Egawa, T. Development of next generation devices amidst global competition due to their huge market potential. *Ultim. Vac. ULVAC* **2012**, *63*, 18–21.

137. Takahashi, K.; Yoshikawa, A.; Sandhu, A. *Wide Bandgap Semiconductors*; Springer: Berlin/Heidelberg, Germany, 2007.

138. Navitassemi. Available online: https://navitassemi.com/gallium-nitride-the-next-generation-of-power/ (accessed on 29 September 2023).

139. Kaminski, N.; Hilt, O. SiC and GaN devices–wide bandgap is not all the same. *IET Circuits Devices Syst.* **2014**, *8*, 227–236. [CrossRef]

140. Mishra, U.K.; Shen, L.; Kazior, T.E.; Wu, Y.-F. GaN-based RF power devices and amplifiers. *IEEE* **2008**, *96*, 287–305. [CrossRef]

141. Meneghesso, G.; Meneghini, M.; Zanoni, E. *GaN-Enabled High Frequency and High Efficiency Power Conversion*; Springer: Berlin/Heidelberg, Germany, 2018.

142. Chou, P.-C.; Chen, S.-H.; Hsieh, T.-E.; Cheng, S.; Del Alamo, J.A.; Chang, E.Y. Evaluation and reliability assessment of GaN-on-Si MIS-HEMT for power switching applications. *Energies* **2017**, *10*, 233. [CrossRef]

143. Flack, T.J.; Pushpakaran, B.N.; Bayne, S.B. GaN technology for power electronic applications: A review. *J. Electron. Mater.* **2016**, *45*, 2673–2682. [CrossRef]

144. Runton, D.W.; Trabert, B.; Shealy, J.B.; Vetury, R. History of GaN: High-power RF GaN from infancy to manufacturable process and beyond. *IEEE Microw. Mag.* **2013**, *14*, 82–93. [CrossRef]

145. Lidow, A. GaN as a displacement technology for Si in power management. In Proceedings of the 2011 ieee energy conversion congress and exposition, Phoenix, AZ, USA, 17–22 September 2011; pp. 1–6.

146. Parikh, P.; Wu, Y.; Shen, L. Commercialization of high 600V GaN-on-Si power HEMTs and diodes. In Proceedings of the 2013 IEEE Energytech, Cleveland, OH, USA, 21–23 May 2013; pp. 1–5.

147. Rohm. Available online: https://www.rohm.com/products/gan-power-devices (accessed on 29 September 2023).

148. Meneghini, M.; De Santi, C.; Abid, I.; Buffolo, M.; Cioni, M.; Khadar, R.A.; Nela, L.; Zagni, N.; Chini, A.; Medjdoub, F. GaN-based power devices: Physics, reliability, and perspectives. *J. Appl. Phys.* **2021**, *130*, 181101. [CrossRef]

149. Oka, T. Recent development of vertical GaN power devices. *Jpn. J. Appl. Phys.* **2019**, *58*, SB0805. [CrossRef]

150. Pu, T.; Younis, U.; Chiu, H.-C.; Xu, K.; Kuo, H.-C.; Liu, X. Review of recent progress on vertical GaN-based PN diodes. *Nanoscale Res. Lett.* **2021**, *16*, 101. [CrossRef]

151. Gu, H.; Tian, F.; Zhang, C.; Xu, K.; Wang, J.; Chen, Y.; Deng, X.; Liu, X. Recovery performance of Ge-doped vertical GaN Schottky barrier diodes. *Nanoscale Res. Lett.* **2019**, *14*, 1–6. [CrossRef]

152. Liu, X.; Liu, Q.; Li, C.; Wang, J.; Yu, W.; Xu, K.; Ao, J.-P. 1.2 kV GaN Schottky barrier diodes on free-standing GaN wafer using a CMOS-compatible contact material. *Jpn. J. Appl. Phys.* **2017**, *56*, 026501. [CrossRef]

153. Zhan, A.; Dang, G.T.; Ren, F.; Cho, H.; Lee, K.-P.; Pearton, S.J.; Chyi, J.-I.; Nee, T.-Y.; Chuo, C.-C. Comparison of GaN pin and Schottky rectifier performance. *IEEE Trans. Electron Devices* **2001**, *48*, 407–411. [CrossRef]

154. Zhang, A.; Johnson, J.; Ren, F.; Han, J.; Polyakov, A.; Smirnov, N.; Govorkov, A.; Redwing, J.M.; Lee, K.; Pearton, S. Lateral Al x Ga 1− x N power rectifiers with 9.7 kV reverse breakdown voltage. *Appl. Phys. Lett.* **2001**, *78*, 823–825. [CrossRef]

155. Millan, J.; Godignon, P.; Perpiñà, X.; Pérez-Tomás, A.; Rebollo, J. A survey of wide bandgap power semiconductor devices. *IEEE Trans. Power Electron.* **2013**, *29*, 2155–2163. [CrossRef]

156. Placidi, M.; Pérez-Tomás, A.; Constant, A.; Rius, G.; Mestres, N.; Millán, J.; Godignon, P. Effects of cap layer on ohmic Ti/Al contacts to Si+ implanted GaN. *Appl. Surf. Sci.* **2009**, *255*, 6057–6060. [CrossRef]

157. Ohta, H.; Kaneda, N.; Horikiri, F.; Narita, Y.; Yoshida, T.; Mishima, T.; Nakamura, T. Vertical GaN pn junction diodes with high breakdown voltages over 4 kV. *IEEE Electron Device Lett.* **2015**, *36*, 1180–1182. [CrossRef]

158. Hu, Z.; Nomoto, K.; Song, B.; Zhu, M.; Qi, M.; Pan, M.; Gao, X.; Protasenko, V.; Jena, D.; Xing, H.G. Near unity ideality factor and Shockley-Read-Hall lifetime in GaN-on-GaN pn diodes with avalanche breakdown. *Appl. Phys. Lett.* **2015**, *107*, 243501. [CrossRef]

159. Kizilyalli, I.C.; Edwards, A.P.; Aktas, O.; Prunty, T.; Bour, D. Vertical power pn diodes based on bulk GaN. *IEEE Trans. Electron Devices* **2014**, *62*, 414–422. [CrossRef]

160. Armstrong, A.; Allerman, A.; Fischer, A.; King, M.; Van Heukelom, M.; Moseley, M.; Kaplar, R.; Wierer, J.; Crawford, M.; Dickerson, J. High voltage and high current density vertical GaN power diodes. *Electron. Lett.* **2016**, *52*, 1170–1171. [CrossRef]

161. Flicker, J.; Kaplar, R. Design optimization of GaN vertical power diodes and comparison to Si and SiC. In Proceedings of the 2017 IEEE 5th Workshop on Wide Bandgap Power Devices and Applications (WiPDA), Albuquerque, NM, USA, 30 October–1 November 2017; pp. 31–38.

162. Liu, A.-C.; Lai, Y.-Y.; Chen, H.-C.; Chiu, A.-P.; Kuo, H.-C. A Brief Overview of the Rapid Progress and Proposed Improvements in GaN Epitaxy and Process for Third-Generation Semiconductors with Wide Bandgap. *Micromachines* **2023**, *14*, 764. [CrossRef]

163. Dang, G.; Zhang, A.; Mshewa, M.; Ren, F.; Chyi, J.-I.; Lee, C.-M.; Chuo, C.; Chi, G.; Han, J.; Chu, S. High breakdown voltage Au/Pt/GaN Schottky diodes. *J. Vac. Sci. Technol. A Vac. Surf. Film.* **2000**, *18*, 1135–1138. [CrossRef]

164. Arslan, E.; Altındal, Ş.; Özçelik, S.; Ozbay, E. Dislocation-governed current-transport mechanism in (Ni/Au)–AlGaN/AlN/GaN heterostructures. *J. Appl. Phys.* **2009**, *105*, 023705. [CrossRef]

165. Applications of Microchip. Available online: https://www.microchip.com/en-us/solutions (accessed on 29 September 2023).

166. Nexgen Power Systems: GaN Semiconductor & Power Devices. Available online: https://nexgenpowersystems.com/technology/vertical-gan-semiconductor (accessed on 29 September 2023).

167. Fu, H.; Fu, K.; Chowdhury, S.; Palacios, T.; Zhao, Y. Vertical GaN power devices: Device principles and fabrication technologies—Part II. *IEEE Trans. Electron Devices* **2021**, *68*, 3212–3222. [CrossRef]

168. Zheng, Z.; Zhang, L.; Song, W.; Chen, T.; Feng, S.; Ng, Y.H.; Sun, J.; Xu, H.; Yang, S.; Wei, J. Threshold voltage instability of enhancement-mode GaN buried p-channel MOSFETs. *IEEE Electron Device Lett.* **2021**, *42*, 1584–1587. [CrossRef]

169. Ma, C.; GaN High-Electron-Mobility Transistor. Encyclopedia. Available online: https://encyclopedia.pub/entry/158 (accessed on 29 September 2023).

170. Hari, N.; Long, T.; Shelton, E. Investigation of gate drive strategies for high voltage GaN HEMTs. *Energy Procedia* **2017**, *117*, 1152–1159. [CrossRef]

171. Zhang, Y.; Li, J.; Wang, J.; Zheng, T.Q.; Jia, P. Dynamic-State Analysis of Inverter Based on Cascode GaN HEMTs for PV Application. *Energies* **2022**, *15*, 7791. [CrossRef]

172. Roberts, J.; Scott, I.H. Power Switching Systems Comprising High Power E-Mode GaN Transistors and Driver Circuitry. US Patent 9,525,413, 20 December 2016.

173. Dong, M.; Elmes, J.; Peper, M.; Batarseh, I.; Shen, Z.J. Investigation on inherently safe gate drive techniques for normally-on wide bandgap power semiconductor switching devices. In Proceedings of the 2009 IEEE Energy Conversion Congress and Exposition, San Jose, CA, USA, 20–24 September 2009; pp. 120–125.

174. Ishibashi, T.; Okamoto, M.; Hiraki, E.; Tanaka, T.; Hashizume, T.; Kachi, T. Resonant gate driver for normally-on GaN high-electron-mobility transistor. In Proceedings of the 2013 IEEE ECCE Asia Downunder, Melbourne, VIC, Australia, 3–6 June 2013; pp. 365–371.

175. Huang, X.; Li, Q.; Liu, Z.; Lee, F.C. Analytical loss model of high voltage GaN HEMT in cascode configuration. *IEEE Trans. Power Electron.* **2013**, *29*, 2208–2219. [CrossRef]

176. EPC. Available online: https://epc-co.com/epc/gallium-nitride/where-is-gan-going (accessed on 29 September 2023).

177. Electronics-Notes. Available online: https://www.electronics-notes.com/articles/electronic_components/fet-field-effect-transistor/gallium-nitride-gan-fet-hemt.php (accessed on 29 September 2023).

178. Morita, T.; Yanagihara, M.; Ishida, H.; Hikita, M.; Kaibara, K.; Matsuo, H.; Uemoto, Y.; Ueda, T.; Tanaka, T.; Ueda, D. 650 V 3.1 mΩcm 2 GaN-based monolithic bidirectional switch using normally-off gate injection transistor. In Proceedings of the 2007 IEEE International Electron Devices Meeting, Washington, DC, USA, 10–12 December 2007; pp. 865–868.

179. Uemoto, Y.; Morita, T.; Ikoshi, A.; Umeda, H.; Matsuo, H.; Shimizu, J.; Hikita, M.; Yanagihara, M.; Ueda, T.; Tanaka, T. GaN monolithic inverter IC using normally-off gate injection transistors with planar isolation on Si substrate. In Proceedings of the 2009 IEEE International Electron Devices Meeting (IEDM), Baltimore, MD, USA, 7–9 December 2009; pp. 1–4.

180. Ueda, T. GaN power devices: Current status and future challenges. *Jpn. J. Appl. Phys.* **2019**, *58*, SC0804. [CrossRef]

181. Hu, J.; Zhang, Y.; Sun, M.; Piedra, D.; Chowdhury, N.; Palacios, T. Materials and processing issues in vertical GaN power electronics. *Mater. Sci. Semicond. Process.* **2018**, *78*, 75–84. [CrossRef]

182. Mizutani, T.; Ohno, Y.; Akita, M.; Kishimoto, S.; Maezawa, K. A study on current collapse in AlGaN/GaN HEMTs induced by bias stress. *IEEE Trans. Electron Devices* **2003**, *50*, 2015–2020. [CrossRef]

183. Yu, C.-H.; Luo, X.-D.; Zhou, W.-Z.; Luo, Q.-Z.; Liu, P.-S. Investigation on the current collapse effect of AlGaN/GaN/InGaN/GaN double-heterojunction HEMTs. *Acta Phys. Sin.* **2012**, *61*, 207301.

184. Meneghini, M.; Vanmeerbeek, P.; Silvestri, R.; Dalcanale, S.; Banerjee, A.; Bisi, D.; Zanoni, E.; Meneghesso, G.; Moens, P. Temperature-dependent dynamic R_{ON} in GaN-Based MIS-HEMTs: Role of surface traps and buffer leakage. *IEEE Trans. Electron Devices* **2015**, *62*, 782–787. [CrossRef]

185. Uren, M.J.; Moreke, J.; Kuball, M. Buffer design to minimize current collapse in GaN/AlGaN HFETs. *IEEE Trans. Electron Devices* **2012**, *59*, 3327–3333. [CrossRef]

186. Blackwell, G.R. Circuit boards. In *The Electronic Packaging Handbook*; CRC Press: Boca Raton, FL, USA, 2017.
187. Islam, R.; Rafin, S.M.S.H.; Mohammed, O.A. Comprehensive Review of Power Electronic Converters in Electric Vehicle Applications. *Forecasting* **2022**, *5*, 22–80. [CrossRef]
188. Rafin, S.M.S.H.; Islam, R.; Mohammed, O.A. Overview of Power Electronic Converters in Electric Vehicle Applications. In Proceedings of the 2023 Fourth International Symposium on 3D Power Electronics Integration and Manufacturing (3D-PEIM), Miami, FL, USA, 1–3 February 2023; pp. 1–7.
189. Keshmiri, N.; Wang, D.; Agrawal, B.; Hou, R.; Emadi, A. Current status and future trends of GaN HEMTs in electrified transportation. *IEEE Access* **2020**, *8*, 70553–70571. [CrossRef]
190. Amano, H.; Baines, Y.; Beam, E.; Borga, M.; Bouchet, T.; Chalker, P.R.; Charles, M.; Chen, K.J.; Chowdhury, N.; Chu, R. The 2018 GaN power electronics roadmap. *J. Phys. D Appl. Phys.* **2018**, *51*, 163001. [CrossRef]
191. Power Electronics. Available online: https://www.powerelectronicsnews.com/ces-2022-gan-technology-for-the-next-future (accessed on 29 September 2023).
192. Dusmez, S.; Fu, L.; Beheshti, M.; Brohlin, P.; Gao, R. Overcurrent Protection in High-Density GaN Power Designs. Available online: https://www.ti.com/lit/an/snoaa15/snoaa15.pdf (accessed on 29 September 2023).
193. Guide, G.A. *Design with Gan Enhancement Mode Hemt*; GaN Systems Inc.: Ottawa, ON, Canada, 2018; Volume 12.
194. Lovati, S. GaN Technology: Challenges and Future Perspectives. Available online: https://www.powerelectronicsnews.com/sic-technology-challenges-and-future-perspectives/ (accessed on 29 September 2023).
195. He, J. Comparison between The ultra-wide band gap semiconductor AlGaN and GaN. *IOP Conf. Ser. Mater. Sci. Eng.* **2020**, *738*, 012009. [CrossRef]
196. Millan, J. A review of WBG power semiconductor devices. In Proceedings of the CAS 2012 (International Semiconductor Conference), Sinaia, Romania, 15–17 October 2012; pp. 57–66.
197. Lo Nigro, R.; Fiorenza, P.; Greco, G.; Schilirò, E.; Roccaforte, F. Structural and Insulating Behaviour of High-Permittivity Binary Oxide Thin Films for SiC and GaN Electronic Devices. *Materials* **2022**, *15*, 830. [CrossRef]
198. Devi, S.; Seyezhai, R. Comparative analysis of Si, SiC and GaN based quasi impedance source inverter. *Mater. Today Proc.* **2022**, *62*, 787–792. [CrossRef]
199. Hassan, A.; Savaria, Y.; Sawan, M. GaN integration technology, an ideal candidate for high-temperature applications: A review. *IEEE Access* **2018**, *6*, 78790–78802. [CrossRef]
200. Ozpineci, B.; Tolbert, L.M. *Comparison of Wide-Bandgap Semiconductors for Power Electronics Applications*; United States Department of Energy: Washington, DC, USA, 2004.
201. Baliga, B.J. *Fundamentals of Power Semiconductor Devices*; Springer Science & Business Media: Berlin/Heidelberg, Germany, 2010.
202. Maier, D.; Alomari, M.; Grandjean, N.; Carlin, J.-F.; Diforte-Poisson, M.-A.; Dua, C.; Chuvilin, A.; Troadec, D.; Gaquière, C.; Kaiser, U. Testing the temperature limits of GaN-based HEMT devices. *IEEE Trans. Device Mater. Reliab.* **2010**, *10*, 427–436. [CrossRef]
203. Shenai, K.; Scott, R.S.; Baliga, B.J. Optimum semiconductors for high-power electronics. *IEEE Trans. Electron Devices* **1989**, *36*, 1811–1823. [CrossRef]
204. Singh, S.; Chaudhary, T.; Khanna, G. Recent advancements in wide band semiconductors (SiC and GaN) technology for future devices. *Si* **2022**, *14*, 5793–5800. [CrossRef]
205. Infineon: SiC MOSFET in Compact SMD Package-IMBG65R022M1H. Available online: https://www.infineon.com/cms/en/product/power/mosfet/Si-carbide/discretes/imbg65r022m1h/ (accessed on 29 September 2023).
206. Meneghini, M.; Meneghesso, G.; Zanoni, E. *Power GaN Devices*; Springer International Publishing: Cham, Switzerland, 2017.
207. Wang, Z.; Zhang, Z.; Wang, S.; Chen, C.; Wang, Z.; Yao, Y. Design and optimization on a novel high-performance ultra-thin barrier AlGaN/GaN power HEMT with local charge compensation trench. *Appl. Sci.* **2019**, *9*, 3054. [CrossRef]
208. Wang, Z.; Li, L. Two-dimensional polarization doping of GaN heterojunction and its potential for realizing lateral p–n junction devices. *Appl. Phys. A* **2022**, *128*, 672. [CrossRef]
209. Wang, Z.; Yang, D.; Cao, J.; Wang, F.; Yao, Y. A novel technology for turn-on voltage reduction of high-performance lateral heterojunction diode with source-gate shorted anode. *Superlattices Microstruct.* **2019**, *125*, 144–150. [CrossRef]
210. Kaess, F.; Mita, S.; Xie, J.; Reddy, P.; Klump, A.; Hernandez-Balderrama, L.H.; Washiyama, S.; Franke, A.; Kirste, R.; Hoffmann, A. Correlation between mobility collapse and carbon impurities in Si-doped GaN grown by low pressure metalorganic chemical vapor deposition. *J. Appl. Phys.* **2016**, *120*, 105701. [CrossRef]
211. Kizilyalli, I.C.; Edwards, A.P.; Nie, H.; Disney, D.; Bour, D. High voltage vertical GaN pn diodes with avalanche capability. *IEEE Trans. Electron Devices* **2013**, *60*, 3067–3070. [CrossRef]
212. Pakes, C.I.; Garrido, J.A.; Kawarada, H. Diamond surface conductivity: Properties, devices, and sensors. *Mrs Bull.* **2014**, *39*, 542–548. [CrossRef]
213. Bose, B.K. *Power Electronics in Renewable Energy Systems and Smart Grid: Technology and Applications*; John Wiley and Sons: Hoboken, NJ, USA, 2019.
214. Adler, M.; Temple, V. Maximum surface and bulk electric fields at breakdown for planar and beveled devices. *IEEE Trans. Electron. Devices* **1978**, *25*, 1266–1270. [CrossRef]
215. Cornu, J. Field distribution near the surface of beveled pn junctions in high-voltage devices. *IEEE Trans. Electron. Devices* **1973**, *20*, 347–352. [CrossRef]

216. Przybilla, J.; Dorn, J.; Barthelmess, R.; Kellner-Werdehausen, U.; Schulze, H.-J.; Niedernostheide, F.-J. Diodes and thyristor—Past, presence and future. In Proceedings of the 2009 13th european conference on Power electronics and applications, Barcelona, Spain, 8–10 September 2009; pp. 1–10.

217. Parameters of Semiconductor Compounds and Heterostructures. Available online: http://www.ioffe.ru/SVA/NSM/Semicond/ (accessed on 29 September 2023).

218. Takaku, T.; Igarashi, S.; Nishimura, T.; Onozawa, Y.; Miyashita, S.; Ikawa, O.; Fujishima, N.; Heinzel, T. 1700V Hybrid Module with Si-IGBT and SiC-SBD for High Efficiency AC690V Application. In Proceedings of the Pcim europe 2015; international exhibition and conference for Power electronics, intelligent motion, renewable energy and energy management, Nuremberg, Germany, 19–20 May 2015; pp. 1–7.

219. Kaneko, S.; Kanai, N.; Hori, M.; Masayoshi, N.; Kakiki, H.; Abe, Y.; Ikeda, Y.; Mochizuki, E. Compact, low loss and high reliable 3.3 kV hybrid power module. In Proceedings of the Pcim Europe 2016; International Exhibition and Conference for Power Electronics, Intelligent Motion, Renewable Energy and Energy Management, 2016; pp. 1–7.

220. Huang, A.Q. Power semiconductor devices for smart grid and renewable energy systems. In *Power Electronics in Renewable Energy Systems and Smart Grid: Technology and Applications*; The Institute of Electrical and Electronics Engineers, Inc.: Piscataway, NJ, USA, 2019; pp. 85–152.

221. Shenai, K. The invention and demonstration of the IGBT [a look back]. *IEEE Power Electron. Mag.* **2015**, *2*, 12–16. [CrossRef]

222. Infineon. Available online: https://www.infineon.com/cms/en/product/power/igbt/igbt-modules/igbt-modules-up-to-4500v-6500v/ (accessed on 29 September 2023).

223. Rahimo, M.; Kopta, A.; Schlapbach, U.; Vobecky, J.; Schnell, R.; Klaka, S. The Bi-mode Insulated Gate Transistor (BiGT) A potential technology for higher power applications. In Proceedings of the 2009 21st International Symposium on Power Semiconductor Devices & IC's, Barcelona, Spain, 14–18 June 2009; pp. 283–286.

224. Voss, S.; Niedernostheide, F.; Schulze, H. Anode design variation in 1200-V trench field-stop reverse-conducting IGBTs. In Proceedings of the 2008 20th International Symposium on Power Semiconductor Devices and IC's, Orlando, FL, USA, 18–22 May 2008; pp. 169–172.

225. Rahimo, M.; Schlapbach, U.; Kopta, A.; Vobecky, J.; Schneider, D.; Baschnagel, A. A high current 3300V module employing reverse conducting IGBTs setting a new benchmark in output power capability. In Proceedings of the 2008 20th International Symposium on Power Semiconductor Devices and IC's, Orlando, FL, USA, 18–22 May 2008; pp. 68–71.

226. Huang, Q.; Amaratunga, G. MOS Controlled Diodes—A new power diode. *Solid-State Electron.* **1995**, *38*, 977–980. [CrossRef]

227. Xu, Z.; Zhang, B.; Huang, A.Q. An analysis and experimental approach to MOS controlled diodes behavior. *IEEE Trans. Power Electron.* **2000**, *15*, 916–922.

228. Xu, Z.; Zhang, B.; Huang, A.Q. Experimental demonstration of the MOS controlled diode (MCD). In Proceedings of the APEC 2000. Fifteenth Annual IEEE Applied Power Electronics Conference and Exposition (Cat. No. 00CH37058), New Orleans, LA, USA, 6–10 February 2000; pp. 1144–1148.

229. Schweizer, M.; Kolar, J.W. Design and implementation of a highly efficient three-level T-type converter for low-voltage applications. *IEEE Trans. Power Electron.* **2012**, *28*, 899–907. [CrossRef]

230. Naito; Takei; Nemoto; Hayashi; Ueno. 1200V reverse blocking IGBT with low loss for matrix converter. In Proceedings of the 2004 Proceedings of the 16th International Symposium on Power Semiconductor Devices and ICs, Kitakyushu, Japan, 24–27 May 2004; pp. 125–128.

231. Kopta, A. High voltage Si based devices for energy efficient power distribution and consumption. In Proceedings of the 2014 IEEE International Electron Devices Meeting, San Francisco, CA, USA, 15–17 December 2014; pp. 2.4.1–2.4.4.

232. Li, Y.; Huang, A.Q.; Lee, F.C. Introducing the emitter turn-off thyristor (ETO). In Proceedings of the Conference Record of 1998 IEEE Industry Applications Conference. Thirty-Third IAS Annual Meeting (Cat. No. 98CH36242), St. Louis, MO, USA, 12–15 October 1998; pp. 860–864.

233. Zhang, B.; Liu, Y.; Zhou, X.; Hawley, J.; Huang, A.Q. The high power and high frequency operation of the emitter turn-off (ETO) thyristor. In Proceedings of the IECON'03. 29th Annual Conference of the IEEE Industrial Electronics Society (IEEE Cat. No. 03CH37468), Roanoke, VA, USA, 2–6 November 2003; pp. 1167–1172.

234. Zhang, B.; Huang, A.Q.; Zhou, X.; Liu, Y.; Atcitty, S. The built-in current sensor and over-current protection of the emitter turn-off (ETO) thyristor. In Proceedings of the 38th IAS Annual Meeting on Conference Record of the Industry Applications Conference, Salt Lake City, UT, USA, 12–16 October 2003; pp. 1264–1269.

235. Wang, H.; Wei, J.; Xie, R.; Liu, C.; Tang, G.; Chen, K.J. Maximizing the performance of 650-V p-GaN gate HEMTs: Dynamic RON characterization and circuit design considerations. *IEEE Trans. Power Electron.* **2016**, *32*, 5539–5549. [CrossRef]

236. Sharma, A.; Lee, S.J.; Jang, Y.J.; Jung, J.P. SiC based technology for high power electronics and packaging applications. *J. Microelectron. Packag. Soc.* **2014**, *21*, 71–78. [CrossRef]

237. Kizilyalli, I.C.; Carlson, E.P.; Cunningham, D.W.; Manser, J.S.; Xu, Y.A.; Liu, A.Y. *Wide Band-Gap Semiconductor Based Power Electronics for Energy Efficiency*; US Department of Energy (USDOE): Washington DC, USA, 2018.

238. Das, S.; Marlino, L.D.; Armstrong, K.O. *Wide Bandgap Semiconductor Opportunities in Power Electronics*; Oak Ridge National Lab.(ORNL): Oak Ridge, TN, USA, 2018.

239. Wang, F.F.; Zhang, Z. Overview of SiC technology: Device, converter, system, and application. *CPSS Trans. Power Electron. Appl.* **2016**, *1*, 13–32. [CrossRef]

240. Market-US. Available online: https://market.us/report/power-electronics-market/request-sample/ (accessed on 29 September 2023).
241. Onsemi: 1200 V, 30 A Si Diode. Available online: https://www.onsemi.com/products/discrete-power-modules/power-modules/igbt-modules/nxh80b120l2q0 (accessed on 29 September 2023).
242. SiC Diodes | FFSM1065A. Available online: https://www.onsemi.com/products/discrete-power-modules/Si-carbide-sic/Si-carbide-sic-diodes/ffsm1065a (accessed on 29 September 2023).
243. Onsemi: SiC MOSFET-NVBG095N065SC1. Available online: https://www.onsemi.com/download/data-sheet/pdf/nvbg095n065sc1-d.pdf (accessed on 29 September 2023).
244. Onsemi: SiC MOSFET-NTH4L028N170M1. Available online: https://www.onsemi.com/download/data-sheet/pdf/nth4l028n170m1-d.pdf (accessed on 29 September 2023).
245. Infineon: GaN CoolGaN™ 600V Enhancement-Mode Power Transistor. Available online: https://www.infineon.com/cms/en/product/power/gan-hemt-gallium-nitride-transistor/iglr60r340d1/ (accessed on 29 September 2023).
246. Infineon: GaN CoolGaN™ 400V Enhancement-MODE Power Transistor. Available online: https://www.infineon.com/cms/en/product/power/gan-hemt-gallium-nitride-transistor/igt40r070d1/ (accessed on 29 September 2023).
247. Infineon: Fast Switching Diode. Available online: https://www.infineon.com/cms/en/product/power/mosfet/n-channel/optimos-and-strongirfet-latest-family-selection-guide/optimos-fast-diode-200v-220-250v-300v/ (accessed on 29 September 2023).
248. Infineon: Fast Switching Emitter Controlled Diode. Available online: https://www.infineon.com/dgdl/IDB18E120_rev2_3G.pdf?fileId=5546d4614815da880148406b0af611e4 (accessed on 29 September 2023).
249. Infineon: 2000 V, 60 A Boost EasyPACK™ 3B CoolSiC™ MOSFET Module-DF4-19MR20W3M1HF_B11. Available online: https://www.infineon.com/cms/en/product/power/mosfet/Si-carbide/modules/df4-19mr20w3m1hf_b11/ (accessed on 29 September 2023).
250. Infineon: HybridPACK™DriveModule-FS380R12A6T4B. Available online: https://www.infineon.com/dgdl/Infineon-FS380R12A6T4B-DataSheet-v03_01-EN.pdf?fileId=5546d4626da6c043016db9e750e268b1 (accessed on 29 September 2023).
251. Infineon: IGBT-Module-F4-50R07W1H3_B11A. Available online: https://www.infineon.com/dgdl/Infineon-F4-50R07W1H3_B11A-DataSheet-v03_00-EN.pdf?fileId=db3a3043405f2978014071e0c83f2079 (accessed on 29 September 2023).
252. ROHM-Gallium-Nitride HEMT-GNE1040TB. Available online: https://www.rohm.com/products/gan-power-devices/gan-hemt/gne1040tb-product#productDetail (accessed on 29 September 2023).
253. ROHM: SiC MOSFET-SCT2H12NY. Available online: https://www.rohm.com/products/sic-power-devices/sic-mosfet/sct2h12ny-product (accessed on 29 September 2023).
254. ROHM: SiC MOSFETs. Available online: https://www.rohm.com/products/sic-power-devices/sic-mosfet (accessed on 29 September 2023).
255. ROHM: Power MOSFET-UT6MA3. Available online: https://www.rohm.com/products/mosfets/small-signal/dual/ut6ma3-product (accessed on 29 September 2023).
256. ROHM: Power MOSFET-QH8M22. Available online: https://www.rohm.com/products/mosfets/small-signal/dual/qh8m22-product#productDetail (accessed on 29 September 2023).
257. ROHM: Ignition IGBT. Available online: https://www.rohm.com/products/igbt/ignition-igbt?page=1&CollectorCurrent100_num=20.030.0#parametricSearch (accessed on 29 September 2023).
258. ST Microelectronics: 600 V Schottky SiC Diode-STPSC2006CW. Available online: https://www.st.com/en/diodes-and-rectifiers/stpsc2006cw.html (accessed on 29 September 2023).
259. ST Microelectronics: 1200 V, 40 A High Surge SiC Power Schottky Diode-STPSC40H12C. Available online: https://www.st.com/en/diodes-and-rectifiers/stpsc40h12c.html (accessed on 29 September 2023).
260. ST Microelectronics: SiC Power MOSFET-SCT1000N170. Available online: https://www.st.com/en/power-transistors/sct1000n170.html (accessed on 29 September 2023).
261. ST Microelectronics: GaN HEMTs-MASTERGAN1. Available online: https://www.st.com/en/power-management/mastergan1.html (accessed on 29 September 2023).
262. ST Microelectronics: GaN HEMTs-MASTERGAN3. Available online: https://www.st.com/en/power-management/mastergan3.html (accessed on 29 September 2023).
263. ST Microelectronics: STPOWER IGBTs Bare die up to 1700V. Available online: https://www.st.com/en/power-transistors/1700v.html (accessed on 29 September 2023).

micromachines

Review

Trap Characterization Techniques for GaN-Based HEMTs: A Critical Review

Xiazhi Zou [1,2], Jiayi Yang [1,2], Qifeng Qiao [3], Xinbo Zou [4], Jiaxiang Chen [4], Yang Shi [1,2] and Kailin Ren [1,2,*]

1   School of Microelectronics, Shanghai University, Shanghai 200444, China; xiazhizou@shu.edu.cn (X.Z.); 1142048504@shu.edu.cn (J.Y.); shiyang@shu.edu.cn (Y.S.)
2   Shanghai Key Laboratory of Chips and Systems for Intelligent Connected Vehicle, Shanghai University, Shanghai 200444, China
3   Shanghai Industrial μTechnology Research Institute, Shanghai 201800, China
4   School of Information Science and Technology, ShanghaiTech University, Shanghai 201210, China; zouxb@shanghaitech.edu.cn (X.Z.); chenjx2@shanghaitech.edu.cn (J.C.)
*   Correspondence: renkailin@shu.edu.cn

Abstract: Gallium nitride (GaN) high-electron-mobility transistors (HEMTs) have been considered promising candidates for power devices due to their superior advantages of high current density, high breakdown voltage, high power density, and high-frequency operations. However, the development of GaN HEMTs has been constrained by stability and reliability issues related to traps. In this article, the locations and energy levels of traps in GaN HEMTs are summarized. Moreover, the characterization techniques for bulk traps and interface traps, whose characteristics and scopes are included as well, are reviewed and highlighted. Finally, the challenges in trap characterization techniques for GaN-based HEMTs are discussed to provide insights into the reliability assessment of GaN-based HEMTs.

Keywords: GaN HEMT; trap; characterization methods; DLTS

Citation: Zou, X.; Yang, J.; Qiao, Q.; Zou, X.; Chen, J.; Shi, Y.; Ren, K. Trap Characterization Techniques for GaN-Based HEMTs: A Critical Review. *Micromachines* **2023**, *14*, 2044. https://doi.org/10.3390/mi14112044

Academic Editors: Zeheng Wang and Jing-Kai Huang

Received: 6 October 2023
Revised: 26 October 2023
Accepted: 27 October 2023
Published: 31 October 2023

1. Introduction

Gallium nitride (GaN) high-electron-mobility transistors (HEMTs) possess superior features such as high breakdown voltage, high electron saturation drift velocity, and low ON-resistance, making them highly promising for power device applications [1,2]. Additionally, due to the polarization effect in GaN-based materials, a high-concentration two-dimensional electron gas (2DEG) channel can be formed at the AlGaN/GaN interface of GaN HEMT devices, significantly enhancing carrier mobility. Consequently, the theoretical figure of merit limit of GaN HEMT is higher than that of Si and SiC-based power electronic devices. In recent years, GaN HEMTs have been intensively studied and widely used in RF amplifiers and power electronics systems [3–5].

Despite the many benefits of GaN-based HEMT devices, the presence of traps in the bulk and at the interface can cause stability and reliability issues such as current collapse [6–11], threshold voltage drift [7,12], deterioration of short channel effect [12], and limited microwave power output [8,13], which seriously limit its large-scale applications. Therefore, trap characterization is crucial for achieving better commercial applications of GaN-based HEMT devices, which can provide more insights into the performances of devices as well as more guidelines for the optimization of the device structure and manufacturing processes to improve both the capability and reliability of GaN HEMTs. In recent decades, several techniques such as low-frequency noise (LFN), frequency dispersion properties, and deep-level transient spectroscopy (DLTS) have been developed to characterize the locations, types, concentrations, energy levels, and capture cross-sections of traps in GaN HEMTs, which are also applicable to novel GaN HEMT structures [14] and GaN diodes [15], p-n junctions [16], etc.

2. Types and Impacts of Traps

Various traps found in GaN HEMTs are classified in this chapter. As illustrated in Figure 1, the main traps in GaN HEMTs can be classified into interface traps and bulk traps according to the locations; the former ones are located mainly between the AlGaN/passivation layer, GaN/Substrate, and AlGaN/GaN heterojunction, while the latter ones are located primarily in the GaN buffer layer and AlGaN barrier layer. In addition, for metal-insulator-semiconductor heterojunction field-effect transistors (MIS-HEMT), interface defects also exist at the interface between AlGaN and insulation.

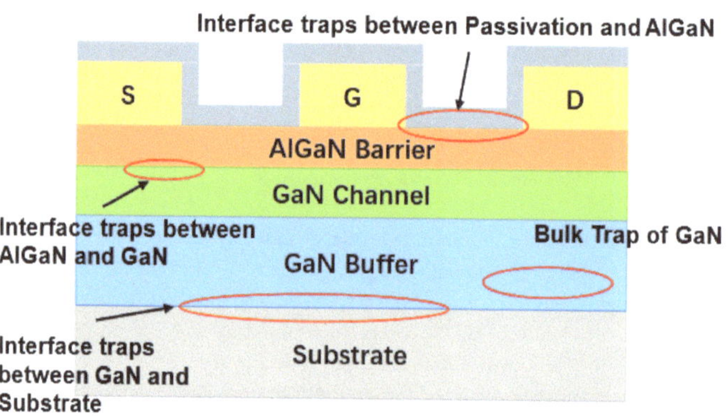

Figure 1. The locations of traps in GaN HEMTs.

The AlGaN/GaN interface and semiconductor/insulator interface traps are mainly caused by dislocations and defects generated during the growth of the material and the manufacturing processes of the device [17,18]. The reasons for the formation of bulk defects in the AlGaN barrier layer and GaN buffer layer are multifaceted. Firstly, buffer traps are introduced because of the high-resistance characteristics exhibited by the GaN buffer layer, which are usually achieved through C or Fe impurity compensation [19–23]. The 2DEG concentration of a device may be affected by the aforementioned traps [24], which, in turn, affects parameters such as current density and threshold voltage [25]. Secondly, V_{Ga}-impurity, V_{N}-impurity, Mg-H complexes, V_{N}-Mg complexes, etc., also form point defects in GaN materials, introducing deep-level traps. In addition, although there has been significant development in GaN-on-GaN homoepitaxial growth and device fabrication [26,27], a considerable portion of devices still use heteroepitaxial substrates, in which large amounts of dislocations and defects are caused by lattice mismatch during the epitaxial growth process, forming deep-level trap states in the bandgap.

The energy levels, positions, and corresponding characterization methods of the traps identified in GaN HEMT devices are shown in Figure 2 [28–66]. From this figure, it can be observed that there are traps located near the energy band of 0.6 eV in SI-GaN/UID-GaN/Si-GaN/Mg GaN, which is said to be attributed mainly to the point defects in GaN in some papers, but there are other reports saying that the source of this type of trap may be caused by V_{N}-impurity, Mg/Si-H complexes, and so on. The 0.7 eV trap near the AlGaN/GaN interface is generally believed to be caused by the spreading defects in the AlGaN/GaN heterostructure. And, beyond that, there are traps located around Ec-0.6 eV at the passivation layer and the semiconductor interface, which may be caused by the high Ga-O components near the passivation layer and semiconductor interface.

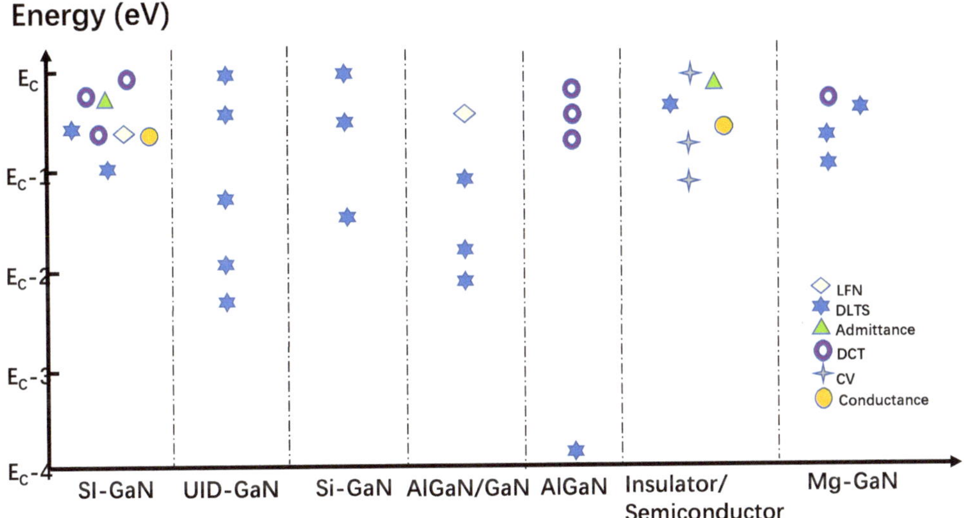

Figure 2. Energy levels and positions of traps in GaN HEMTs.

From the perspective of characterization methods, DCT, LFN, and low-frequency output admittance methods are used more frequently to characterize traps in semiconductor bodies, while other methods such as C-V and dispersion of conductance output are mainly used to characterize the traps at semiconductor interfaces. It is also illustrated that DLTS can characterize traps on the surface and in vivo based on their modes. In addition, it can be observed that DCT is usually used to characterize shallower traps, while DLTS can characterize deeper-level traps. DCT, a method that is used to characterize shallower traps, is different from DLTS since DLTS is usually used to characterize deeper traps.

From the perspective of trap location, DCT, DLTS, and C-V can identify the trap location through different voltage biases, and the conductivity method can measure semiconductor/insulator interface traps in MOS structures. As for LFN and transconductance methods, although they cannot identify trap locations, they can be used together with other characterization methods to comprehensively analyze traps.

3. Characterization Methods of Bulk Traps

3.1. Drain Current Transient

The drain current transient (DCT) test involves applying large positive V_{ds} bias, large negative V_{gs} bias, or both to measure the change in I_{DS} [48–56]. The transient current I_{DS} can be expressed as

$$I_{DS}(t) = \sum \Delta I_i \exp\left(-\frac{t}{\tau_i}\right) + I_\infty, \tag{1}$$

where ΔI_i is the amplitude, τ_i is the time constant of the trap, and I_∞ is the current which is at a steady state [51]. The Bayesian deconvolution method can be used to obtain the time constant of traps (τ_n) and the energy level and cross-section of the trap can be derived from Arrhenius plots:

$$\ln(\tau_n T^2) = -\frac{E_a}{k_B T} + \ln(\sigma_n \gamma_n), \tag{2}$$

where σ_n is the electron capture cross-section, γ_n contains the density and thermal velocity of electrons, E_a is the trap activation energy, and k_B is the Boltzmann constant [51].

The type of trap can be determined by the peak in the derivative spectrum of DCT where a positive peak represents the existence of an electron trap, while a negative peak shows that there exists a hole trap.

The process of extracting trap parameters with a leakage current response is illustrated in Figure 3. The change of I_{DS} at different temperatures during trap launch, the time constant of the trap extracted through the Bayesian convolution method, and the Arrhenius plot extracted from the transient curve are shown in Figure 3a, Figure 3b, and Figure 3c, respectively.

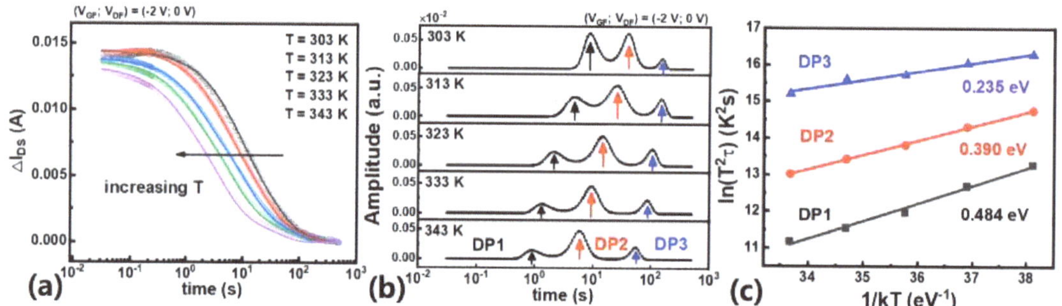

Figure 3. (Color online.) (**a**) The actual detrapping transient curves with different temperatures. (**b**) The time constant spectra at different temperatures. (**c**) The Arrhenius plots and the corresponding energy levels. Reprinted from [51], with the permission of AIP Publishing.

The physical location of the traps can still be determined by other means, although the inherent spatial sensitivity cannot be provided by this method. Different filling pulse conditions with different combinations of the gate-source voltage (V_{gs}) and the gate-source voltage (V_{ds}) stress conditions can be used to fill traps in different areas of GaN HEMTs. For instance, filling pulses with a strong negative $V_{gs} < V_{th}$ can cause electrons to fill surface channel regions located beneath the gate area or defects in both the gate electrode and drain regions [67]. The channel capture can be highlighted by applying a strong positive V_{ds} bias and a strong negative V_{gs} bias due to the increase in the electron tunneling in the drain direction. A strong positive V_{ds} bias with $V_{gs} > V_{th}$ can allow electrons to be captured in the barrier layer or buffer layer between the gate and drain, as large positive V_{gs} can scatter hot electrons out of the channel [68]. Therefore, the physical location of traps can be distinguished through DCT tests on a device performed by stress conditions.

The DCT technique is simple and can locate trap positions, but the trap density cannot be quantitatively measured and traps can only be detected with energy levels below 1 eV. Therefore, other approaches are necessary to characterize deep-level traps.

3.2. Low-Frequency Leakage Noise

Low-frequency leakage noise (LFN) is a noise signal generated in the low-frequency range, usually below a few hundred Hz, which reflects the charge and energy-level distribution inside a device. In GaN HEMT devices, the presence of traps affects the device's leakage current and conductivity [69]. A small current noise is generated by traps when a small signal voltage is applied to the device. This noise can be measured by means of low-frequency leakage noise measurement techniques [39,56,60,61]. By analyzing the noise signals at different frequencies, parameter energy levels and capture cross-sections can be obtained by analyzing the noise signals at different frequencies and extracting the time constants of the G-R. The relationship between output noise and frequency is illustrated in Figure 4a. The measured output drain noise spectral density must be multiplied by the frequency to distinguish G-R noise from other measurement noise sources, as shown in Figure 4b. The cutoff frequency of traps can be extracted at different temperatures and then the trap parameters can be extracted by using the Arrhenius equation in Figure 4c.

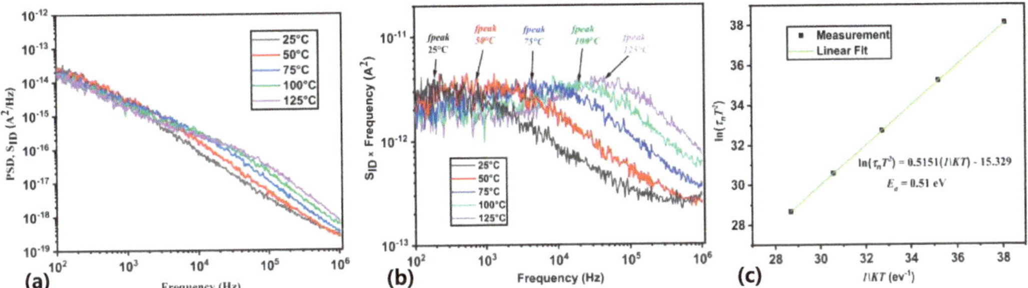

Figure 4. (Color online.) (**a**) Output noise PSD versus frequency. (**b**) Output noise PSD multiplied by frequency measured. (**c**) Extracted Arrhenius plot using LFN measurement [39].

Compared to DLTS and DCT, LFN has more difficulty in locating traps and lacks quantitative measurements of trap density, but it does not require a large reverse bias voltage to be applied to a device to degrade the de-trapping performance, and it is also capable of detecting traps in small area devices.

3.3. Low-Frequency Output Admittance Measurements

The characterization technique for low-frequency output admittance measurements characterizes traps by measuring the characteristics of their S/Y parameters as a function of frequency [39,57–59,65,70]. Then, the measured parameters can be calculated as equivalent Y_{22} parameters. Due to the influence of traps, the Y_{22} parameter obtained will reach its peak at a certain frequency. This peak will shift towards a higher frequency as the temperature increases. The emission time constant of the trap can be extracted from the peak frequency (f_{peak}) using Equation (3). The parameters of the trap can then be obtained using the Arrhenius equation.

$$f_{\text{peak}} = f_{\text{Imag}[Y_{22}]} = \frac{1}{2\pi\tau_n},$$

(3)

The relationship between the imaginary part of the measured Y_{22} parameter and the frequency is demonstrated in Figure 5a, and the Arrhenius plot of this measurement is shown in Figure 5b.

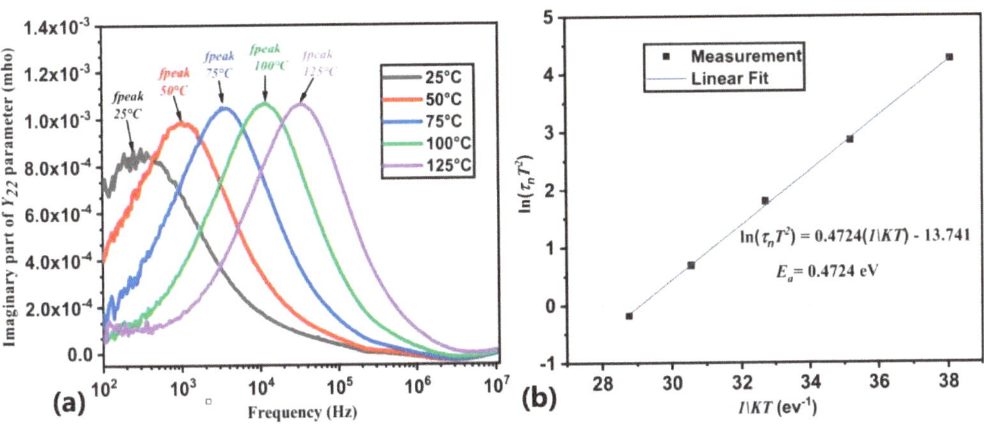

Figure 5. (Color online.) (**a**) Imaginary part of the measured Y_{22}-parameter vs. frequency. (**b**) Extracted Arrhenius plot [39].

In practical applications, the appropriate parameter to represent the dispersion of HEMT traps should be chosen based on specific requirements. Generally, if an analysis of the influence of HEMT traps on the entire system is required, the S parameter should be used as it provides the relative response between input and output ports. On the other hand, if a deeper understanding of the characteristics of HEMT traps is required, the Y parameter may be more suitable as it provides the internal response of the device, including the relationship between voltage and current.

Pulse effects such as voltage stabilization time or unstable temperature are avoided and wide dynamic range and measurement speed are provided in this method [70]. However, only energy levels and capture cross-sections for traps can be obtained through this method.

3.4. DLTS

DLTS has the advantages of being sensitive to measurement, having a wide range of detectable defect energy levels, being able to simultaneously measure both majority and minority carrier traps, and being able to determine trap positions. The processes involved in trap emission and capture during DLTS testing are summarized in Figure 6. In the steady-state condition, as indicated in Figure 6a, the energy level E_T is not occupied by electrons. The Fermi level is forced to shift towards the conduction band when applying a filling pulse bias voltage (V_f), as indicated in Figure 6b, which attracts electrons and consequently weakens the built-in electric field. The trap levels within the depletion region are situated below the Fermi level. Charges captured by traps within the depletion region with energy levels above the Fermi level will be emitted by applying a measurement voltage (V_m) that is more negative than the filling pulse voltage V_f, as shown in Figure 6c. Taking the contribution of thermal emission into account, the trap emission constant can be obtained by measuring the change in capacitance through the DLTS from time t_1 to t_2 after the pulse [71]. As for the determination of the type of trap, the method of identifying the type of peak in the derivative of capacitance versus the time plot can be applied, where a positive peak represents the trap type as an electron and a negative peak represents the trap type as a hole. Taking an n-type semiconductor as an example, the trap density can be calculated from Equation (4).

$$N_T = \frac{\Delta C_{\max}}{C_0} \frac{2N_D r^{\frac{r}{r-1}}}{r-1},$$

(4)

where r = t_2/t_1 and C_0 is the steady-state capacitance value.

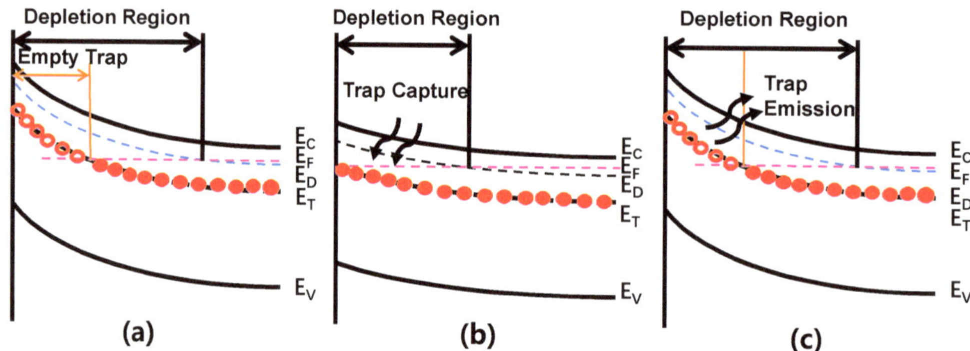

Figure 6. (Color online.) Schottky diode energy-band diagrams: (**a**) steady-state; (**b**) trap filling state; (**c**) trap emission state.

Capacitor DLTS (C-DLTS) first applies a filling pulse voltage to the device being tested and then measures the change in capacitance after the pulse to characterize the trap [35,72–75]. Traps in the AlGaN barrier layer, GaN channel layer, and buffer layer can be effectively distinguished through C-DLTS. The main principle is based on the state of the 2DEG whereby the gate capacitance primarily comes from the GaN channel layer and buffer layer, while the contribution of the barrier layer to the total capacitance is very small when the 2DEG is depleted, and the depletion region is mainly confined to the AlGaN barrier layer when the 2DEG is accumulated.

It should be noted that the transient values of C-DLTS are usually less than 10% of the total depletion capacitance. Therefore, a sufficient area must be produced to enable experimental resolution ΔC. Furthermore, it is necessary to consider the signal-to-noise ratio of DLTS devices.

Constant drain-current DLTS (CID-DLTS) is applied to obtain specific trap parameters beneath the gate by adjusting the V_{GS} to maintain a constant drain current (I_{DS}), as depicted in Figure 7b. In the gate-controlled mode, the gate voltage is pulsed to V_{fill} in order to populate the deep levels, and the I_{DS} must be kept constant in order to measure the transient response of V_{GS}. The merits of CID-DLTS are that it can detect traps at very low concentrations and capture trap responses with high sensitivity as well as in very short time scales within the device [37,46].

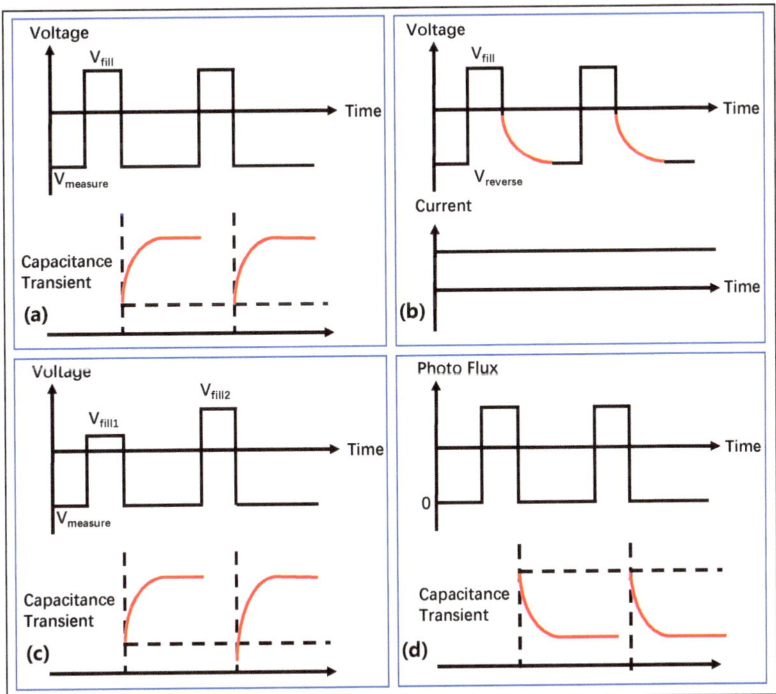

Figure 7. (Color online.) Measurement conditions and curves for (**a**) C-DLTS, (**b**) CID-DLTS, (**c**) D-DLTS, and (**d**) O-DLTS.

Double correlation DLTS (D-DLTS) replaces one amplitude pulse in C-DLTS by using pulses with different amplitudes, as shown in Figure 7c. Although this method makes the experiment and data processing more complex, it facilitates the observation of defect behavior within the space–charge region. Moreover, the variation of defects with depth can be analyzed by changing the pulse amplitude and rate window [76,77].

In DLTS technology, there is also a commonly used drain current DLTS (I-DLTS), whose testing method is highly similar to that of DCT technology, except that DCT observes long-term changes in I_{DS}, while I-DLTS selects a small interval with significant current changes for analysis.

Several of the above DLTS techniques are suitable for detecting majority traps, while the detection of minority traps is difficult. This is due to the very stringent test conditions for minority traps, which require a few traps to be filled and majority traps to be emptied. The optical DLTS (ODLTS) technique has been proposed and validated in order to better characterize minority traps.

ODLTS is a method that uses light pulses as injection pulses to excite electrons and holes, causing carriers to be captured from the trap. Once the light pulse is switched off, the carriers are detrapped so that the capacitance of the device can be measured to obtain the trap parameters. As shown in Figure 7d, photo-generated carriers are generated and captured by traps when a light pulse is applied to a device. After the end of the optical pulse, these captured carriers are de-captured, causing a gradual change in capacitance. Trap parameters can be extracted from capacitance changes at different temperatures after going through the above steps. One of the superiorities of ODLTS is that it overcomes the limitations of DLTS in studying minority traps, with high sensitivity to ultra-deep-level traps [42].

So far, the conventional DLTS method described has the drawback of poor energy resolution. Laplace DLTS (L-DLTS) is an isothermal technique in which the capacitance transients at a fixed temperature are digitized and averaged, and then the defect emission rate is obtained through numerical methods equivalent to the Laplace inverse transform [78]. Compared with traditional DLTS, the main advantage of L-DLTS is a significant improvement in energy resolution, which provides a more precise detection for traps. However, a better signal-to-noise ratio is required in L-DLTS than in conventional DLTS, which makes it less sensitive by a factor of 5.

4. Characterization Methods of Interface Traps

4.1. Constant Capacitance Deep-Level Transient Spectroscopy

Constant capacitance DLTS (CC-DLTS) employs a customized feedback control circuit to maintain a consistent capacitance and regulate voltage to measure voltage transients resulting from trap discharge [66,79,80]. Significant advantages of this method over the traditional DLTS are evident as it ensures a constant depletion of capacitance and SCR width throughout the entire transient process. This feature is particularly beneficial when studying the interface traps in MIS-HEMTs since it keeps the Fermi level constant during the transient response process. Figure 8 illustrates the process and principle of CC-DLTS for measuring MIS HEMT interface traps.

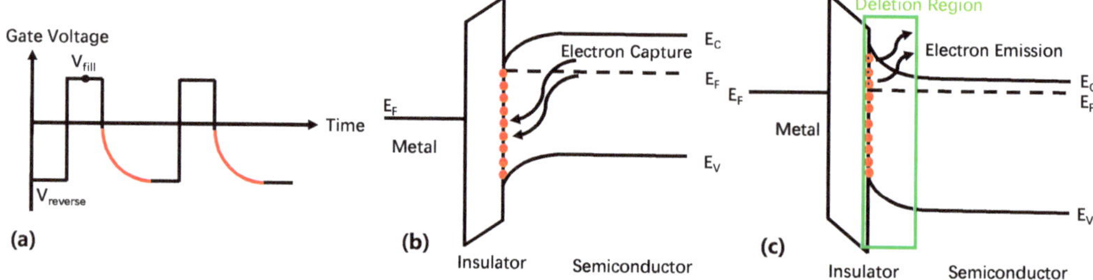

Figure 8. (Color online.) (**a**) The process of CC-DLTS measuring MIS HEMT interface traps. Energy-band diagrams for an MIS HEMT on an n-type semiconductor for (**b**) pulsed accumulation bias and (**c**) nonequilibrium depletion bias.

As shown in Figure 8a, the MIS-HEMT is biased at a negative pressure V_{reverse}, and a depletion capacitor is established at V_{reverse}. Following that, a pulsed filling voltage referred to as V_{fill} is employed to transition the device into an accumulation state. Throughout this phase, electrons are introduced and occupy the interface/oxide states, as demonstrated in Figure 8b. After the filling pulse, the device returns to the depletion state, and the traps emit electrons. The depletion bias is used to maintain a constant capacitance, which means maintaining an almost-fixed SCR. At this condition, the change in bias voltage is reflected by the release process of interface traps, and the energy and density of the traps can be extracted from $V_{\text{ds}}(t)$. Equation (5) allows for the calculation of the interface state density (N_{it}).

$$N_{\text{it}} = \int_{E_{\text{FR}}}^{E_{\text{FP}}} N_{\text{SS}}(E) \left(1 - \exp\left(\frac{-t_P}{\tau_C(E)}\right)\right) dE, \tag{5}$$

It involves the density of detected interface states ($N_{\text{SS}}(E)$), the Fermi level at the gate bias of UR (E_{FP}), the Fermi level at the gate bias of UP (E_{FR}), and the capture time constants ($\tau_C(E)$) associated with the interface states [79].

Nevertheless, CC-DLTS has a slow response due to the influence of the feedback circuit, but it is still highly sensitive and suitable for measuring interface traps [81–84].

4.2. Quasi-Static C-V Measurement

Quasi-static C-V (QSCV) testing is the process of testing the C-V curve of a device under quasi-static and high-frequency conditions. The C-V curve obtained at high frequencies is generally considered to be the ideal curve without interface traps, whereas the C-V curve changes in response to interface traps at low frequencies. The density of interface traps can be determined by comparing the C-V curves measured at quasi-static and high frequencies (HFCV) using Equation (6) [85,86].

$$D_{\text{it}} = \int_{V < V_{\text{th}}}^{V_G} (C_{\text{QSCV}} - C_{\text{HFCV}}) dV / e, \tag{6}$$

Figure 9 demonstrates the CV testing of MIS diodes, with 10 kHz selected as the high frequency and 1 Hz as the lowest quasi-static state.

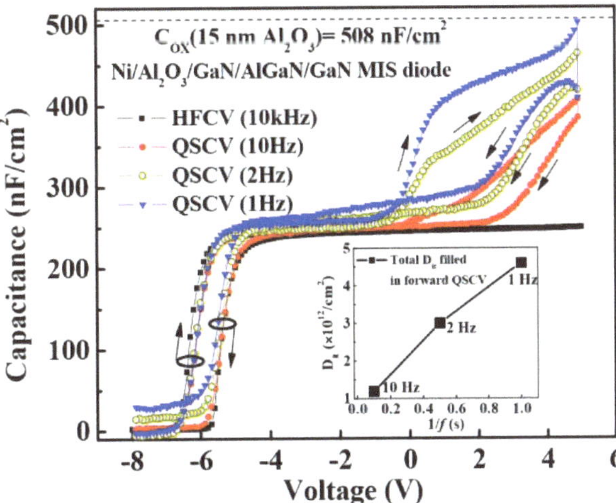

Figure 9. (Color online.) CV characteristics measured at 10, 2, 1, and 10 kHz of Ni/Al$_2$O$_3$/GaN/AlGaN/GaN MIS diodes. Reproduced with permission from [85]. Copyright 2012, Wiley.

Quasi-static C-V measurement can provide not only the interface trap charge density but also the determination of the trap energy level and capture cross-section.

4.3. Dispersion of Conductance Output

The dispersion of conductance output is a highly sensitive technique for characterizing interface defect density, which is capable of detecting interface defects on orders of 10^{10} cm^{-2}/eV or lower [63,64,87].

The equivalent circuit for measuring MOS interface traps using the dispersion of conductance output method is shown in Figure 10a, where C_{OX} is the oxide layer capacitance, C_S is the semiconductor capacitance, and C_{it} is the interface trap capacitance. The charge loss caused by the trap capture emission is represented by R_{it}. By circuit transformation, Figure 10b can be obtained, where C_P and G_P are represented by the Equations (7) and (8):

$$C_P = C_S + \frac{C_{it}}{1+(\omega\tau_{it})^2},\tag{7}$$

$$\frac{G_P}{\omega} = \frac{qD_{it}}{2\omega\tau_{it}}\ln\left[1+(\omega\tau_{it})^2\right],\tag{8}$$

Here, $\tau_{it} = C_{it}R_{it}$. By plotting the G_P/ω-ω curve, the trap density D_{it} and the trap time constant can be extracted from the peak of the curve and the frequency corresponding to the peak, respectively.

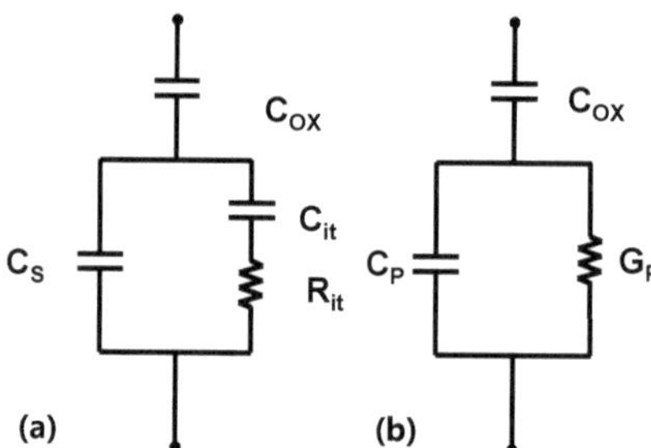

Figure 10. (**a**) The equivalent circuit of MOS. (**b**) Simplified circuit of (**a**).

The conductivity method is highly sensitive, but requires operation over a wide frequency range and is therefore slow to measure [87].

4.4. Single-Pulse Charge Pump

To characterize the interface trap density, a technique known as single-pulse charge pump (SPCP), a method of applying a pulse voltage to gates, can be applied on devices. The SPCP measurement process involves applying a high enough pulse to the gate when the gate voltage is small [88,89]. As the rising edge of the pulse advances, channel electrons begin to accumulate and, subsequently, a portion is captured by the interface trap. At the beginning of the falling edge, the trapped electrons cannot respond quickly due to the long trap time constant. As a result, a difference in current between rise time and fall time occurs, which determines the interface trap density, as shown in Equations (9) and (10):

$$Q_{it} = \int \left(I_{tcp,RISE} - I_{tcp,FALL}\right)dt,\tag{9}$$

$$Q_{it} = qN_{it},\tag{10}$$

where $I_{tcp,RISE}$ and $I_{tcp,FALL}$ are CP current during rise time and fall time, respectively.

SPCP is less susceptible to interference from gate leakage and well characterized for traps with recovery times within 100 μs [89]. However, except for the value of the interface trap density, other parameters cannot be determined.

5. Conclusions and Outlook

The characterization of GaN HEMT traps is a complex and challenging task that requires advanced experimental techniques and sophisticated theoretical models. Although considerable advancements have been made in this domain, numerous unanswered questions and technical challenges remain, necessitating further exploration to comprehensively understand the characteristics and dynamics of these traps.

A significant obstacle arises from the restrictions imposed by existing measurement techniques. This review focuses on the principles and processes of characterizing traps using electrical, optical, and junction capacitance methods, which have already been widely used to probe traps in GaN HEMTs. However, these techniques have their own drawbacks, such as low sensitivity, poor spatial resolution, and limited frequency range. Furthermore, some of these techniques require special sample preparations or expensive equipment, which can limit their accessibility and practicality. The scope and characteristics of each characterization technique are summarized in Table 1.

Table 1. The range of applications and characteristics of characterization methods.

Methods	Range of Application	Sensitivity	Speed	Non-Destructiveness	Characteristics
DCT	Bulk trap	High	Low	No	Easily interferes with leakage current
LFN	Bulk trap	High	Fast	Yes	Easily interferes with noise
Low-frequency output admittance	Bulk trap	Low	Fast	Yes	Wide range
C-DLTS	Bulk trap	High	Low	No	Wide range
CID-DLTS	Bulk trap	High	Low	No	Complex
D-DLTS	Bulk trap in the SCR	High	Low	No	Complex
ODLTS	Bulk trap	High	Low	No	Measures minority carrier traps
CC-DLTS	Interface trap with high concentrations	High	Low	No	Complex
QSCV	Interface trap with high concentrations	Low	Fast	Yes	Energy levels and capture cross-sections cannot be obtained
Dispersion of conductance	Interface trap	Most high	Low	Yes	Wide frequency range
SPCP	Interface trap	High	Fast	Yes	Energy levels and cross-sections unknown

Another challenge is related to the characterization of interface traps. The characterization of GaN HEMT interface traps is mostly applicable to MIS HEMT, while the interface traps of conventional HEMT structures are difficult to measure due to the presence of MS junctions.

To overcome these challenges, further optimizations and innovations in characterization techniques are needed. The characterization of GaN HEMT traps will continue to be a vibrant field, with many exciting opportunities, such as the new measurement techniques that can provide higher sensitivity and resolution a and wider frequency range while reducing the cost and complexity of further research and developments. Moreover, by integrating multiple measurement techniques and theoretical models, complementary and consistent information about traps can be obtained.

In conclusion, the characterization of GaN HEMT traps is a challenging and rewarding task that is critical for improving device performance and reliability. With continuous advancements in measurement techniques and theoretical models, it can be anticipated that profound understandings of the characteristics and dynamics of these traps can be further unlocked so that the full potential of traps for future applications can be realized. This knowledge will enable us to enhance the stability and reliability of GaN HEMTs, broaden their range of applications, and unleash their complete potential for future utilization.

Author Contributions: Conceptualization, X.Z. (Xiazhi Zou) and K.R.; writing—original draft preparation, X.Z. (Xiazhi Zou); writing—review and editing, K.R., J.Y. and Y.S.; supervision, Q.Q., X.Z. (Xinbo Zou) and J.C. All authors have read and agreed to the published version of the manuscript.

Funding: This work was supported by the Shanghai Natural Science Foundation under grant 23ZR1422500 and the National Natural Science Foundation of China under grant 62204150.

Data Availability Statement: Data sharing not applicable.

Conflicts of Interest: The authors declare no conflict of interest.

References

1. Ikeda, N.; Niiyama, Y.; Kambayashi, H.; Sato, Y.; Nomura, T.; Kato, S.; Yoshida, S. GaN Power Transistors on Si Substrates for Switching Applications. *Proc. IEEE* **2010**, *98*, 1151–1161. [CrossRef]
2. Lee, F.C.; Li, Q. High-Frequency Integrated Point-of-Load Converters: Overview. *IEEE Trans. Power Electron.* **2013**, *28*, 4127–4136. [CrossRef]
3. Kumar, V.; Lu, W.; Khan, F.A.; Schwindt, R.; Kuliev, A.; Simin, G.; Yang, J.; Khan, M.A.; Adesida, I. High performance 0.25 μm gate-length AlGaN/GaN HEMTs on sapphire with transconductance of over 400 mS/mm. *Electron. Lett.* **2002**, *38*, 252–253. [CrossRef]
4. Li, C.; Li, Z.; Peng, D.; Ni, J.; Pan, L.; Zhang, D.; Dong, X.; Kong, Y. Improvement of breakdown and current collapse characteristics of GaN HEMT with a polarization-graded AlGaN buffer. *Semicond. Sci. Technol.* **2015**, *30*, 035007. [CrossRef]
5. Kang, M.; Lee, M.; Choi, G.; Hwang, I.; Cha, H.; Seo, K. High-performance normally off AlGaN/GaN-on-Si HEMTs with partially recessed SiNx MIS structure. In Proceedings of the 9th International Workshop on Nitride Semiconductors, Orlando, FL, USA, 1 August 2017.
6. Klein, P.B.; Freitas, J.J.; Binari, S.C.; Wickenden, A.E. Observation of deep traps responsible for current collapse in GaN metal–semiconductor field-effect transistors. *Appl. Phys. Lett.* **1999**, *75*, 4016–4018. [CrossRef]
7. Meneghesso, G.; Chini, A.; Zanoni, E.; Manfredi, M.; Pavesi, M.; Pavesi, B.; Pavesi, C. Diagnosis of trapping phenomena in GaN MESFETs. In Proceedings of the Electron Devices Meeting, 2000. IEDM Technical Digest. International, San Francisco, CA, USA, 10–13 December 2000.
8. Binari, S.C.; Ikossi, K.; Roussos, J.A.; Kruppa, W.; Park, D.; Dietrich, H.B.; Koleske, D.D.; Wickenden, A.E.; Henry, R.L. Trapping effects and microwave power performance in AlGaN/GaN HEMTs. *IEEE Trans. Electron Devices* **2001**, *48*, 465–471. [CrossRef]
9. Klein, P.B.; Binari, S.C.; Ikossi, K.; Wickenden, A.E.; Koleske, D.D.; Henry, R.L.; Katzer, D.S. Investigation of traps producing current collapse in AlGaN/GaN high electron mobility transistors. *Electron. Lett.* **2001**, *37*, 661. [CrossRef]
10. Klein, P.B.; Binari, S.C.; Ikossi, K.; Wickenden, A.E.; Koleske, D.D.; Henry, R.L.; Katzer, D.S. Current collapse and the role of carbon in AlGaN/GaN high electron mobility transistors grown by metalorganic vapor-phase epitaxy. *Appl. Phys. Lett.* **2001**, *79*, 3527–3529. [CrossRef]
11. Vetury, R.; Zhang, N.Q.; Keller, S.; Mishra, U.K. The impact of surface states on the DC and RF characteristics of AlGaN/GaN HFETs. *IEEE Trans. Electron Devices* **2001**, *48*, 560–566. [CrossRef]
12. Hasumi, Y.; Kodera, K. Simulation of the surface trap effect on the gate lag in GaAs MESFETs. *Electron. Commun. Jpn.* **2002**, *85*, 18–26. [CrossRef]
13. Binari, S.C.; Klein, P.B.; Kazior, T.E. Trapping effects in GaN and SiC microwave FETs. *Proc. IEEE* **2002**, *90*, 1048–1058. [CrossRef]
14. Wang, Z.; Zhang, Z.; Wang, S.; Chen, C.; Wang, Z.; Yao, Y. Design and Optimization on a Novel High-Performance Ultra-Thin Barrier AlGaN/GaN Power HEMT With Local Charge Compensation Trench. *Appl. Sci.* **2019**, *9*, 3054. [CrossRef]
15. Wang, Z.; Li, L. Two-dimensional polarization doping of GaN heterojunction and its potential for realizing lateral p–n junction devices. *Appl. Phys. A* **2022**, *8*, 672. [CrossRef]
16. Wang, Z.; Yang, D.; Cao, J.; Wang, F.; Yao, Y. A novel technology for turn-on voltage reduction of high-performance lateral heterojunction diode with source-gate shorted anode. *Superlattices Microstruct.* **2019**, *125*, 144–150. [CrossRef]
17. Meneghini, M.; Rossetto, I.; Bisi, D.; Stocco, A.; Pantellini, A.; Lanzieri, C.; Nanni, A.; Meneghesso, G.; Zanoni, E. Buffer Traps in Fe-Doped AlGaN/GaN HEMTs: Investigation of the Physical Properties Based on Pulsed and Transient Measurements. *IEEE Trans. Electron Devices* **2014**, *61*, 4070–4077. [CrossRef]
18. Guo, W.L.; Chen, Y.F.; Li, S.Y.; Lei, L.B.; Chang, Q. Defect studies of GaN-based HEMT devices: A review. *Chin. J. Luminesc.* **2017**, *38*, 760–767.
19. Bergsten, J.; Thorsell, M.; Adolph, D.; Chen, J.; Kordina, O.; Sveinbjörnsson, E.; Rorsman, N. Electron trapping in extended defects in microwave AlGaN/GaN HEMTs with carbon-doped buffers. *IEEE Trans. Electron Devices* **2018**, *65*, 2446–2453. [CrossRef]
20. Villamin, M.E.; Kondo, T.; Iwata, N. Effect of C- and Fe-doped GaN buffer on AlGaN/GaN high electron mobility transistor performance on GaN substrate using side-gate modulation. *Jpn. J. Appl. Phys.* **2021**, *60*, SBBF01. [CrossRef]

21. Bisi, D.; Stocco, A.; Rossetto, I.; Meneghini, M.; Rampazzo, F.; Chini, A.; Soci, F.; Pantellini, A.; Lanzieri, C.; Gamarra, P.; et al. Effects of buffer compensation strategies on the electrical performance and RF reliability of AlGaN/GaN HEMTs. *Microelectron. Reliab.* **2015**, *55*, 1662–1666. [CrossRef]

22. Meneghini, M.; Bisi, D.; Rossetto, I.; Santi, C.D.; Stocco, A.; Hilt, O.; Bahat-Treidel, E.; Wuerfl, J.; Rampazzo, F.; Meneghesso, G.; et al. Trapping processes related to iron and carbon doping in AlGaN/GaN power HEMTs. *Proc. SPIE* **2015**, *9363*, 936320.

23. Joshi, V.; Gupta, S.D.; Chaudhuri, R.R.; Shrivastava, M. Interplay between surface and buffer traps in governing breakdown characteristics of AlGaN/GaN HEMTs-Part II. *IEEE Trans. Electron Devices* **2020**, *68*, 80–87. [CrossRef]

24. Wang, F.Z.; Chen, W.J.; Wang, Y.; Sun, R.Z.; Ding, G.J.; Yu, P.; Wang, X.H.; Feng, Q.; Xu, W.J.; Xu, X.R.; et al. Analytically Modeling the Effect of Buffer Charge on the 2DEG Density in AlGaN/GaN HEMT. *IEEE Trans. Electron Devices* **2023**, 1–8. [CrossRef]

25. Wang, F.; Chen, W.; Sun, R.; Wang, Z.; Zhang, B. An analytical model on the gate control capability in p-gan gate algan/gan high-electron-mobility transistors considering buffer acceptor traps. *J. Phys. D Appl. Phys.* **2020**, *54*, 095107. [CrossRef]

26. Wu, M.; Zhang, M.; Yang, L.; Hou, B.; Yu, Q.; Li, S.M.; Shi, C.Z.; Zhao, W.; Lu, H.; Chen, W.W.; et al. First Demonstration of State-of-the-art GaN HEMTs for Power and RF Applications on A Unified Platform with Free-standing GaN Substrate and Fe/C Co-doped Buffer. In Proceedings of the 2022 International Electron Devices Meeting (IEDM), San Francisco, CA, USA, 3–7 December 2022.

27. Nomoto, K.; Hu, Z.; Song, B.; Zhu, M.; Qi, M.; Yan, R.; Protasenko, V.; Imhoff, E.; Kuo, J.; Kaneda, T.; et al. GaN-on-GaN p-n power diodes with 3.48 kV and 0.95 mΩ-cm^2: A record high figure-of-merit of 12.8 GW/cm^2. In Proceedings of the 2015 IEEE International Electron Devices Meeting (IEDM), San Francisco, CA, USA, 7–9 December 2015.

28. Lee, W.I.; Huang, T.C.; Guo, J.D.; Feng, M.S. Effects of column III alkyl sources on deep levels in GaN grown by organometallic vapor phase epitaxy. *Appl. Phys. Lett.* **1995**, *67*, 1721–1723. [CrossRef]

29. Soh, C.B.; Chua, S.J.; Lim, H.F.; Chi, D.; Liu, W.; Tripathy, S. Identification of deep levels in GaN associated with dislocations. *J. Phys. Condens. Matter* **2004**, *16*, 6305–6315. [CrossRef]

30. Yutaka, T.; Yujiro, Y.; Shibata, T.; Yamaguchi, S.; Ueda, H.; Uesugi, T.; Kachi, T. Hole traps in n-GaN detected by minority carrier transient spectroscopy. *Phys. Status Solidi* **2011**, *8*, 433–436.

31. Hacke, P.; Nakayama, H.; Detchprohm, T.; Hiramatsu, K.; Sawaki, N. Deep levels in the upper band-gap region of lightly Mg-doped GaN. *Appl. Phys. Lett.* **1996**, *68*, 1362–1364. [CrossRef]

32. Chung, H.M.; Chuang, W.C.; Pan, Y.C.; Tsai, M.C.; Lee, M.C.; Chen, W.K.; Chiang, C.I.; Lin, C.H.; Chang, H. Electrical characterization of isoelectronic in-doping effects in GaN films grown by metalorganic vapor phase epitaxy. *Appl. Phys. Lett.* **2000**, *76*, 897–899. [CrossRef]

33. Umana-Membreno, G.A.; Dell, J.M.; Hessler, T.P.; Nener, B.D.; Parish, G.; Faraone, L.; Mishra, U.K. 60Co gamma-irradiation-induced defects in n-GaN. *Appl. Phys. Lett.* **2002**, *80*, 4354–4356. [CrossRef]

34. Cho, H.K.; Kim, C.S.; Hong, C.H. Electron capture behaviors of deep level traps in unintentionally doped and intentionally doped n-type GaN. *J. Appl. Phys.* **2003**, *94*, 1485–1489. [CrossRef]

35. Hacke, P.; Detchprohm, T.; Hiramatsu, K.; Speck, J.S.; Mishra, U.K.; DenBaars, S.P. Analysis of deep levels in n-type GaN by transient capacitance methods. *J. Appl. Phys.* **1994**, *76*, 304–309. [CrossRef]

36. Hierro, A.; Ringel, S.A.; Hansen, M.; Speck, J.S.; Mishra, U.K.; DenBaars, S.P. Hydrogen passivation of deep levels in n-GaN. *Appl. Phys. Lett.* **2000**, *77*, 1499–1501. [CrossRef]

37. Sasikumar, A.; Arehart, A.R.; Cardwell, D.W.; Jackson, C.M.; Sun, W.; Zhang, Z.; Ringel, S.A. Deep trap-induced dynamic on-resistance degradation in GaN-on-Si power MISHEMTs. *Microelectron. Reliab.* **2016**, *61*, 39–44. [CrossRef]

38. Calleja, E.; Sanchez, F.J.; Basak, D.; Sánchez-García, M.A.; Muñoz, E.; Izpura, I.; Calle, F.; Tijero, J.M.G.; Sánchez-Rojas, J.L.; Beaumont, B.; et al. Yellow luminescence and related deep states in undoped GaN. *Phys. Rev. B* **1997**, *55*, 4689–4694. [CrossRef]

39. Bouslama, M.; Gillet, V.; Chang, C.; Nallatamby, J.C.; Sommet, R.; Prigent, M.; Quéré, R.; Lambert, B. Dynamic Performance and Characterization of Traps Using Different Measurements Techniques for the New AlGaN/GaN HEMT of 0.15-µm Ultrashort Gate Length. *IEEE Trans. Microw. Theory Tech.* **2019**, *67*, 2475–2482. [CrossRef]

40. Benvegnu, A.; Bisi, D.; Laurent, S.; Meneghini, M.; Meneghesso, G.; Muraro, J.L.; Barataud, D.; Zanoni, E.; Quere, R. Trap Investigation under Class AB Operation in AlGaN/GaN HEMTs Based on Output-Admittance Frequency Dispersion, Pulsed and Transient Measurements. In Proceedings of the IEEE Microwave Integrated Circuits Conference, Paris, France, 7–8 September 2015.

41. Hernandez, C.; López-López, M.; Casallas-Moreno, Y.L.; Rangel-Kuoppa, V.; Cardona, D.; Hu, Y.; Kudriatsev, Y.; Zambrano-Serrano, M.A.; Gallardo-Hernandez, S. Study of the Heavily p-Type Doping of Cubic GaN with Mg. *Sci. Rep.* **2020**, *10*, 16858.

42. Chen, J.; Huang, W.; Qu, H.; Zhang, Y.; Zhou, J.; Zou, X. Study of Minority Carrier Traps in p-GaN Gate HEMT by Optical Deep Level Transient Spectroscopy. *Appl. Phys. Lett.* **2022**, *120*, 211105. [CrossRef]

43. Armstrong, A.; Chakraborty, A.; Speck, J.S.; DenBaars, S.P.; Mishra, U.K.; Ringel, S.A. Characterization and Discrimination of AlGaN- and GaN-Related Deep Levels in AlGaN/GaN Heterostructures. *AIP Conf. Proc.* **2007**, *893*, 223–224.

44. Fang, Z.Q.; Look, D.C.; Kim, D.H.; Adesida, I. Traps in AlGaN/GaN/SiC heterostructures studied by deep level transient spectroscopy. *Appl. Phys. Lett.* **2005**, *87*, 182103. [CrossRef]

45. Nakano, Y.; Irokawa, Y.; Takeguchi, M. Deep-Level Optical Spectroscopy Investigation of Band Gap States in AlGaN/GaN Hetero-Interfaces. *Appl. Phys. Express* **2008**, *1*, 091101. [CrossRef]

46. Arehart, A.R.; Sasikumar, A.; Via, G.D.; Winningham, B.; Poling, B.; Heller, E.; Ringel, S.A. Spatially-discriminating trap characterization methods for HEMTs and their application to RF-stressed AlGaN/GaN HEMTs. In Proceedings of the IEEE Microwave Symposium Digest, San Francisco, CA, USA, 6–8 December 2010; pp. 394–397.

47. Cho, H.K.; Khan, F.A.; Adesida, I.; Fang, Z.Q.; Look, D.C. Deep level characteristics in n-GaN with inductively coupled plasma damage. *J. Phys. D Appl. Phys.* **2008**, *41*, 155314. [CrossRef]

48. Dupouy, E.; Raja, P.V.; Gaillard, F.; Sommet, R.; Nallatamby, J.C. Trap Characterization in InAlN/GaN and AlN/GaN based HEMTs with Fe- and C-doped Buffers. In Proceedings of the 2021 16th European Microwave Integrated Circuits Conference (EuMIC), London, UK, 3–4 April 2022.

49. Bouslama, M.; Raja, P.V.; Gaillard, F.; Sommet, R.; Nallatamby, J.C. Investigation of Electron Trapping in AlGaN/GaN HEMT with Fe-doped Buffer through DCT Characterization and TCAD Device Simulations. *AIP Adv.* **2021**, *11*, 025018. [CrossRef]

50. Raja, P.V.; Dupouy, E.; Bouslama, M.; Sommet, R.; Nallatamby, J.C. Estimation of Trapping Induced Dynamic Reduction in 2DEG Density of GaN-Based HEMTs by Gate-Lag DCT Technique. *IEEE Trans. Electron Devices* **2022**, *69*, 4864–4869. [CrossRef]

51. Pan, S.; Feng, S.; Li, X.; Bai, K.; Lu, X.; Li, Y.; Zhang, J.; Zhou, L. Characterization of Hole Traps in Reverse-Biased Schottky-type p-GaN Gate HEMTs by Current-Transient Method. *Appl. Phys. Lett.* **2022**, *121*, 153501. [CrossRef]

52. Pan, S.; Feng, S.; Li, X.; Bai, K.; Lu, X.; Zhu, J.; Zhang, J.; Zhou, L. Identification of Traps in p-GaN Gate HEMTs During OFF-State Stress by Current Transient Method. *IEEE Trans. Electron Devices* **2022**, *69*, 4877–4882. [CrossRef]

53. Tapajna, M.; Simms, R.J.T.; Faqir, M.; Kuball, M.; Pei, Y.; Mishra, U.K. Identification of Electronic Traps in AlGaN/GaN HEMTs Using UV Light-Assisted Trapping Analysis. In Proceedings of the 010 IEEE International Reliability Physics Symposium, Anaheim, CA, USA, 2–6 May 2010.

54. Meneghesso, G.; Verzellesi, G.; Pierobon, R.; Rampazzo, F.; Chini, A.; Mishra, U.K.; Canali, C.; Zanoni, E. Surface-Related Drain Current Dispersion Effects in AlGaN-GaN HEMTs. *IEEE Trans. Electron Devices* **2004**, *51*, 1554–1561. [CrossRef]

55. Joh, J.; Alamo, J. A Current-Transient Methodology for Trap Analysis for GaN High Electron Mobility Transistors. *IEEE Trans. Electron Devices* **2011**, *58*, 132–140. [CrossRef]

56. Tartarin, J.G.; Karboyan, S.; Olivie, F.; Astre, G.; Bary, L.; Lambert, B. I-DLTS, Electrical Lag and Low Frequency Noise Measurements of Trapping Effects in AlGaN/GaN HEMT for Reliability Studies. In Proceedings of the 2011 6th European Microwave Integrated Circuit Conference, Manchester, UK, 10–11 October 2011.

57. Subramani, N.K.; Couvidat, J.; Hajjar, A.A.; Nallatamby, J.; Sommet, R.; Quéré, R. Identification of GaN Buffer Traps in Microwave Power AlGaN/GaN HEMTs Through Low Frequency S-Parameters Measurements and TCAD-Based Physical Device Simulations. *IEEE J. Electron Devices Soc.* **2017**, *5*, 175–181. [CrossRef]

58. Gustafsson, S.; Chen, J.; Forsberg, U.; Thorsell, N.; Janzén, E.; Rorsman, N. Dispersive Effects in Microwave AlGaN/AlN/GaN HEMTs with Carbon-Doped Buffer. *IEEE Trans. Electron Devices* **2015**, *62*, 2162–2169. [CrossRef]

59. Raja, P.V.; Subramani, N.K.; Gaillard, F.; Bouslama, M.; Sommet, R.; Nallatamby, J.C. Identification of Buffer and Surface Traps in Fe-Doped AlGaN/GaN HEMTs Using Y21 Frequency Dispersion Properties. *Electronics* **2021**, *10*, 3096. [CrossRef]

60. Bouslama, M.; Hajjar, A.A.; Laurent, S.; Subramani, N.K.; Nallatamby, J.C.; Prigent, M. Low Frequency Drain Noise Characterization of Different Technologies of GaN HEMTs: Investigation of Trapping Mechanism. In Proceedings of the International Workshop on Integrated Nonlinear Microwave and Millimetre-wave Circuits (INMMIC), Brive La Gaillarde, France, 10–11 December 2018.

61. Tartarin, J.G.; Soubercaze-Pun, G.; Grondin, J.J.; Bary, L.; Mimila-Arroyo, J.; Chevallier, J. Generation-Recombination Defects In AlGaN/GaN HEMT on SiC Substrate, Evidenced By Low Frequency Noise Measurements and SIMS Characterization. *AIP Conf. Proc.* **2007**, *922*, 163–166.

62. Polyakov, A.Y.; Smirnov, N.B.; Govorkov, A.V.; Shlensky, A.A.; Pearton, S.J. Influence of high-temperature annealing on the properties of Fe doped semi-insulating GaN structures. *J. Appl. Phys.* **2004**, *95*, 5591–5596. [CrossRef]

63. Silvestri, M.; Uren, M.J.; Kuball, M. Iron-induced deep-level acceptor center in GaN/AlGaN high electron mobility transistors: Energy level and cross section. *Appl. Phys. Lett.* **2013**, *102*, 073501. [CrossRef]

64. Ma, X.H.; Zhu, J.J.; Liao, X.Y.; Yue, T.; Chen, W.W.; Hao, Y. Quantitative characterization of interface traps in Al_2O_3/AlGaN/GaN metal-oxide-semiconductor high-electron-mobility transistors by dynamic capacitance dispersion technique. *Appl. Phys. Lett.* **2013**, *103*, 033510. [CrossRef]

65. Potier, C.; Martin, A.; Campovecchio, M.; Laurent, S.; Quere, R.; Jacquet, J.C.; Jardel, O.; Piotrowicz, S.; Delage, S. Trap characterization of microwave GaN HEMTs based on frequency dispersion of the output-admittance. In Proceedings of the European Microwave Integrated Circuit Conference, Rome, Italy, 6–9 October 2014; pp. 464–467.

66. Huang, S.; Wang, X.; Liu, X.; Kang, X.W.; Fan, J.; Yang, S.; Yin, H.; Wei, K.; Zheng, Y.; Wang, X.; et al. Revealing the Positive Bias Temperature Instability in Normally-OFF AlGaN/GaN MIS-HFETs by Constant-Capacitance DLTS. In Proceedings of the 2019 31st International Symposium on Power Semiconductor Devices and ICs (ISPSD), Shanghai, China, 19–23 May 2019.

67. Meneghesso, G.; Meneghini, M.; Bisi, D.; Rossetto, I.; Cester, A.; Mishra, U.K.; Zanoni, E. Trapping phenomena in AlGaN/GaN HEMTs: A study based on pulsed and transient measurements. *Semicond. Sci. Technol.* **2013**, *28*, 014005. [CrossRef]

68. Meneghini, M.; Bisi, D.; Marcon, D.; Stoffels, S.; Hove, M.V.; Wu, T.L.; Decoutere, S.; Meneghesso, G.; Zanoni, E. Trapping in GaN-based metal-insulator-semiconductor transistors: Role of high drain bias and hot electrons. *Appl. Phys. Lett.* **2014**, *104*, 143505. [CrossRef]

69. Subramani, N.K.; Couvidat, J.; Hajjar, A.A.; Nallatamby, J.; Floriot, D.; Prigent, M.; Quéré, R. Low-Frequency Noise Characterization in GaN HEMTs: Investigation of Deep Levels and Their Physical Properties. *IEEE Electron Device Lett.* **2017**, *38*, 1109–1112. [CrossRef]

70. Potier, C.; Jacquet, J.; Dua, C.; Martin, A.; Campovecchio, M.; Oualli, M.; Jardel, O.; Piotrowicz, S.; Laurent, S.; Aubry, R.; et al. Highlighting trapping phenomena in microwave GaN HEMTs by low-frequency S-parameters. *Int. J. Microw. Wirel. Technol.* **2015**, *7*, 287–296. [CrossRef]

71. Matteo, M.; Carlo, D.S.; Idriss, A.; Matteo, B.; Marcello, C.; Riyaz, A.K.; Luca, N.; Nicolò, Z.; Alessandro, C.; Farid, M.; et al. GaN-based power devices: Physics, reliability, and perspectives. *J. Appl. Phys.* **2021**, *130*, 181101.

72. Yang, S.; Huang, S.; Wei, J.; Zheng, Z.; Wang, Y.; He, J.; Chen, K.V. Identification of Trap States in p-GaN Layer of a p-GaN/AlGaN/GaN Power HEMT Structure by Deep-Level Transient Spectroscopy. *IEEE Electron Device Lett.* **2020**, *41*, 685–688. [CrossRef]

73. Gassoumi, M.; Bluet, J.M.; Guillot, G.; Gaquiere, C.; Maaref, H. Characterization of deep levels in high electron mobility transistor by conductance deep level transient spectroscopy. *Mater. Sci. Eng. C* **2008**, *28*, 787–790. [CrossRef]

74. Salah, S.; Saadaoui, M.; Gassoumi, M.; Maaref, H.; Christophe, G. Anomaly and defects characterization by I-V and current deep level transient spectroscopy of Al0.25Ga0.75N/GaN/SiC high electron-mobility transistors. *J. Appl. Phys.* **2012**, *111*, 073713.

75. Lang, D.V. Deep-level transient spectroscopy: A new method to characterize traps in semiconductors. *J. Appl. Phys.* **1974**, *45*, 3023–3032. [CrossRef]

76. Lefèvre, H.; Schulz, M. Double correlation technique (DDLTS) for the analysis of deep level profiles in semiconductors. *Appl. Phys. A* **1977**, *12*, 45–53. [CrossRef]

77. Kosai, K. External generation of gate delays in a boxcar integrator—Application to deep level transient spectroscopy. *Rev. Sci. Instrum.* **1982**, *53*, 210–213. [CrossRef]

78. Auret, F.D.; Meyer, W.E.; Diale, M.; Rensburg, P.; Song, S.F.; Temst, K. Electrical Characterization of Metastable Defects Introduced in GaN by Eu-Ion Implantation. *Mater. Sci. Forum* **2011**, *804*, 679–680. [CrossRef]

79. Huang, S.; Wang, X.; Liu, X.; Zhao, R.; Shi, W.; Zhang, Y.; Fan, J.; Yin, H.; Wei, K.; Zheng, Y.; et al. Capture and emission mechanisms of defect states at interface between nitride semiconductor and gate oxides in GaN-based metal-oxide-semiconductor power transistors. *J. Appl. Phys.* **2019**, *126*, 164505. [CrossRef]

80. Johnson, N.M. Measurement of semiconductor–insulator interface states by constant-capacitance deep-level transient spectroscopy. *J. Vac. Sci. Technol.* **1982**, *21*, 303–314. [CrossRef]

81. Li, M.F.; Sah, C.T. New techniques of capacitance-voltage measurements of semiconductor junctions. *Solid-State Electron.* **1982**, *25*, 95–99. [CrossRef]

82. Dejule, R.Y.; Haase, M.A.; Ruby, D.S.; Stillman, G.E. Constant capacitance DLTS circuit for measuring high purity semiconductors. *Solid State Electron.* **1985**, *28*, 639–641. [CrossRef]

83. Kolev, P. An improved feedback circuit for constant-capacitance voltage transient measurements. *Solid-State Electron.* **1992**, *35*, 387–389. [CrossRef]

84. Li, M.F.; Sah, C.T. A new method for the determination of dopant and trap concentration profiles in semiconductors. *IEEE Trans. Electron Devices* **1982**, *29*, 306–315.

85. Huang, S.; Shu, Y.; Roberto, J.; Chen, K.V. Characterization of Vth-instability in Al$_2$O$_3$/GaN/AlGaN/GaN MIS-HEMTs by quasi-static C-V measurement. *Phys. Status Solid* **2012**, *9*, 923–926. [CrossRef]

86. Gray, P.V.; Brown, D.M. Density of SiO$_2$-Si interface states. *Appl. Phys. Lett.* **1966**, *8*, 31–33. [CrossRef]

87. Nicollian, E.H.; Goetzberger, A. The Si-SiO$_2$ interface-electrical properties as determined by the metal-insulator-silicon conductance technique. *Bell Syst. Tech. J.* **1967**, *46*, 1055–1133. [CrossRef]

88. Alghamdi, S.; Si, M.; Bae, H.; Zhou, H.; Ye, P.D. Single Pulse Charge Pumping Measurements on GaN MOS-HEMTs: Fast and Reliable Extraction of Interface Traps Density. *IEEE Trans. Electron Devices* **2020**, *67*, 444–448. [CrossRef]

89. Lin, L.; Ji, Z.; Zhang, J.F.; Zhang, W.D.; Kaczer, B.; Gendt, S.D.; Groeseneken, G. A Single Pulse Charge Pumping Technique for Fast Measurements of Interface States. *IEEE Trans. Electron Devices* **2011**, *58*, 1490–1498. [CrossRef]

MDPI AG
Grosspeteranlage 5
4052 Basel
Switzerland
Tel.: +41 61 683 77 34

Micromachines Editorial Office
E-mail: micromachines@mdpi.com
www.mdpi.com/journal/micromachines